Subdivision Surface Modeling Technology

Wenhe Liao · Hao Liu · Tao Li

Subdivision Surface Modeling Technology

Higher Education Press

Springer

Wenhe Liao
College of Mechanical and Electrical Engineering
Nanjing University of Aeronautics and Astronautics
Nanjing, Jiangsu
China

Tao Li
Suzhou University of Science and Technology
Suzhou, Jiangsu
China

Hao Liu
College of Mechanical and Electrical Engineering
Nanjing University of Aeronautics and Astronautics
Nanjing, Jiangsu
China

ISBN 978-981-10-9892-5 ISBN 978-981-10-3515-9 (eBook)
DOI 10.1007/978-981-10-3515-9

Jointly published with Higher Education Press, Beijing, China
ISBN: 978-7-04-047514-2 Higher Education Press

The print edition is not for sale in China Mainland. Customers from China Mainland please order the print book from: Higher Education Press.

Printed on acid-free paper

This Springer imprint is published by Springer Nature
The registered company is Springer Nature Singapore Pte Ltd.
The registered company address is: 152 Beach Road, #21-01/04 Gateway East, Singapore 189721, Singapore

Foreword

Subdivision surface is a popular modeling technique in the field of computer-aided design (CAD) and computer graphics (CG) for its strong modeling capabilities for meshes of any topology. This book makes a comprehensive introduction to subdivision modeling technologies, the focus of which lies in not only fundamental theories but also practical applications. In theory aspect, this book seeks to make readers understand the contacts between spline surfaces and subdivision surfaces and makes the readers master the analysis techniques of subdivision surfaces. In application aspect, it introduces some typical modeling techniques, such as interpolation, fitting, fairing, intersection, trimming and interactive edit. By studying this book, readers can grasp the main technologies of subdivision surface modeling and use them in software development. This knowledge also benefits understanding of CAD/CG software operations.

Due to flexible topology adaptivity and strong modeling capability, subdivision surface modeling technology has developed quickly in the field of CAD, CG, and geometric modeling since its appearance during the 1970s. Many famous 3D modeling software, such as 3DMax, Maya, and Meshlab, has involved subdivision surface as a modeling tool. Subdivision modeling technology has been successfully applied in making characters of games and special effects of movies. As the saying goes, "Give a man a fish; you have fed him for today. Teach a man to fish; and you have fed him for a lifetime." On the one hand, the book has done a detailed exposition to the basic theory of subdivision surfaces and strives to make readers to achieve the mastery. On the other hand, although the contents of the book are limited, we make a remarks about the main topic at the end of each chapter and list the closely related references for readers to self-improve. We believe that by learning through this book, readers will have a capability of researching and developing with subdivision surfaces independently.

The book was planned by Prof. Wenhe Liao, and most materials came from doctorial dissertations supervised by him. Associate Professor Hao Liu complied this book and wrote Chaps. 1–6 and Sects. 10.1 and 10.2. Dr. Tao Li arranged the rest of the book. He wrote Chap. 8, and Sects. 7.3, 10.3 and revised Sects. 7.1, 7.2 and Chaps. 9, 11. Chapter 9 was taken from Gang He's doctorial dissertation, and

Dr. He revised the English manuscript of this chapter. Sections 7.1 and 7.2 and Chap. 11 came from Xiangyu Zhang's doctorial dissertation, and Dr. Zhang revised the corresponding English manuscript. Graduate Wei Fan made a lot of work for the final proof. The authors thank Dr. He, Dr. Zhang and Graduate Wei Fan for their contributions to this book.

This book is suitable for graduate students, teachers, and technical personals majoring in CAGD, CAD/CG, and other related fields as a reference book on surface modeling. Due to the limitation of our knowledge, there are inevitably some drawbacks in this book. If any flaw found, we are grateful for your contact with us (liuhao-01@nuaa.edu.cn).

Table of Symbols

Notation

Our general approach to notation is to accord to traditional convention meanwhile precisely to express our intentions. Consequently, we try to use traditional notation as far as possible. At the same time, some special notation is introduced; for example, $\boldsymbol{M}[i,j]$ denotes the entry of a matrix $\boldsymbol{M}$ in the ith row and in the jth row. The highlights of this notation are the following:

- Function application is denoted using parentheses (), for example $\boldsymbol{p}(u), \boldsymbol{p}(u,v)$; a combinational number is also be denoted using parentheses (), for example $\binom{5}{3} = \frac{5!}{3!(5-3)!}$. Diploid or triple is also denoted using parentheses (), for example $K = (V, E, F)$.
- Vectors and matrices are created by enclosing their members in square brackets, for example $\boldsymbol{U} = [u_0.u_1, u_2, \cdots]$, $\boldsymbol{M} = \begin{bmatrix} v & e & f \\ 0 & 0 & e \\ 0 & e & 0 \end{bmatrix}$. These members are scalars. Conversely, the ith entry of the vector $\boldsymbol{U}$ is denoted by $\boldsymbol{U}[i]$. The entry in the ith row and in the jth row is denoted by $\boldsymbol{M}[i,j]$. If members of a vector are also vectors, we especially denote the vector as $\overrightarrow{\bullet}$. For example $\overrightarrow{\boldsymbol{E}} = \begin{bmatrix} \boldsymbol{e}_0 \\ \boldsymbol{e}_1 \\ \vdots \\ \boldsymbol{e}_{n-1} \end{bmatrix}$. When a vector denotes a coordinate of a point, we also directly use name of components. For example for $\boldsymbol{R} = [x,y,z]$, $\boldsymbol{R}[x]$ denote the x component. We also use a vector to denote a form of a Lave tiling, for example [4,6,12] Lave tiling.
- Sets are created by enclosing their members in curly brackets $\{\}$. We arrange that indices of members of a vector, matrix, or a set start from 0.

- The expression $\frac{\partial^s \partial^t \boldsymbol{p}(\boldsymbol{x},\boldsymbol{y})}{\partial^s \boldsymbol{x}[i] \partial^t \boldsymbol{y}[j]}$ denotes the sth partial derivative with respect to $\boldsymbol{x}[i]$ and the tth partial derivative with respect to $\boldsymbol{y}[i]$ of the function $\boldsymbol{p}(\boldsymbol{x},\boldsymbol{y})$. For convenience, we also use $f_u(u,v)$ to denote the partial derivative of $f(u,v)$ with respect to u; $f'(u)$ denote derivative of the function $f(u)$ with respect to the variable u.

We also follow several important stylistic rules when assigning variable names. We assign any variable name to be denoted by italics, for example $\boldsymbol{U}$, $\boldsymbol{M}$, a, b, λ. Bold italics denote vectors or matrices, while italics denote scalar variables or names of geometric shapes. Often, a same letter probably has both a bold italic version and an italic version denoting different meanings. For example, M denotes a mesh, while $\boldsymbol{M}$ denotes a matrix.

Roman

Notation of points and vertices does most probably cause confusions. The highlights of these familiar letter notations are the following:

- $\boldsymbol{p}$, $\boldsymbol{q}$, point on continuous curves or surfaces.
- $\boldsymbol{V}$, $\boldsymbol{E}$, $\boldsymbol{F}$, vertex of polygon or mesh or grid. Note that $\boldsymbol{V}$ also denotes a knot vector for spline surfaces;
- $\boldsymbol{v}$, $\boldsymbol{e}$, $\boldsymbol{f}$, vertex of subpolygon or submesh.
- $\vec{\boldsymbol{V}}$, $\vec{\boldsymbol{E}}$, $\vec{\boldsymbol{F}}$, $\vec{\boldsymbol{e}}$, $\vec{\boldsymbol{f}}$, vector formed by vertices.
- $\tilde{\boldsymbol{v}}$, $\tilde{\boldsymbol{e}}$, $\tilde{\boldsymbol{f}}$, vertex after Fourier transformation for $\boldsymbol{v}$, $\boldsymbol{e}$, $\boldsymbol{f}$.

For other some familiar notations, we give their meanings as the following:

- i, j, integer indices
- u, v, continuous parameter variables
- $\boldsymbol{U} = [u_0, u_1, \ldots]$, $\boldsymbol{V} = [v_0, v_1, \ldots]$, knot vector for spline curves or knot vectors for spline surfaces.
- r, s, t, k, l, temporary variables; k usually used as level of subdivision; degree of polynomial; degree of continuity; size of a generation vector. r usually used as multiplicity of a knot u_i in a knot vector $\boldsymbol{U}$ or variable for integer index in a sequence;
- $\boldsymbol{s}(u,v)$ or $\boldsymbol{s}(s,t)$, a part of a subdivision surface
- C^k, G^k, k degree derivative continuity and k degree geometric continuity
- m, n, size of grid, mesh, polygon, matrix, or vector; n also denotes valence of vertex.
- M, M^k, mesh and mesh on kth level of subdivision;
- $N_{i,k}(u)$, $N_i(u)$, B-spline basic function
- d_i, e_i, parameters of vertices or edges in polygons or meshes
- $f()$, $g()$, $p()$, $q()$, $h()$, scalar functions
- d, differential operator
- T, subdivision operator

- x, y, z continuous domain variables. Usually denote coordinates of points or vertices
- $\boldsymbol{s}_i$, $\boldsymbol{t}_i$. generation vector for a grid. It is a unit vector
- D^k, generation vector group formed by generation vectors. It is a set.
- G_D^k, grid formed the vector group D^k
- $\boldsymbol{i} = (i_0,i_1)$, or $\boldsymbol{i} = (i_0,i_1,i_2)$, integer coordinates in a grid
- $\boldsymbol{x}$, $\boldsymbol{y}$, real coordinates in a grid
- $N_{D^k}(\boldsymbol{x})$, box spline basic function
- $C(z)$, generating function for $N_D(\boldsymbol{x})$:
- $X, \overset{s}{X}, \overset{s+}{X}, \overset{s+}{\overline{X}}$, parameter mesh of a manifold patch and sth side of X, extension of $\overset{s}{X}$, rectangular mesh mapped from $\overset{s+}{X}$.
- $c_{i,j}^s$, a chart in the parameter mesh $\overset{s}{X}$,
- w, weight for vertex in a mesh
- T_C, the non-uniform subdivision operator
- T, the skirt-removed operator
- T_{RC}, non-uniform skirt-removed scheme
- T_{CT}, non-uniform subdivision operator constructing subdivision surface interpolating mesh corner vertices
- $C(\bullet)$ denotes the operator taking the continuity degree
- $\boldsymbol{E}$, energy of curve or surface
- $\boldsymbol{M}$, subdivision matrix
- $\boldsymbol{K}$, picking matrices
- S, a usual name for a surface
- $\boldsymbol{S}(\cdot,\cdot)$, $\boldsymbol{S}(\cdot)$ or $\boldsymbol{S}$, a surface equation or any a point on a surface or a mapping from parameter region to a space or a set formed by all points in the surface S.

Greek Letters

α, β, $\boldsymbol{\alpha}$, $\boldsymbol{\beta}$, coefficients or vector formed by coefficients

γ, aperture factor

φ, ϕ, functions or mappings

$\mu(y)$, a mapping constructing the basic curve in parameter regions

κ slope of line

Θ^s A set formed by all related vertices of $\boldsymbol{x}$ in X^s

ε, temporary variables, usually denote very little real number

ζ, a given vector function that represent imposed loads

ξ, eigenvector

$\boldsymbol{\Xi}$ denotes a matrix formed by eigenvectors

Σ, plane

Ω, parameter region

Miscellaneous

D
Z, Z^2, $Z^2/2$, set of integers, set of integer pairs. Set of integer pairs divided by 2
R, R^2, real number space, and two-dimension real number space
$\otimes$, convolution operator of two functions or Kronecker product of matrices
k, curvature

Functions

$supp(D^k)$, support region of a vector group D^k
$O(supp(D^k))$ or $O(n^k)$, an open set which is the inner region of $supp(D^k)$ or a polynomial of n^k
$span(D^k)$, space spanned by D^k
$U(x, \varepsilon)$ a ε neighborhood *of* $\boldsymbol{p}$
edges(M) set of a mesh M
$max\{\bullet\}$, the maximum element of a set denoted by •
$min\{\bullet\}$, the minimum element of a set denoted by •
$|\bullet|$, the valence of a vertex or element number of a set or a vector
$a\%b$, the remainder after a divided by b

Contents

Chapter 1
Introduction

Surface modeling is a fundamental realm of CAD and greatly affects the development of CAD compared with NURBS. Subdivision is a subsequently emerged surface modeling technique. It can be regarded as a bridge between continuous modeling and discrete modeling. Subdivision modeling has a very wide prospect. Just as what DeRose has predicted [1], subdivision will largely supplant B-splines in many application domains in the coming years. This chapter firstly describes the surface modeling. And then, a review for achievements of subdivision surfaces is provided. Lastly, some surveys and books on subdivision modeling are recommended, and meanwhile, the main idea of this book is presented.

1.1 Surface Modeling

Surface modeling generally refers to free-form surface modeling. Different from the analytic surface, the shape of which is determined by an equation, the shape of a free-form surface can vary according to designers' intention. Surface modeling is derived from the aeronautics industry because shapes of airplanes are complex and it is difficult to express these shapes by using analytical surfaces. So far, it has become one of the most important fundamental technologies of CAD/CAM and the most important part of CAGD. It is also the fundamental technologies of many other fields, such as computer graphics, computer animation, computer vision.

Surface modeling mainly researches the representation, design, analysis, and rendering of surfaces in computer graphic systems [2]. These technologies on design, rendering, and analysis depend on representation methods of surfaces. For example, for the point interpolating technique, its implementing algorithms are probably different under different representation methods. Consequently, after a new surface representation appears, people will usually research the technologies of design, analysis, and rendering of the new representation.

W. Liao et al., *Subdivision Surface Modeling Technology*,
DOI 10.1007/978-981-10-3515-9_1

From the middle ages of last century to present, parameter spline method, Coons method, Bézier method, B-spline method, and NURBS method [3–5] have been adopted as main methods of the surface representations. Nowadays, NURBS method is a method that is mostly used. Bézier method and B-spline method can be regarded as special cases of NURBS method. Subdivision method is different from NURBS method. Subdivision surfaces do not have expressions, and they are defined by limits of mesh sequences. T-spline method is another surface representation method that can be regarded as an extension of B-spline method. Due to achievements of operating abilities and memories of PCs, the polygon mesh has become an important shape representation method. A complex shape can be accurately represented by a polygon mesh with hundreds or thousands of polygons.

The polygon mesh is a discrete representation, whereas the NURBS surface and the T-spline surface are continuous representations. A polygon mesh means that their face elements are polygons, for example, triangles, quadrangles, pentagons. The subdivision surface is a representation between the discrete representation and the continuous representation. Because control meshes of spline surfaces and subdivision surfaces are polygon meshes, the polygon mesh is usually mentioned in this book. In this book, *polygon mesh* is simply called as *mesh*. When we use polygon meshes to represent surfaces, especially in the case of meshes with dense vertices, *polygon mesh* is also called as the *mesh surface*. The geometric model expressed by a polygon mesh is called as the *polygon model* or the *mesh model*.

Surface design is a comprehensive topic. We discuss it in two parts: modeling methods and operation methods. Approximation, interpolation, and fitting are three fundamental modeling methods. Figure 1.1 gives explanations for these three modeling methods. There are also other modeling methods, such as skinning, sweeping, extrusion, revolution.

Intersection, trimming, and union are fundamental surface operations that are called Boolean operations. Other surface operations, such as deformation, blending, fairing, reconstruction, simplification, conversion, offset and multiresolution analysis, have been increasingly focused on since the 80s of last century [5, 6].

Surface analysis involves the classical differential geometric properties of surfaces, such as continuity, curvature, principle directions. Surface analysis estimates

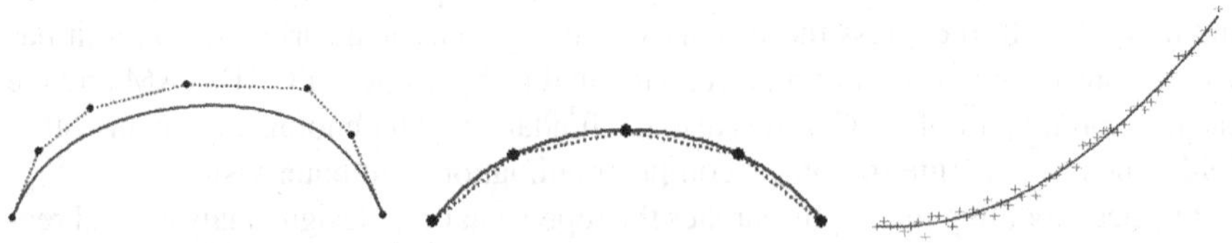

(a) Approximation; (b) Interpolation; (c) Fitting

Fig. 1.1 Three fundamental modeling methods

the qualities of surfaces and abilities of modeling methods. It can guide modifications of surface shapes and constructions of modeling methods.

Surface rendering involves drawing, color, light, texture, *etc.*. Hardware rendering is an important research area of subdivision surfaces. In order to rapidly render subdivision surfaces, an efficient method is to subdivide control meshes using hardware.

1.2 Concept of Subdivision Surfaces

Although NURBS has become the industry standard of data exchange of geometric information in computers and is widely applied in industry and animation modeling, it still has some disadvantages. Just as what DeRose [7] has pointed out, a single NURBS surface cannot express a complex surface with arbitrary topology, for example, surfaces in human animation modeling. Patchwork of trimmed NURBS is the most commonly adopted way to model complex smooth surfaces. However, this way does have at least two difficulties: (1) Trimming is expensive and prone to numerical error; (2) it is difficult to maintain smoothness, or even approximate smoothness, at the seams of the patchwork as the model is animated.

Subdivision surfaces have the potential to overcome these problems: They do not require trimming, and smoothness of the model is automatically guaranteed, even as the model animates. They have advantages of mesh surfaces and spline surfaces.

What is subdivision? Subdivision densifies vertices of meshes by a set of given rules. The process of densifying vertices of meshes is also called refinement. A mesh sequence is obtained when we recursively subdivide a mesh. The limit of the mesh sequence is called a subdivision surface if the mesh sequence is convergent. Any mesh in the sequence is called a subdivision mesh. If a mesh has the same topology as that of a subdivision mesh, we call that the mesh has subdivision connectivity. A set of subdivision rules is called a subdivision scheme. In applications, we usually replace the subdivision surface by using the subdivision mesh on a certain subdivision level. Figure 1.2 gives a construction process of a subdivision surface. In this case, the pipeline is expressed by an entire surface without trimmings and joins.

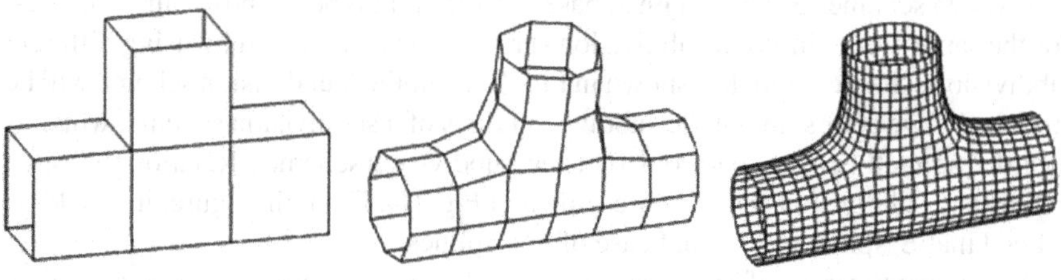

(a) Initial mesh; (b) Subdivide once; (c) Subdivide twice

Fig. 1.2 Construction process of a subdivision surface

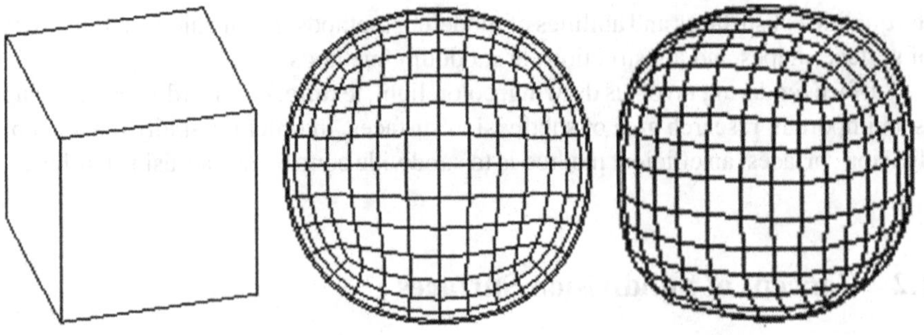

(a) Initial mesh; (b) C-C subdivision scheme; (c) Doo-Sabin subdivision scheme

Fig. 1.3 Different subdivision schemes produce different subdivision surfaces for the same initial mesh

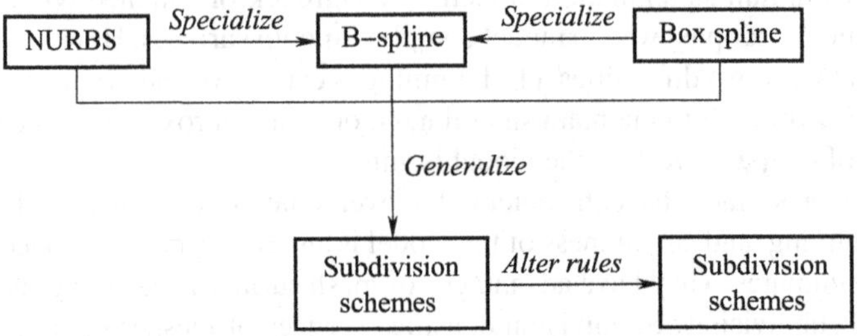

Fig. 1.4 Relations between splines and subdivision schemes

Subdivision is closely related to box spline theories. How is it related to box splines? A box spline surface has a control mesh. For convenience, we call the subdivision of control meshes of spline surfaces as refinement. Box spline theories define refinement rules for control meshes. The limit surface is the box spline surface when we refine a control mesh. We obtain a subdivision scheme if we generalize these refinement rules to arbitrary meshes. It is one of the subdivision scheme construction methods to generalize refinement rules to arbitrary topology meshes. Different subdivision schemes can be obtained based on different types of box spline surfaces. For the same mesh, different subdivision surfaces can be obtained by using different subdivision schemes, which is shown in Fig. 1.3. Another subdivision scheme will be produced if we alter some of the subdivision rules of a subdivision scheme, which is regarded as another method to construct new subdivision schemes. Relations between splines and subdivision schemes are shown in Fig. 1.4. From the figure, it should be noticed that B-spline is a special case of box splines.

As a generalization of spline surfaces, subdivision surfaces can be regarded as bridges between continuous surfaces and discrete surfaces. On the one hand, we can replace the subdivision surface by using the subdivision mesh on a certain subdivision level. On the other hand, we can analyze geometric properties of subdivision surfaces in the view of continuous surfaces. Generally, subdivision elegantly addresses many issues that are confronted in computer graphics [8]:

Arbitrary Topology: Subdivision generalizes classical spline surface approaches to arbitrary topology. This implies that there is no need for trimming curves or awkward constraint management between patches.
Scalability: Because of its recursive structure, subdivision naturally accommodates level-of-detail rendering and adaptive approximation with error bounds. The result let us be able to make the best use of limited hardware resources, such as those found on low-end PCs.
Uniformity of Representation: Many of traditional modeling methods use either polygon meshes or spline patches. Subdivision spans the spectrum between these two extremes. Surfaces can behave as if they are made of patches, or they can be treated as if consisting of many small polygons.
Numerical Stability: The meshes produced by subdivision have many of the nice properties finite element solvers require. As a result, subdivision representations are also highly suitable for many numerical simulation tasks which are important in engineering and computer animation settings.
Code Simplicity: Subdivision is simple to implement and very efficient to execute.

1.3 Development of Subdivision Surfaces

Though subdivision surfaces have prevailed since the late 1990s, the basic ideas behind subdivision are very old indeed and can be traced as far back as the early 1950s when De Rham G. used "corner cutting" to describe smooth curves. However, the application in geometric modeling starts with the proposal of Chaikin [9], who devised a method of generating smooth curves for plotting in the middle 1970s. In the limit, Chaikin's algorithm produced uniform quadratic B-spline curves. In 1978, Catmull and Clark [10], and Doo and Sabin [11] generalized refinement rules of control meshes of biquadratic and bicubic B-spline surfaces to meshes of arbitrary topology, which indicates that subdivision formally becomes one of the surface modeling methods. The generalization of biquadratic B-spline surfaces is called Doo–Sabin subdivision surfaces, and their subdivision rules are called Doo–Sabin subdivision scheme. Similarly, we have Catmull–Clark subdivision scheme that is usually simply called as C-C subdivision scheme.

The number of mesh vertices exponentially increases when a mesh is subdivided. For example, there are totally m vertices on a mesh. If the mesh is subdivided k times and a vertex becomes n vertices after a subdivision step, there are approximately mn^k vertices on the mesh. Limited to computers' abilities of operation and memories, subdivision has not been focused on until the 90s of last century. Before the 90s of last century, most researchers mainly have paid attention to NURBS surfaces. The theoretical system of NURBS became mature. However, subdivision techniques also constantly grew in this period. The growth was embedded in two aspects. One is construction of subdivision schemes and the other is analysis of subdivision surfaces.

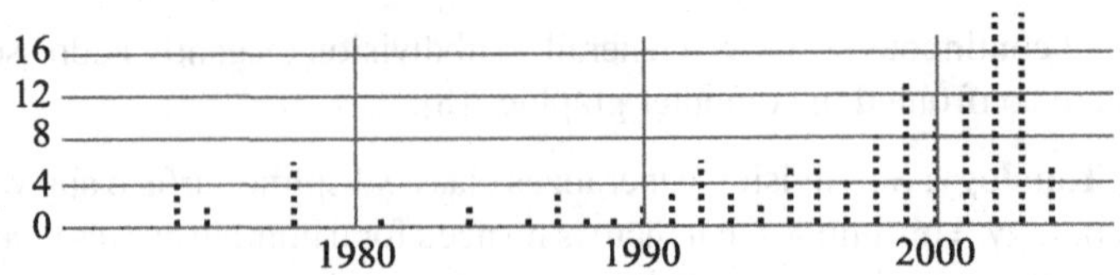

Fig. 1.5 Subdivision papers, plotted by year

For construction of subdivision schemes, there are also two big ideas. One is called Loop subdivision scheme and the other is the four-point scheme. Loop subdivision scheme was described in Loop's Masters thesis [12] in 1987, and it was defined over a grid of triangles. The four-point scheme was presented by Dyn, Levin, and Gregory [13] in 1987. It was a subdivision scheme of curves. That is, we subdivide a polygon and then obtain a curve by using the subdivision scheme. Different from the previous approximating subdivision schemes, it was an interpolating scheme. The generalization of the four-point scheme began in 1990 [14]. The subdivision scheme is called Butterfly scheme that is also defined over a triangular grid.

For the analysis of subdivision surfaces, we pay attention to convergence of subdivision mesh sequences and continuity of subdivision surfaces. For simplicity, we call them convergence and continuity of subdivision schemes, respectively. In this period, the main method of subdivision surface analysis is the eigenanalysis of subdivision matrices. In 1978, Doo and Sabin [11] gave the famous eigenanalysis method—discrete Fourier transformation. In later eigenanalysis analysis, Fourier transformation was always used without exception. As the development and application of Doo and Sabin's idea, Ball and Story considered a more general form of the algorithm of Doo and Sabin [15]. Loop also gave eigenanalysis of his Loop subdivision surfaces when constructing the subdivision scheme. Ball et al.'s work and Loop's work show that the eigenanalysis of subdivision matrices could be used explicitly in the original design of a scheme.

After 1990, especially in the late 1990s and early 2000s, subdivision modeling techniques have developed rapidly. The opinion is reflected by the statistics in [16]. The statistics is given in Fig. 1.5. We summarize these achievements in the following ten aspects:

(1) Subdivision surface analysis. A representative work was given by Reif [17] in 1995. He identified that we could not ensure that the subdivision surfaces were G^1 continuous only by using properties of subdivision matrices. Consequently, he constructed the characteristic map and gave sufficient and necessary conditions for G^1 continuity of subdivision surfaces. By using the theory, Reif [18] showed that the attempts to make a G^2 variant without flat points for a C-C surface will not succeed, which evoked many discussions on how to construct G^2 continuous subdivision surfaces in later years. General criteria to construct G^k continuous subdivision surfaces were given by Zorin in his Phd thesis [19] in 1998. By using these criteria, he designed an algorithm to verify G^1 continuity of subdivision surfaces. Prautzsch [20] generalizes results in Reif [17] and gave the sufficient and necessary conditions

for G^k continuity of subdivision surfaces. As a shape analysis method, Peter [21, 22] considered the differential geometric properties of subdivision surfaces, such as the fundamental forms, the Weingarten map, the principal curvatures, the principal directions in 2004. By relating the shape properties to the subdivision matrices, they obtained some conditions for the construction of high-quality subdivision schemes.

(2) New subdivision schemes. Methods to construct subdivision schemes are classified into two types: One is to design new subdivision schemes, and the other is to alter old subdivision schemes to improve continuity of subdivision surfaces, especially to obtain G^2 continuous subdivision surfaces, in light of subdivision surface analysis. For the first type, Kobbelt [23] presented an interpolating subdivision scheme over quadrilateral grids. The scheme is a generalization of product of the four-point scheme for curve subdivisions. Peters and Reif [24] presented the "simplest" scheme that was also midpoint subdivision scheme. The midpoint subdivision surface was in fact a generalization of a box spline surface. Sederberg [25] gave the non-uniform subdivision schemes by generalizing non-uniform quadratic and cubic B-spline surfaces. Kobbelt [26] described a $\sqrt{3}$ subdivision scheme defined over a grid of triangles. Compared with the Loop scheme, the number of faces of subdivision meshes produced by a $\sqrt{3}$ subdivision scheme increases slowly. When a subdivision step is executed, the number of faces has an increase of a $\sqrt{3}$ multiple. The theme on the increase rate leads to Velho's 4–8 scheme [27]. Based on 4–8 scheme, Li [28] gave a $\sqrt{2}$ subdivision scheme. Dyn [29] introduced a hexagon subdivision scheme that constructed interpolating convexity-preserving subdivision surfaces. Different from Dyn's hexagon subdivision scheme, Zhang's hexagon subdivision scheme [30] was an approximating scheme. For the second type, Zorin [8, 31] gave new subdivision rules for Butterfly subdivision scheme. The improved Butterfly subdivision scheme can construct C^1 continuous subdivision surfaces. By using the sufficient and necessary conditions for G^k continuity of subdivision surfaces, Prautzsch and Umlauf [32–34] altered the subdivision rules of C-C subdivision scheme, Loop subdivision scheme, and Butterfly subdivision scheme. After altering the subdivision rules, C-C subdivision surfaces and Loop subdivision surfaces are G^2 continuous and Butterfly subdivision surfaces are G^1 continuous. It is a character of the period that new subdivision schemes bloomed up, which was due to two reasons. One is that the relations between box splines and subdivision are opened out. Because every box spline has a set of refinement rules, we can generalize these refinement rules to arbitrary meshes and obtain new subdivision schemes. Of course, each such scheme would have to have its extraordinary point rules invented. The other is that the construction of subdivision schemes has the directions of subdivision surface analysis. Subdivision surface analysis is helpful to invent extraordinary point rules of subdivision schemes.

(3) Classifications of subdivision schemes. It is an important task to classify these subdivision schemes since there are so many subdivision schemes. In fact, we have used some classification methods in above discussions, such as interpolating subdivision and approximating subdivision, Zorin et al. [8] enumerated some basic classification methods. The topic was profoundly researched after 2000. Ivrissimtzis et al. [35] gave a classification system for subdivision schemes by using similarity

transformations of grids. The classification is a generalization of Alexa' classification [36]. The classification shows that subdivision schemes with low incensement ratio of mesh elements (vertices, edges, and faces) come at the expense of symmetry and uniformity. It is a natural idea to relate classification and unification of subdivision schemes. Stam [37] presented a class of subdivision surfaces which generalized uniform tensor product B-spline surfaces of any bi-degree to meshes of arbitrary topology. In the class, Doo–Sabin subdivision scheme and C-C subdivision scheme become two special cases. Zorin [38] gave an analogous work almost at the same time. Similar to the generalization of B-spline surfaces of any bi-degree, Oswald [39] presented a new family for a$\sqrt{3}$ subdivision. After inserting new vertices into meshes, there are smooth iterative steps. When iterative times increase, the resulting surfaces become smoother at regular vertices. As a further generalization of the new family, they also introduced a wider class of composite subdivision schemes suitable for arbitrary topologies and topology rules of subdivisions.

(4) Parameter evaluation. After taking a pair of parameters (u, v), can we compute the coordinates of $\boldsymbol{S}(u, v)$? Parameter evaluation can answer this question. Except for coordinates of points, parameter evaluation methods can usually compute partial derivatives of subdivision surfaces. Consequently, we can compute some differential geometric variables, such as normals, curvature, and principle directions, if these variables exist. Stam's method [40] is perfect in theory, and it can evaluate subdivision surfaces generalized from spline surfaces [41]. Based on Stam's method, Wang [42] gave a parameter evaluation method for non-uniform C-C subdivision scheme. Different from Stam's method, Lai [43] gave another evaluation method that employed less eigenbasis functions. Halstead et al. [44] gave formulas to compute limit positions and limit normals of mesh vertices of subdivision surfaces. Though their methods cannot be regarded as parameter evaluation methods, those formulas are convenient to compute limit properties of mesh vertices.

(5) Interpolation. The interpolating methods based on subdivision surfaces can be classified into two categories: interpolation of vertices and interpolation of curves. The interpolation of vertices falls into two topics: interpolating subdivision and approximating subdivision. The famous interpolating subdivision scheme—Butterfly scheme appeared in 1990. Two sets of improved rules of Butterfly scheme were, respectively, given by Zorin [8, 31] and Prautzsch and Umlauf [20, 32]. Kobbelt's interpolating subdivision scheme was a generalization of the four-point scheme and was given in 1996 [23]. Except for these schemes generalized from the four-point scheme, there are some schemes obtained by altering approximating schemes. Labisk and Greiner [45] gave an interpolating $\sqrt{3}$ subdivision scheme. Li et al. [46] gave an interpolating $\sqrt{2}$ subdivision scheme. Zhang [47] gave push-back interpolating subdivision schemes for C-C subdivision scheme, Doo–Sabin subdivision scheme, and Loop subdivision scheme. For algorithms based on approximating subdivision schemes, Nasri [48] noticed that Doo–Sabin subdivision surfaces interpolate the center of each face of meshes. Consequently, control vertices can be computed by linear systems. Based on the idea, Nasri [49] presented a method to construct Doo–Sabin subdivision surfaces interpolating given points and normals. Similar to Nasri's method, Li et al. [50] discussed the method of interpolation of points and

normals based on C-C subdivision scheme. Zheng et al.'s method [51] was similar to Li's method while Zheng considered the fairing of C-C subdivisions. In order to improve the fairing of subdivision surfaces, Zheng adjusted control vertices by using local optimization models. By using shape similarity, Lai [52] also gave an interpolating method of C-C subdivision surfaces. Helatead's fairing interpolating method [44] was very representative, and he used physics energy model to compute control vertices. Helatead's method was a global method. It was an interesting phenomenon that interpolations of points are always related to interpolations of normals when people construct interpolating surfaces using approximating subdivisions. For the interpolation of curves, the representative algorithm was Levin's combined subdivision schemes [53] and the algorithm was also applied in trimming of subdivision surfaces, filling of holes, and blending of surfaces [54–56]. By using polygonal complexes, Nasri [57] researched the curve interpolations of Doo–Sabin subdivision surfaces and C-C subdivision surfaces. Zhang et al.'s method [58] can be regarded as developments of the polygonal complex method. In Zhang's method [58], polygonal complexes are called "symmetric zonal meshes."

(6) Fitting. Fitting is a key of reverse engineering. It is usually called the surface reconstruction that fits unstructured triangle meshes or points clouds by using subdivision surfaces. Fitting can mainly be classified into two categories: local parameterization method [59] and circularly adjusting-vertices method [60, 61]. The first method constructs harmonic maps between a coarse polygon model and the fitted original surface. The process of constructing harmonic maps is also the process of parameterization. By using harmonic maps, we sample the original surfaces. These samples form a mesh with subdivision connectivity. That is, these meshes are not obtained by subdivisions while they have topologies of subdivision meshes. The second method adjusts vertices of meshes after each subdivision step in order that the shape of subdivision meshes or limit surfaces is as close as possible to the shape of original surfaces. Ma's method [62] does not fall into the above two categories. After parameterizing the original surface, he computes control vertices by the least square method. Since shapes in the real world are not all smooth and they probably have some sharp characters, such as creases, darts, sharp points, some special subdivision rules have to be designed for existing subdivision schemes in order to truly fit these sharp characters [63].

(7) Multiresolution analysis of subdivision surfaces. For a mesh M^k with subdivision connectivity, how should another mesh M^{k-1} with the subdivision connectivity in a lower subdivision level be chosen to approximate the shape? How do we restore the shape of M^k from M^{k-1}? Multiresolution analysis of subdivision surfaces asks the question. Wavelet is an important tool of multiresolution analysis of subdivision surfaces. Wavelet transformation of subdivision surfaces is originally explored by Lounsbery et al. [64]. Based on lifting scheme for B-spline wavelets, Martin [65, 66] presented a construction method of lifted biorthogonal wavelets for meshes with C-C subdivision connectivity. Their algorithm can be executed in linear time. Not using wavelet transformations, Zorin [67] described a multiresolution representation for meshes based on subdivision. Based on the representation, they built a scalable interactive multiresolution editing system.

(8) Boolean operations. Intersection and trimming of surfaces are the basis of Boolean operations. Litke [54] researched the trimming of subdivision surfaces by using the combined subdivision scheme. The combined subdivision scheme ensured the surfaces after trimming accurately interpolate trimming curves. Using the union operation as an example, Biermann [68] discussed the Boolean operation of subdivision surfaces defined on triangle meshes. In Biermann's method, it is important to construct multiresolution meshes for resulting surfaces of union operations. Hui [69] presented an algorithm for blending of subdivision surfaces. Zhou [70] also researched the intersection and trimming operations of Loop subdivision surfaces. Generally, it is the key of Boolean operations to control the precision of operation results.

(9) Rendering. How do we rapidly render a subdivision surface? This is a problem on evaluation of subdivision surfaces [71]. The evaluation methods of subdivision surfaces can be classified into two categories: software evaluation and hardware evaluation. Recursive subdivision is the most direct software evaluation method. Since ordinary recursive subdivision results in exponential increase of elements of meshes, adaptive subdivisions are considered by many researchers. Different from ordinary recursive subdivisions, adaptive subdivision results in such resulting meshes: There are higher subdivision levels on regions with larger curvatures. Kobbelt [23] is one of the researchers that first used the adaptive subdivision. The adaptive subdivision scheme is given for quadrilateral subdivisions. The Y-split technique in the adaptive subdivision is a famous tackle to repair holes between regions with different subdivision levels. Kobbelt [26] also presented an adaptive subdivision method for a $\sqrt{3}$ subdivision scheme. The adaptive subdivision method still focused on eliminating gaps between regions with different subdivision levels. Most literature on adaptive subdivisions appeared in the 2000s and late of 1990s. Li [72] presented a survey for adaptive subdivisions. He considered that there were three topics for adaptive subdivisions: criteria to determine regions that should have high subdivision levels, methods to eliminate gaps between regions with different subdivision levels, and rules to compute coordinates of vertices. By combining C-C subdivision with T-spline method, Sederberg [73] presented T-NURCC subdivision scheme that can also be regarded as an adaptive subdivision scheme. The topic on the hardware evaluation focuses on balancing the workload between CPU and GPU (graphics processing unit). These algorithms take advantage of parallel execution streams in programmable graphics hardware. This is an interesting topic after 2000. In 2000, Bischoff et al. [74] proposed a hardware solution for Loop subdivision surface rendering. The pretabulated basis function composition method explored by Bolz [71] is a representative hardware rendering method for C-C subdivision surfaces. Unfortunately, the tabulated evaluation limits flexibility and can increase downstream complexity, and it is probably troublesome to produce some sharp characters. By contrast, Shiue's GPU subdivision kernel [75] generated the subdivision mesh at different levels on the GPU so that all evaluation work rested with GPU shaders.

(10) Applications. It is a milestone that the animation short film *Geri's game* succeeded in 1998. Its dramatis personae model is made by using subdivision techniques [7]. So far, some leading animation software, such as Maya, 3D-Max, has used the

subdivision as a main modeling method. The 16th part of the MPEG4 standard—DAFX(Animation Framework extension)—also introduces the subdivision as a main modeling method. The first AFX version was released in the beginning of 2003. In 2006, a new AFX version was released.

Generally, every topic about surface modeling is almost referred in the researches of subdivision surfaces after 2000. Except for the above topics, there are also some other topics, such as mixed subdivision [16, 70], surface deformation [76], surface conversion [77, 78], surface offset [79–81].

1.4 Idea of This Book

So far, subdivision surface modeling techniques have formed a perfect system from theories to applications. Furthermore, applications of subdivision surfaces have reached great success in the animation field. Subdivision surfaces have become data exchange standards of 3D animations. Practices show that the subdivision is a powerful modeling tool and has the powerful potential. Consequently, it is a significant task to sum up existing researching results and generalize them. However, these achievements on subdivision surfaces are abundant and profound. It is impossible for this book to involve all these achievements. Consequently, this book aims at gathering achievements in aspect of applications. Some contents that the authors think are difficult, such as conditions for G^k continuity [19, 20, 32–34], variational subdivision [82, 83], Loudbery's wavelet transformation [64], hardware rendering [71, 74, 75], will not be introduced in this book. We hope that this book can lead beginners to the subdivision modeling field.

This book is not the first work that summarizes achievements of subdivision modeling. There have been some surveys and books on subdivision before this book. Literatures [8, 16, 86] are all-around surveys. The survey [8] summarizes achievements before 1998 based on the following topics: relations between subdivision and B-splines, analysis of convergence and continuity of subdivision surfaces. Discussions of the survey [8] are very detailed and can be considered as explanations for original literatures. The survey [86] summarizes achievements of subdivision surfaces before 2004. The survey puts emphasis on relations between subdivisions and refinements of B-splines. Some common issues on subdivision surface modeling are also addressed. Several key topics, such as subdivision scheme construction, subdivision surface property analysis, parametric evaluation, and subdivision surface fitting, are discussed. Some other important topics are also summarized for potential future research and development. Compared with the above surveys, The survey [16] contains the most comprehensive contents. He summarized achievements before 2003 and reviewed those new ideas and new methods after 1995. Except for reviews of new subdivision schemes and classification methods, he still referred to multivariate subdivision and face-valued subdivision and discusses non-uniform subdivision, non-stationary subdivision, mixed subdivision etc., as new ideas. For subdivision surface analysis, shape tuning and the parameter evaluation, he gives advantages and disadvantages of some main methods. For applications of subdivision surfaces, he discusses finite element, data compression, and reverse engineering. The survey [87] gathers these achievements of subdivision surfaces before 2006 in two parts: fundamental theories and application techniques. Compared with the above surveys, he still discusses hardware rendering techniques and applications in animations.

The literature [1] is a monograph on subdivision surfaces and discusses achievements before 2001. This book mainly discusses the fundamental theories of subdivisions. It points out that subdivision may be viewed as the synthesis of two previously distinct approaches to modeling shape: functions and fractals. Construction of subdivision schemes is a main content of the book. Analysis of convergence and continuity of subdivision surfaces are also discussed as a "hot" topic in that period. There are other some monographs [6, 88, 89] on surface modeling that introduce some knowledge of subdivision surfaces. However, only several classical subdivision schemes and basic principles of subdivision surface analysis are discussed in those books.

The outline of this book is designed according to the idea of [87] and consists of roughly two parts: fundamental theories and applied techniques. The first part includes relations between subdivision and splines, introductions of some main subdivision schemes, and theories of subdivision surface analysis. Different from other surveys and books, this book directly starts from the recursive definition of B-spline and hopes that give readers a concise cognition of spline theories. The method to construct subdivision schemes from box splines is a main content of this part. The second part includes n-side patches, optimization modeling interpolation, fitting, deformations, intersection and trimming, mesh editing. Those technologies are necessary to construct geometric models using subdivision surfaces.

Remarks

This chapter gives a review for the development of subdivision surfaces. This review focuses on the stationary subdivision scheme though there are also some literatures on the non-stationary subdivision [82, 83, 90]. Stationary subdivision schemes currently are the most frequently applied subdivision schemes. Discussions in this book will also focus on stationary subdivision schemes. As far as expressions of geometric shapes are concerned, there are three forms: univariate, bivariate, and trivariate which are, respectively, fits for curves, surfaces, and volumes. There are probably expressions of higher dimensions [16]. However, our discussions are also limited to bivariate subdivision schemes, i.e., subdivision schemes to construct surfaces. Surfaces are the most frequently applied geometric shapes. A majority of researches on subdivision modeling are researches of subdivision surfaces. Topics involved by researches of subdivision surfaces are very extensive. These topics discussed in this book are only several ones. However, we try to enumerate other topics and literatures so that the readers can know them. For meshes, there are manifold meshes and non-manifold meshes. This book also refers to two-manifold meshes because a single surface is a two-manifold in the differential geometric.

In this chapter, there are probably some concepts that are not known by beginners. We have given simple descriptions for some elementary concepts such as **mesh, subdivision scheme, triangle subdivision**. Strict definitions will be given in later chapters. It is a good idea for a beginner to directly read the second chapter if he or she does not have a clear understanding of those elementary concepts. After reading the second chapter, you may read the introduction again. We want to give readers a comprehensive cognition for subdivision surfaces by the introduction.

Note: In some literatures, C^k continuity of subdivision surfaces is discussed, while G^k continuity of subdivision surfaces is discussed in other literatures. In the view of derivations, C^k continuity and G^k continuity are different. That is to say, C^k continuity is a concept related to expressions of curves and surfaces, while G^k continuity is related to geometric variables. However, it is easy to know that a surface must be G^k continuous if it is C^k continuous. Consequently, for geometric shapes, we do not use the concept of G^k continuity except special requirements. For functions whose values are scalar, we do use the concept of C^k continuity.

Exercises

1. What is the polygon mesh? Why is a polygon mesh able to represent a complex shape?
2. What is the subdivision surface? Why is a single subdivision surface able to represent a complex shape but why a single NURBS surface hasn't the ability?
3. Why do most of the polygons of a mesh have the similar shape after the mesh is subdivided several times?

Chapter 2
Splines and Subdivision

Most subdivision schemes are derived from refinement methods of control meshes of spline curves and surfaces. This chapter mainly discusses various refinement methods for control meshes of spline curves and surfaces. Firstly, definitions and basic properties of spline functions are introduced; secondly, refinement rules of spline functions are deduced based on their definitions and basic properties; lastly, refinement rules of control meshes of spline curves and surfaces are deduced based on refinement rules of spline functions. In next chapter, we will generalize these refinement rules of control meshes to arbitrary 2-manifold meshes to construct subdivision surfaces.

2.1 B-Splines

There are many definitions for B-splines. The recursive definition is the most commonly used one in computational field. Its discovery is attributed to de-Boor, Cox, and Mansfield [3, 4].

Definition 2.1 Given a non-decreasing sequence of the real u-axis:

$$U = [\cdots, u_{-2}, u_{-1}, u_0, u_1, u_2, \cdots], \text{ where } u_i \leqslant u_{i+1}.$$

B-splines (or B-spline basic functions) can be defined as:

$$\begin{cases} N_{i,0}(u) = \begin{cases} 1, & if\ u_i \leqslant u < u_{i+1} \\ 0, & otherwise \end{cases} \\ N_{i,k}(u) = \dfrac{u - u_i}{u_{i+k} - u_i} N_{i,k-1}(u) + \dfrac{u_{i+k+1} - u}{u_{i+k+1} - u_{i+1}} N_{i+1,k-1}(u) \\ assume\ 0/0 = 0. \end{cases} \tag{2.1}$$

W. Liao et al., *Subdivision Surface Modeling Technology*,
DOI 10.1007/978-981-10-3515-9_2

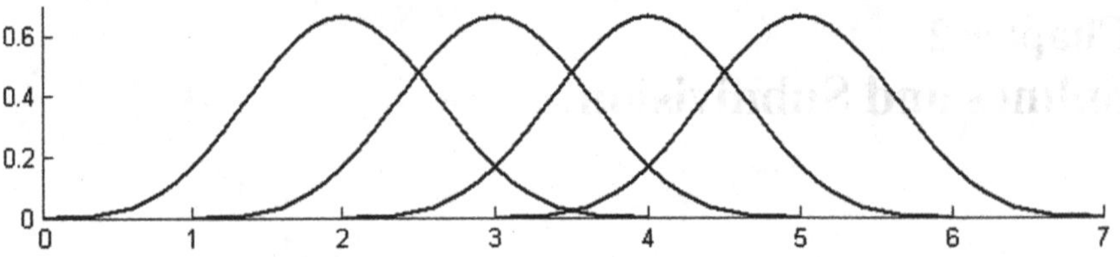

Fig. 2.1 Uniform cubic B-splines

In the above definition, the u_i are called knots and $\boldsymbol{U}$ is called the knot vector. The half-open interval, $[u_i, u_{i+1})$, is called the i-th knot span. $N_{i,k}(u)$ is the i-th B-spline basic function of k degree (order $k+1$). From the recursive Formula (2.1), we can find:

• To define a set of B-spline basic functions, we should firstly define a knot vector $\boldsymbol{U}$, and the degree k.

• To compute $N_{i,k}(u), k+2$ knots, $u_i, u_{i+1}, \cdots, u_{i+k+1}$, have to be used. $[u_i, u_{i+k+1}]$ is called the support span of $N_{i,k}(u)$. The first subscript of $N_{i,k}(u)$ is equal to the subscript of the left end knot of the support interval. The subscript of $N_{i,k}(u)$ indicates its position on the u-axis, as shown in Fig. 2.1.

• The knot vector $\boldsymbol{U} = [u_0, u_1, \cdots, u_{n-k-1}, \ldots, u_n]$ can define $n-k$ B-spline basic functions of k-degree: $N_{i,k}(u)$. The last one is $N_{n-k-1,k}(u)$.

• The knot span $[u_i, u_{i+1}]$ can have zero length, since the knots do not need to be distinct.

• The recursive Formula (2.1) can yield the quotient 0/0, and we define this quotient as zero. If knots of $\boldsymbol{U}$ are uniformly distributed on the u-axis, i.e.,

$$u_{i+1} - u_i = constant, \quad i \in Z.$$

We call $N_{i,k}(u)$ a uniform B-spline. Figure 2.1 gives the shapes of uniform cubic B-spline basic functions. From Formula (2.1), we can find that $\boldsymbol{U}$ and $t\boldsymbol{U}$ define the basic functions with same shapes.

We now investigate how to obtain the value or expression of $N_{i,k}(u)$ from Formula (2.1). Obviously,

$$N_{i,1}(u) = \frac{u - u_i}{u_{i+1} - u_i} N_{i,0}(u) + \frac{u_{i+2} - u}{u_{i+2} - u_{i+1}} N_{i+1,0}(u).$$

Based on $N_{i,0}(u) = \begin{cases} 1, & if\ u_i \leqslant u < u_{i+1} \\ 0, & otherwise \end{cases}$, we can get

$$N_{i,1} = \begin{cases} (u - u_i)/(u_{i+1} - u_i), & if\ u_i \leqslant u < u_{i+1} \\ (u_{i+2} - u)/(u_{i+2} - u_{i+1}), & if\ u_{i+1} \leqslant u < u_{i+2} \\ 0, & otherwise. \end{cases}$$

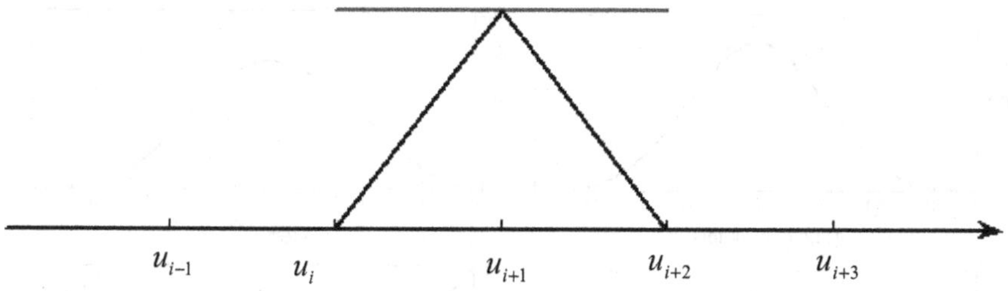

Fig. 2.2 Recursion from zero-degree B-spline to one-degree B-spline

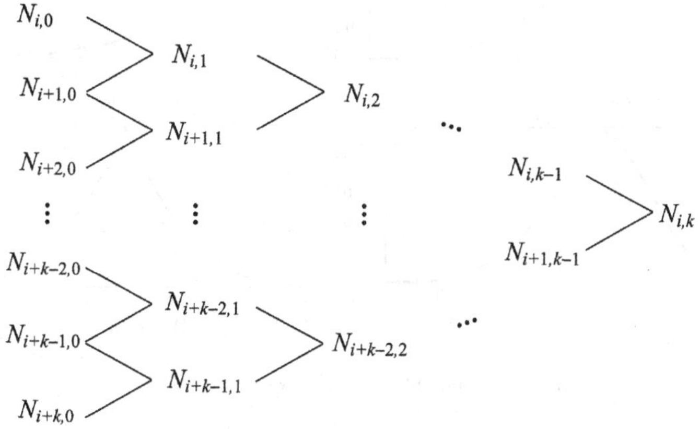

Fig. 2.3 Recursion of $N_{i,k}(u)$

Figure 2.2 shows the recursive process.

To obtain $N_{i,k}(u)$, we may apply the recursive method shown in Fig. 2.3.

From the recursive process of $N_{i,k}(u)$, we can find $N_{i,k}(u)$ is a piecewise polynomial function. It is defined on the whole real span, and

$$N_{i,k}(u) = \begin{cases} > 0 & u \in [u_i, u_{i+k+1}) \\ = 0 & u \notin [u_i, u_{i+k+1}) \end{cases} \quad (k \geqslant 0).$$

In the recursive process, when there are multiknots in the knot vector $\boldsymbol{U}$, i.e.,

$$u_i = \cdots = u_{i+r-1} \quad (r > 0), \tag{2.2}$$

quotient 0/0 may appear. We assume the quotient is zero. Multiknots affect the continuity degree of $N_{i,k}(u)$. For a knot u_i, if Eq. 2.2 exists, we call the knot u_i have the multiplicity r. The continuity degree of $N_{i,k}(u)$ is $k - r$ at the knot u_i. When u_0 has different multiplicities, the shapes of $N_{0,3}(u)$ are shown in Fig. 2.4.

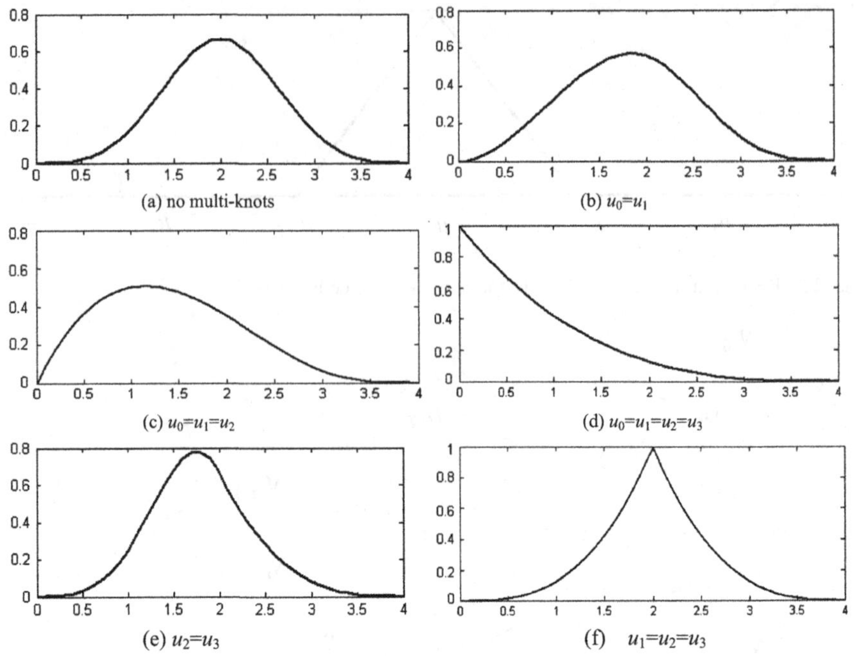

Fig. 2.4 Multiplicity of u_0 effects continuity degree of $N_{0,3}(u)$

From the above discussion, we know that B-splines have the following properties:

(1) Recursion: $N_{i,k}(u)=\dfrac{u-u_i}{u_{i+k}-u_i}N_{i,k-1}(u)+\dfrac{u_{i+k+1}-u}{u_{i+k+1}-u_{i+1}}N_{i+1,k-1}(u)$, $(k>0)$

(2) Positivity: $N_{i,k}(u)\geqslant 0$

(3) Local support: $N_{i,k}(u)=0,\ u\in(-\infty,u_i)\cup[u_{i+k+1},+\infty)$

(4) Partition of unity: $\sum_i N_{i,k}(u)=1$

(5) Continuity: $N_{i,k}(u)$ is $(k-r)$ times continuously differentiable, where r is the maximum multiplicity of its knots in the support span.

Let us investigate the property of partition of unity. Assume $u\in[u_i,u_{i+1})$, according to the local support property,

$$N_{j,k}(k)=0 \text{ where } j=\ldots,i-k-2,i-k-1;\ j=i+1,i+2,\ldots$$

Consequently,

$$\sum_j N_{j,k}(u)=\sum_{j=i-k}^{i} N_{j,k}(u)=1. \tag{2.3}$$

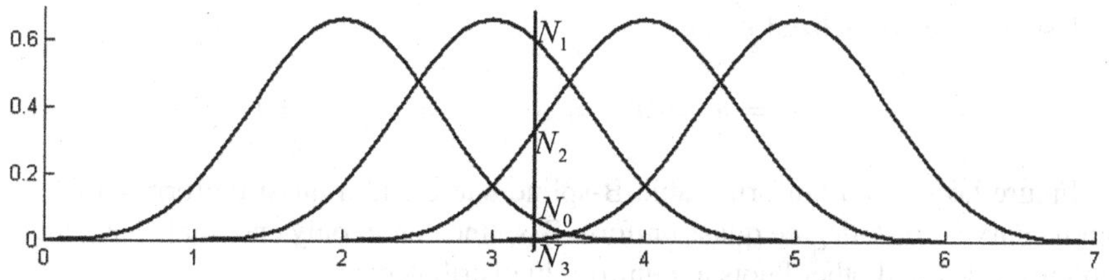

Fig. 2.5 Partition of unity

When $U = [u_0, u_1, \ldots, u_{n-k-1}, \ldots, u_n]$, there must be $k + 1$ nonzero curve segments in the interval $[u_i, u_{i+1})$. Only in this case can we have Eq. (2.3). The property will be explained in Fig. 2.5.

Except for the recursive definition, the convolution definition and the truncated power definition are always used in theory fields. Starting from the convolution definition, the subdivision rules of uniform B-splines are easy to be obtained. Uniform B-splines can be considered as simple examples of box splines when the convolution definition is used, which will be found in Sects. 2.3 and 2.4.

2.2 B-Spline Curves and Surfaces

2.2.1 B-Spline Curves and Their Properties

A B-spline curve can be defined as:

$$\boldsymbol{p}(u) = \sum_{i=0}^{n} \boldsymbol{P}_i N_{i,k}(u), \quad u \in [u_k, u_{n+1}], \tag{2.4}$$

where the knot vector $\boldsymbol{U} = [u_0, u_1, \ldots, u_{n+k+1}]$. $\boldsymbol{P}_i (i = 0, 1, \ldots, n)$ are called control points of the B-spline curve $\boldsymbol{p}(u)$, and the polygon $\boldsymbol{P}_0 \boldsymbol{P}_1 \ldots \boldsymbol{P}_n$ is called control polygon of the B-spline curve $\boldsymbol{p}(u)$. To make ends of the curve $\boldsymbol{p}(u_k)$ and $\boldsymbol{p}(u_{n+1})$ identical to ends of the control polygon, we need that:

$$u_0 = \cdots = u_k, \quad u_{n+1} = \cdots = u_{n+k+1}.$$

In this case,

$$\boldsymbol{p}(u_k) = \boldsymbol{P}_0, \ \boldsymbol{p}'(u_k) = k \frac{\boldsymbol{P}_1 - \boldsymbol{P}_0}{u_{k+1} - u_1},$$
$$\boldsymbol{p}(u_{n+1}) = \boldsymbol{P}_n, \ \boldsymbol{p}'(u_k) = k \frac{\boldsymbol{P}_n - \boldsymbol{P}_{n-1}}{u_{n+k} - u_n}.$$

Usually, we normalized $\boldsymbol{U}$, i.e.,

$$u_0 = \cdots = u_k = 0, \quad u_{n+1} = \cdots = u_{n+k+1} = 1.$$

Figure 2.6 gives a uniform cubic B-spline curve and a quasi-uniform cubic B-spline curve. For k-degree quasi-uniform B-spline curve, only end knots have multiplicity $k+1$ and other knots are uniformly distributed.

$\boldsymbol{p}(u)$ is a segment polynomial curve since $N_{i,k}(u)$ is a segment polynomial. According to the local support property of $N_{i,k}(u)$, the curve segment

$$\boldsymbol{p}_i(u) = \sum_{j=i-k}^{i} \boldsymbol{P}_j N_{j,k}(u), \quad u \in [u_i, u_{i+1}) \tag{2.5}$$

is a polynomial curve segment and $\boldsymbol{p}_{i+1}(u)$ is another, as shown in Fig. 2.6.

Based on the continuity property of $N_{i,k}(u)$, $\boldsymbol{p}(u)$ is C^{k-1} continuous in a knot interval $[u_i, u_{i+1})$. If the multiplicity of knot u_i is r, $\boldsymbol{p}(u)$ is C^{k-r} at the knot u_i.

Based on the local support property of $N_{i,k}(u)$, if the position of the control point $\boldsymbol{P}_i$ is altered, only $k+1$ curve segments, $\boldsymbol{p}_i(u)$, $\boldsymbol{p}_{i+1}(u), \ldots, \boldsymbol{p}_{i+k}(u)$, will change their shapes, as shown in Fig. 2.7.

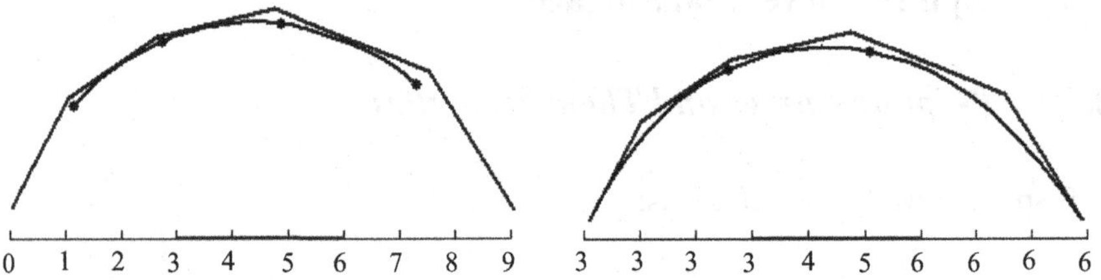

(a) Uniform knots (b) End knots with multiplicity 4

Fig. 2.6 Cubic B-spline curve

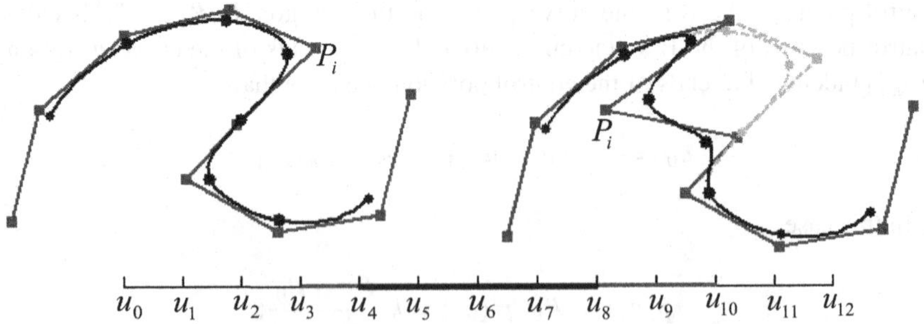

(a) Curve before pulling $\boldsymbol{P}_i$ (b) Curve after pulling $\boldsymbol{P}_i$

Fig. 2.7 Change of shape of cubic curve after control point $\boldsymbol{P}_i$ is pulled

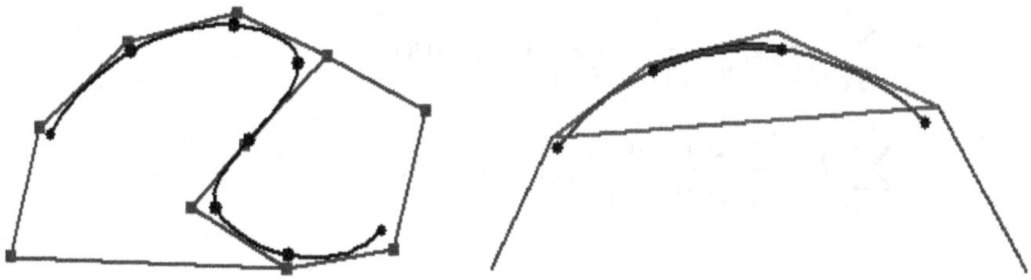

(a) $p(u)$ in convex hull of $\{P_i|i=0,\ldots,n\}$ (b) $p_i(u)$ in convex hull of $\{P_{i-k},\ldots,P_i\}$

Fig. 2.8 Cubic curve in convex hull formed by its control point set

According to the positivity property and the local support property of $N_{i,k}(u)$, we obtain the convex hull property of B-spline curves, namely, $\boldsymbol{p}(u)$ is in the convex hull formed by $\{\boldsymbol{P}_i\}$, and $\boldsymbol{P}_i(u)$ is in the convex hull formed by $\{\boldsymbol{P}_{i-k},\ldots,\boldsymbol{P}_i\}$, as shown in Fig. 2.8.

According to the recursive property of B-spline basic functions, we can obtain the de-Boor algorithm which calculates points on B-spline curves. For the curve (2.4), let $u \in [u_i, u_{i+1}] \subset [u_k, u_{n+1}]$, then,

$$\boldsymbol{p}(u) = \sum_{j=i-k}^{i} \boldsymbol{P}_j N_{j,k}(u) = \cdots = \sum_{j=i-k}^{i-s} \boldsymbol{P}_j^s N_{j+s,k-s}(u) = \cdots = \boldsymbol{P}_{j-k}^k, \tag{2.6}$$

where

$$\boldsymbol{P}_j^s = \begin{cases} \boldsymbol{P}_j, & s = 0 \\ (1-\alpha_j^s)\boldsymbol{P}_j^{s-1} + \alpha_j^s \boldsymbol{P}_{j+1}^{s-1}, & s > 0 \end{cases} \tag{2.7}$$

$$j = i-k,\ldots,i-s;\ s = 1,2,\ldots,k$$

$$\alpha_j^s = \frac{u-u_{j+s}}{u_{j+k+1}-u_{j+s}}$$

Assume $0/0 = 0$.

In fact,

$$\boldsymbol{p}(u) = \sum_{j=i-k}^{i} \boldsymbol{P}_j N_{j,k}(u) = \sum_{j=i-k}^{i} \boldsymbol{P}_j$$

$$\left[\frac{u-u_j}{u_{j+k}-u_j} N_{j,k-1}(u) + \frac{u_{j+k+1}-u}{u_{j+k+1}-u_{j+1}} N_{j+1,k-1}(u)\right]$$

$$= \sum_{j=i-k}^{i} \boldsymbol{P}_j \frac{u-u_j}{u_{j+k}-u_j} N_{j,k-1}(u) + \sum_{j=i-k}^{i} \boldsymbol{P}_j \frac{u_{j+k+1}-u}{u_{j+k+1}-u_{j+1}} N_{j+1,k-1}(u)$$

$$= \sum_{j=i-k-1}^{i-1} \boldsymbol{P}_{j+1} \frac{u - u_{j+1}}{u_{j+k+1} - u_{j+1}} N_{j+1,k-1}(u) +$$
$$\sum_{j=i-k}^{i} \boldsymbol{P}_j \frac{u_{j+k+1} - u}{u_{j+k+1} - u_{j+1}} N_{j+1,k-1}(u)(*).$$

Note that $u \in [u_i, u_{i+1}]$, and $N_{i-k,k-1}(u) = 0$, $N_{i+1,k-1}(u) = 0$. Consequently,

$$(*) = \sum_{j=i-k}^{i-1} \frac{u - u_{j+1}}{u_{j+k+1} - u_{j+1}} \boldsymbol{P}_{j+1} N_{j+1,k-1}(u) +$$
$$\sum_{j=i-k}^{i-1} \frac{u_{j+k+1} - u}{u_{j+k+1} - u_{j+1}} \boldsymbol{P}_j N_{j+1,k-1}(u)$$
$$= \sum_{j=i-k}^{i-1} \left(\frac{u_{j+k+1} - u}{u_{j+k+1} - u_{j+1}} \boldsymbol{P}_j + \frac{u - u_{j+1}}{u_{j+k+1} - u_{j+1}} \boldsymbol{P}_{j+1} \right) N_{j+1,k-1}(u)$$
$$= \sum_{j=i-k}^{i-1} \boldsymbol{P}_j^1 N_{j+1,k-1}(u).$$

The recursive process expressed in Formulas (2.6) and (2.7) can be depicted by Fig. 2.9. Figure 2.10 gives the process of computing a point $\boldsymbol{p}(u)$ on a cubic B-spline curve.

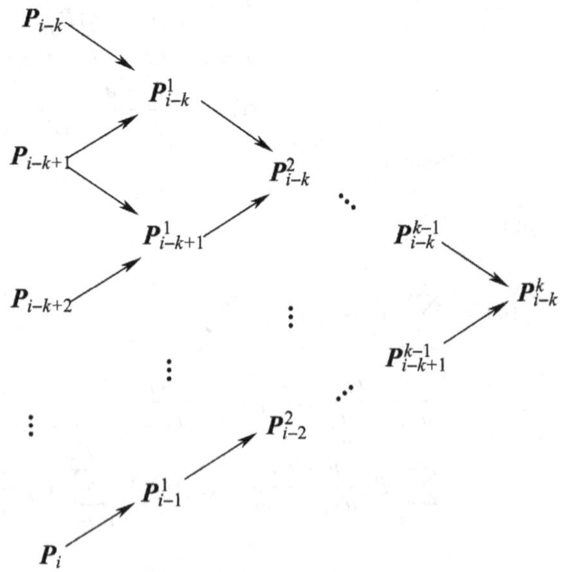

Fig. 2.9 Recursive process of de-Boor algorithm

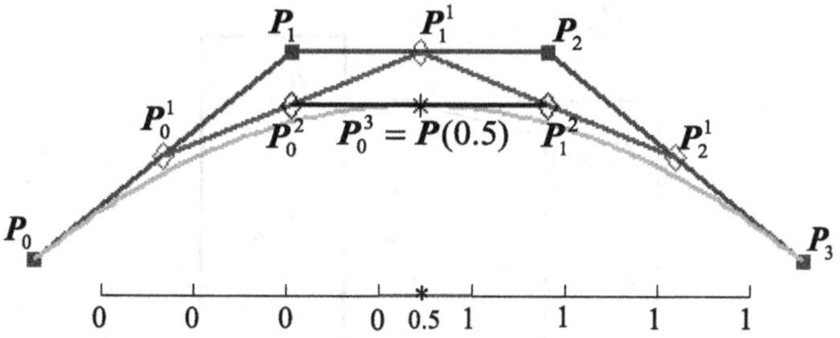

Fig. 2.10 Calculating a point on cubic B-spline curve with de-Boor algorithm

The knot insertion algorithm is an important technique of B-spline curves and surfaces. Its discovery is attributed to Boehm [3, 4]. For the curve expressed by Eq. (2.5), we insert a knot in its definition interval:

$$u \in [u_i, u_{i+1}] \subset [u_k, u_{n+1}].$$

So a new knot vector can be obtained:

$$\overline{\boldsymbol{U}} = [u_0, u_1, \ldots, u_i, u, u_{i+1}, \ldots, u_{n+k+1}]$$

Readjusting numbers of knots in the knot vector, we have

$$\overline{\boldsymbol{U}} = [\overline{u}_0, \overline{u}_1, \ldots, \overline{u}_i, \overline{u}_{i+1}, \overline{u}_{i+2}, \ldots, \overline{u}_{n+k+2}].$$

The new vector $\overline{\boldsymbol{U}}$ defines a set of new B-spline basic functions: $\overline{N}_{j,k}(u), j = 0, 1, \ldots, n+1$. The B-spline curve can be re-expressed by the set of new B-spline basic functions and new control points $\overline{\boldsymbol{P}}_j, j = 0, 1, \ldots, n+1$:

$$\boldsymbol{p}(u) = \sum_{j=0}^{n+1} \overline{\boldsymbol{P}}_j \overline{N}_{j,k}(u), \quad u \in [\overline{u}_k, \overline{u}_{n+2}].$$

$\overline{\boldsymbol{P}}_j (j = 0, 1, \ldots, n+1)$ can be calculated by the following formulas:

$$\overline{\boldsymbol{P}}_j = \boldsymbol{P}_j, j = 0, 1, \ldots, i-k$$
$$\overline{\boldsymbol{P}}_{j+1} = \boldsymbol{P}_j^1 = (1-\alpha_j^1)\boldsymbol{P}_j + \alpha_j^1 \boldsymbol{P}_{j+1}, j = i-k, i-k+1, \ldots, i-r-1$$
$$\alpha_j^1 = \frac{u - u_{j+1}}{u_{j+k+1} - u_{j+1}}, \quad \text{let } 0/0 = 0$$
$$\overline{\boldsymbol{P}}_j = \boldsymbol{P}_j, j = i-r+1, \ldots, n+1.$$

In above formulas, r is the multiplicity of u in the original knot vector $\boldsymbol{U}$. If $u_i < u < u_{i+1}, r = 0$. When $0 < r < k, u = u_i = u_{i-1} = \cdots = u_{i-r+1}$. The nature of the algorithm is the first recursion of the de-Boor algorithm when the de-Boor

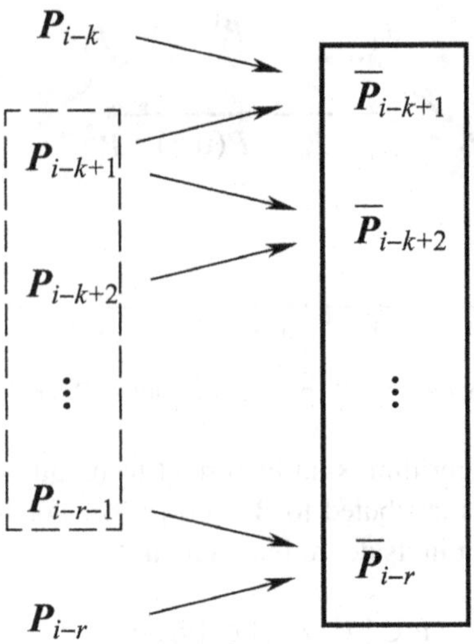

Fig. 2.11 Vertex replacement in knot insertion

Fig. 2.12 Example of knot insertion

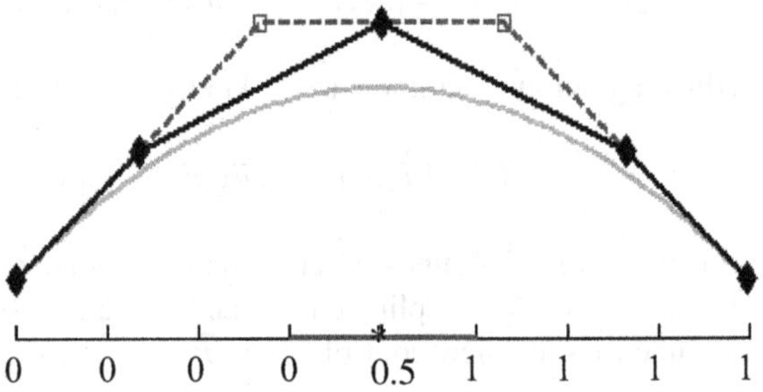

algorithm is applied to calculate a point $\boldsymbol{p}(u)$ of a B-spline curve. Figure 2.11 gives the process of inserting a knot. After those new vertices $\overline{\boldsymbol{P}}_{i-k+1}, \ldots, \overline{\boldsymbol{P}}_{i-r}$ replace old vertices $\boldsymbol{P}_{i-k+1}, \ldots, \boldsymbol{P}_{i-r-1}$, a knot insertion is completed. Figure 2.12 gives an example of inserting a knot to a cubic B-spline curve.

2.2.2 B-Spline Surfaces and Their Properties

A B-spline surface can be represented as:

$$\boldsymbol{p}(u, v) = \sum_{i=0}^{m} \sum_{j=0}^{n} \boldsymbol{P}_{i,j} N_{i,k}(u) N_{j,l}(v),\ (u, v) \in [u_k, u_{m+1}] \times [v_l, v_{n+1}], \tag{2.8}$$

where $\boldsymbol{U} = [u_0, u_1, \ldots, u_{m+k+1}]$, $\boldsymbol{V} = [v_0, v_1, \ldots, v_{n+l+1}]$, k and l are knot vectors and degrees of basic functions on the two parameter directions u, v, respectively.

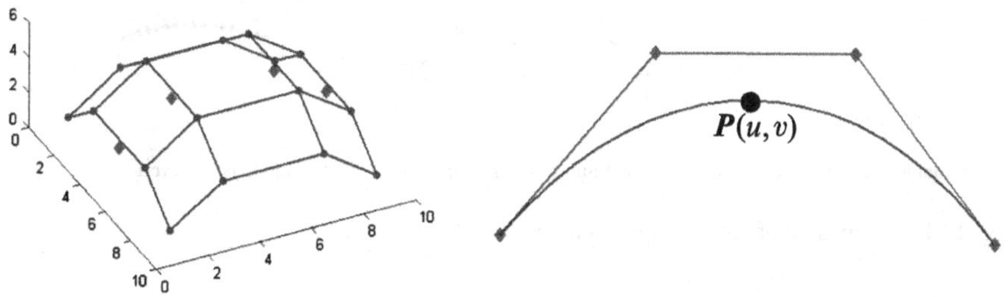

(a) De-Boor algorithm on u parameter (b) De-Boor algorithm on v parameter

Fig. 2.13 Apply de-Boor algorithm on bi-cubic B-spline surface

B-spline surfaces are tensor surfaces, i.e., a B-spline surface can be constructed by applying operators φ, ψ on a point array formed by control points. The two operators are equally applied on u, v, the two parameter directions. Consequently,

$$\boldsymbol{p}(u,v) = (\psi \cdot \varphi)\boldsymbol{P} = [N_{0,k}, \cdots, N_{m,k}] \begin{bmatrix} \boldsymbol{P}_{0,0} & \cdots & \boldsymbol{P}_{0,n} \\ \vdots & \cdots & \vdots \\ \boldsymbol{P}_{m,0} & \cdots & \boldsymbol{P}_{m,n} \end{bmatrix} \begin{bmatrix} N_{0,l} \\ \vdots \\ N_{n,l} \end{bmatrix}.$$

A main advantage of tensor surfaces is that many problems about tensor surfaces can be converted to problems about curves, which give great convenience to theory analysis and programming computation. With the property of tensor production, we can generalize the de-Boor algorithm and the knot insertion algorithm from B-spline curves to B-spline surfaces.

Let $(u, v) \in [u_i, u_{i+1}] \times [v_j, v_{j+1}]$ and pick a parameter direction, such as u. Now, we apply the de-Boor algorithm on every column control points. In fact, we easily find that we can only apply curve de-Boor algorithm on $l+1$ columns: $j-l, \ldots, j$. So, there will be $l+1$ points. On the v parameter direction, after applying the de-Boor algorithm to the $l+1$ points, we get the point $\boldsymbol{p}(u, v)$ we need. The calculating process is shown in Fig. 2.13.

For knot insertion, if we insert a knot u for the $\boldsymbol{U}$ knot vector, we need to apply the curve knot insertion algorithm to every column control points. So a new point array is formed, and it is the vertex array needed by us after inserting the knot u. The calculation is similar when we insert a knot v for the $\boldsymbol{V}$ knot vector.

2.3 Knot Insertion Algorithm and Refinement of B-Spline Curves and Surfaces

We have pointed out that refinement is subdivision in Chap. 1. Particularly, the subdivision of control meshes of spline surfaces is called refinement. In this chapter, we furthermore point out that the process of doubling knot intervals of B-spline basis

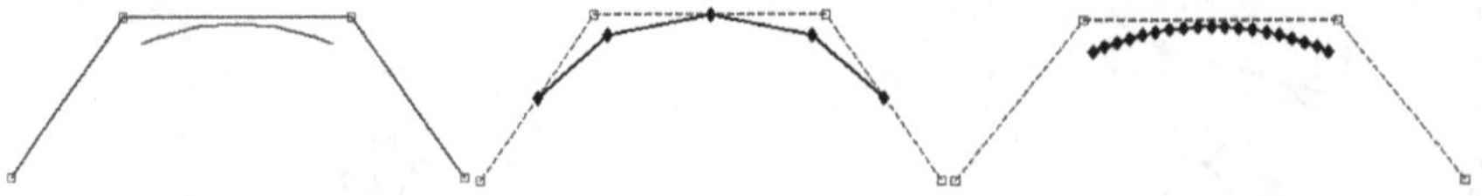

(a) Initial Mesh and curve (b) Subdivision once (c) Subdivision several times

Fig. 2.14 Subdivision of cubic B-spline curve

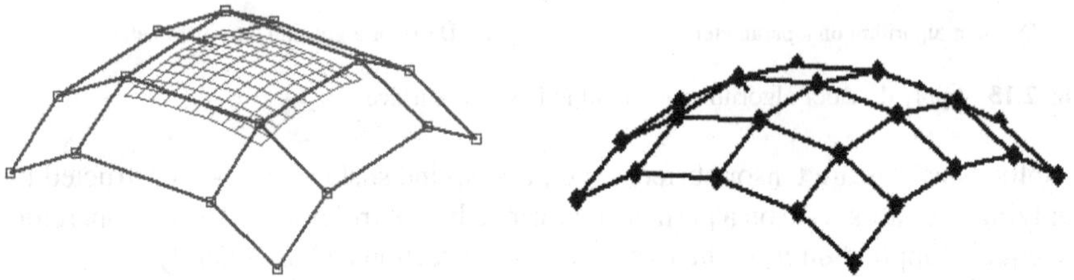

(a) Initial Mesh and surface (b) Subdivision once

Fig. 2.15 Subdivision of cubic B-spline surface

functions and then obtaining new spline basis functions or new control polygons or control meshes is called refinement. After a B-spline curve or surface is refined, the new control polygon or mesh defines the same curve or surface as the one defined by the former control polygon or mesh. When we continuously refine these polygons or meshes, we can obtain a sequence of polygons or meshes that converge to the B-spline curve or surface defined by it. Figures 2.14 and 2.15 give the process of refinement of a cubic B-spline curve.

Sederberg et al. [25] give non-uniform subdivision schemes. When we apply these subdivision schemes on topology rectangle meshes, we can obtain B-spline surfaces. These subdivision schemes can be derived from knot-doubling algorithms of B-spline curves and surfaces.

2.3.1 Subdivision of B-Spline Curves

Assume that there is no multiplicity knot in knot vector $\boldsymbol{U} = [\ldots, u_{-1}, u_0, u_1, \ldots]$. Now, we insert a middle point for every knot interval and obtain a new knot vector: $\overline{\boldsymbol{U}} = [\ldots, \overline{u}_{-2}, \overline{u}_{-1}, \overline{u}_0, \overline{u}_1, \overline{u}_2, \ldots]$. Consider quadric B-spline curve:

$$p(u) = \sum_{i=-\infty}^{+\infty} \boldsymbol{P}_i N_{i,2}(u).$$

Its shape is shown in Fig. 2.16a.

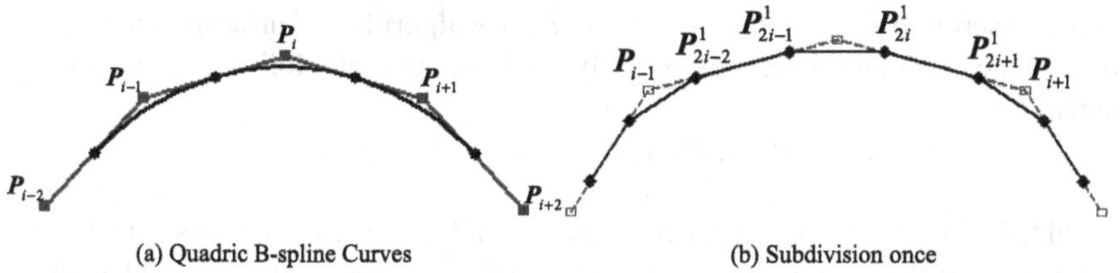

Fig. 2.16 Quadric B-spline curves and their subdivision

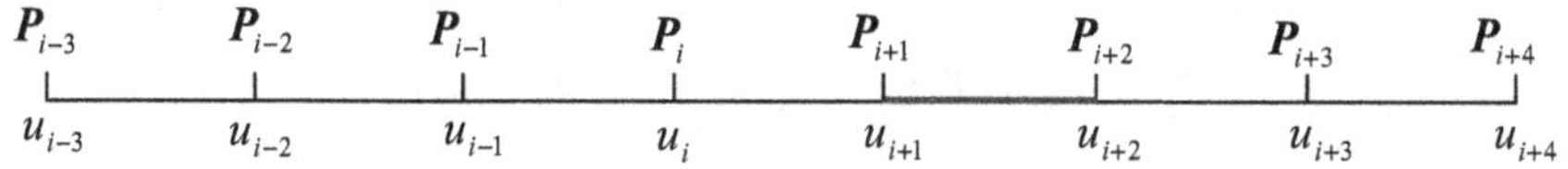

Fig. 2.17 Correspondence between knots and control vertices

From Fig. 2.16a, we know that every control vertex $\boldsymbol{P}_i$ corresponds to a curve segment. Set the length of parameter interval of the curve segment as d_i, and parameter of every vertex can be given as $\boldsymbol{P}_i \to d_i$.

After doubling knots, we can get:

$$\begin{aligned} \boldsymbol{P}^1_{2i} &= \frac{(d_i + 2d_{i+1})\boldsymbol{P}_i + d_i\boldsymbol{P}_{i+1}}{2(d_i + d_{i+1})} \\ \boldsymbol{P}^1_{2i+1} &= \frac{d_{i+1}\boldsymbol{P}_i + (2d_i + d_{i+1})\boldsymbol{P}_{i+1}}{2(d_i + d_{i+1})}. \end{aligned} \tag{2.9}$$

As shown in Fig. 2.16b, in order to deduce Formula (2.9), let knots and control vertices correspond as shown in Fig. 2.17:

In this case,

$$d_i = u_{i+2} - u_{i+1}. \tag{2.10}$$

Now consider that how $\boldsymbol{P}_{i-1}$, $\boldsymbol{P}_i$, $\boldsymbol{P}_{i+1}$ are replaced in the process of subdivision. It is easy to know that $\boldsymbol{P}_{i-1}$, $\boldsymbol{P}_i$, $\boldsymbol{P}_{i+1}$ correspond to the curve segment:

$$\boldsymbol{p}_i(u) = \sum_{j=i-1}^{i+1} \boldsymbol{P}_j N_{j,2}(u), \quad u \in [u_{i+1}, u_{i+2}],$$

where $\boldsymbol{U}_i = [u_{i-1}, u_i, u_{i+1}, u_{i+2}, u_{i+3}, u_{i+4}]$. Since only considering curve segment $\boldsymbol{p}_i(u)$, we can know that new control vertices are only linear combinations of $\boldsymbol{P}_{i-1}$, $\boldsymbol{P}_i$, $\boldsymbol{P}_{i+1}$. Consequently, we insert knots for all knot intervals in $\boldsymbol{U}_i$ by following steps:

Step 1: Insert $u = (u_{i-1} + u_i)/2 \in [u_{i-1}, u_i]$. By algorithm of inserting knots (2.7), to calculate new control vertices, we only need to control the following vertices and knots:

$$\boldsymbol{P}_{i-3}, \boldsymbol{P}_{i-2}, \boldsymbol{P}_{i-1}; \quad u_{i-3}, u_{i-2}, u_{i-1}, u_i, u_{i+1}.$$

So, added vertices are not a linear combination of $\boldsymbol{P}_{i-1}, \boldsymbol{P}_i, \boldsymbol{P}_{i+1}$. Consequently, we may not insert the knot $u = (u_{i-1} + u_i)/2 \in [u_{i-1}, u_i]$ and may not calculate these new added vertices, either.

Step 2: Insert $u = (u_i + u_{i+1})/2 \in [u_i \; u_{i+1}]$. To calculate new control vertices, we only use the following control vertices and knots:

$$\boldsymbol{P}_{i-2}, \boldsymbol{P}_{i-1}, \boldsymbol{P}_i; \quad u_{i-2}, u_{i-1}, u_i, u_{i+1}, u_{i+2}$$

New added vertices are:

$$\begin{aligned}
\boldsymbol{Q}_{i-2}^1 &= \frac{u_{i+1} - u_i}{2(u_{i+1} - u_{i-1})}\boldsymbol{P}_{i-2} + \frac{u_{i+1} + u_i - 2u_{i-1}}{2(u_{i+1} - u_{i-1})}\boldsymbol{P}_{i-1} \\
\boldsymbol{Q}_{i-1}^1 &= \frac{2u_{i+2} - u_{i+1} - u_i}{2(u_{i+2} - u_i)}\boldsymbol{P}_{i-1} + \frac{u_{i+1} - u_i}{2(u_{i+2} - u_i)}\boldsymbol{P}_i.
\end{aligned}$$

Correspondence between vertices and knots is

$\boldsymbol{P}_{i-2}$	$\boldsymbol{Q}_{i-2}^1$	$\boldsymbol{Q}_{i-1}^1$	$\boldsymbol{P}_i$	$\boldsymbol{P}_{i+1}$	$\boldsymbol{P}_{i+2}$	$\boldsymbol{P}_{i+3}$	$\boldsymbol{P}_{i+4}$
u_{i-2}	u_{i-1}	u_i	$(u_i + u_{i+1})/2$	u_{i+1}	u_{i+2}	u_{i+3}	u_{i+4}

Step 3: Insert $u = (u_{i+1} + u_{i+2})/2 \in [u_{i+1}, u_{i+2}]$. To calculate new control vertices, we only use the following control vertices and knots:

$$\boldsymbol{Q}_{i-1}^1, \boldsymbol{P}_i, \boldsymbol{P}_{i+1}; \quad u_i, (u_i + u_{i+1})/2, u_{i+1}, u_{i+2}, u_{i+3}$$

$$\begin{aligned}
\boldsymbol{Q}_{i-1}^2 &= \frac{u_{i+2} - u_{i+1}}{2u_{i+2} - u_{i+1} - u_i}\boldsymbol{Q}_{i-1}^1 + \frac{u_{i+2} - u_i}{2u_{i+2} - u_{i+1} - u_i}\boldsymbol{P}_i \\
&= \frac{u_{i+2} - u_{i+1}}{2(u_{i+2} - u_i)}\boldsymbol{P}_{i-1} + \frac{u_{i+2} + u_{i+1} - 2u_i}{2(u_{i+2} - u_i)}\boldsymbol{P}_i \\
\boldsymbol{Q}_i^2 &= \frac{2u_{i+3} - u_{i+2} - u_{i+1}}{2(u_{i+3} - u_{i+1})}\boldsymbol{P}_i + \frac{u_{i+2} - u_{i+1}}{2(u_{i+3} - u_{i+1})}\boldsymbol{P}_{i+1}.
\end{aligned}$$

Correspondence between vertices and knots is

$\boldsymbol{Q}_{i-2}^1$	$\boldsymbol{Q}_{i-1}^1$	$\boldsymbol{Q}_{i-1}^2$	$\boldsymbol{Q}_i^2$	$\boldsymbol{P}_{i+1}$	$\boldsymbol{P}_{i+2}$	$\boldsymbol{P}_{i+3}$	$\boldsymbol{P}_{i+4}$
u_{i-1}	u_i	$(u_i + u_{i+1})/2$	u_{i+1}	$(u_{i+1} + u_{i+2})/2$	u_{i+2}	u_{i+3}	u_{i+4}

Step 4: Insert $u = (u_{i+2} + u_{i+3})/2 \in [u_{i+2}, u_{i+3}]$. To calculate new control vertices, we only use control vertices and knots:

$$\boldsymbol{Q}_i^2, \boldsymbol{P}_{i+1}, \boldsymbol{P}_{i+2}; u_{i+1}, (u_{i+1}+u_{i+2})/2, u_{i+2}, u_{i+3}, u_{i+4}$$

$$\begin{aligned}\boldsymbol{Q}_i^3 &= \frac{u_{i+3}-u_{i+2}}{2u_{i+3}-u_{i+2}-u_{i+1}}\boldsymbol{Q}_i^2 + \frac{u_{i+3}-u_{i+1}}{2u_{i+3}-u_{i+2}-u_{i+1}}\boldsymbol{P}_{i+1} \\ &= \frac{u_{i+3}-u_{i+2}}{2(u_{i+3}-u_{i+1})}\boldsymbol{P}_i + \frac{u_{i+3}+u_{i+2}-2u_{i+1}}{2(u_{i+3}-u_{i+1})}\boldsymbol{P}_{i+1} \\ \boldsymbol{Q}_{i+1}^3 &= \frac{2u_{i+4}-u_{i+3}-u_{i+2}}{2(u_{i+4}-u_{i+2})}\boldsymbol{P}_{i+1} + \frac{u_{i+3}-u_{i+2}}{2(u_{i+4}-u_{i+2})}\boldsymbol{P}_{i+2}.\end{aligned}$$

Correspondence between vertices and knots is

$\boldsymbol{Q}_{i-2}^1$	$\boldsymbol{Q}_{i-1}^1$	$\boldsymbol{Q}_{i-1}^2$	$\boldsymbol{Q}_i^2$	$\boldsymbol{Q}_i^3$	$\boldsymbol{Q}_{i+1}^3$	$\boldsymbol{P}_{i+2}$	$\boldsymbol{P}_{i+3}$
u_{i-1}	u_i	$(u_i+u_{i+1})/2$	u_{i+1}	$(u_{i+1}+u_{i+2})/2$	u_{i+2}	$(u_{i+2}+u_{i+3})/2$	u_{i+3}

Now, we can find that $\boldsymbol{P}_{i-1}, \boldsymbol{P}_i, \boldsymbol{P}_{i+1}$ have already been replaced by $\boldsymbol{Q}_{i-1}^1$, $\boldsymbol{Q}_{i-1}^2$, $\boldsymbol{Q}_i^2$, $\boldsymbol{Q}_i^3$. Based on Formula (2.10) and Fig. 2.15b, we can get

$$\begin{aligned}\boldsymbol{P}_{2i-2}^1 &= \boldsymbol{Q}_{i-1}^1 = \frac{(d_{i-1}+2d_i)\boldsymbol{P}_{i-1}+d_{i-1}\boldsymbol{P}_i}{2(d_{i-1}+d_i)}, \\ \boldsymbol{P}_{2i-1}^1 &= \boldsymbol{Q}_{i-1}^2 = \frac{d_i\boldsymbol{P}_{i-1}+(2d_{i-1}+d_i)\boldsymbol{P}_i}{2(d_{i-1}+d_i)} \\ \boldsymbol{P}_{2i}^1 &= \boldsymbol{Q}_i^2 = \frac{(d_i+2d_{i+1})\boldsymbol{P}_i+d_i\boldsymbol{P}_{i+1}}{2(d_i+d_{i+1})}, \\ \boldsymbol{P}_{2i+1}^1 &= \boldsymbol{Q}_i^3 = \frac{d_{i+1}\boldsymbol{P}_i+(2d_i+d_{i+1})\boldsymbol{P}_{i+1}}{2(d_i+d_{i+1})}.\end{aligned}$$

Consequently, we get Formula (2.9). These subdivided new vertices have the following parameters:

$$\boldsymbol{P}_{2i-1}^1 \to d_i/2, \quad \boldsymbol{P}_{2i}^1 \to d_i/2.$$

Similarly, we consider cubic B-spline curve:

$$\boldsymbol{p}(u) = \sum_{i=-\infty}^{+\infty} \boldsymbol{P}_i N_{i,3}(u).$$

where $\boldsymbol{U} = [\ldots, u_{-1}, u_0, u_1, \ldots]$. Assume that its shape is shown in Fig. 2.18a

From Fig. 2.18a, we know that each edge of the control polygon corresponds to a curve segment. Write the length of parameter interval of the curve segment as d_i, and the parameter of each edge can be given as $\boldsymbol{P}_i\boldsymbol{P}_{i+1} \to d_i$.

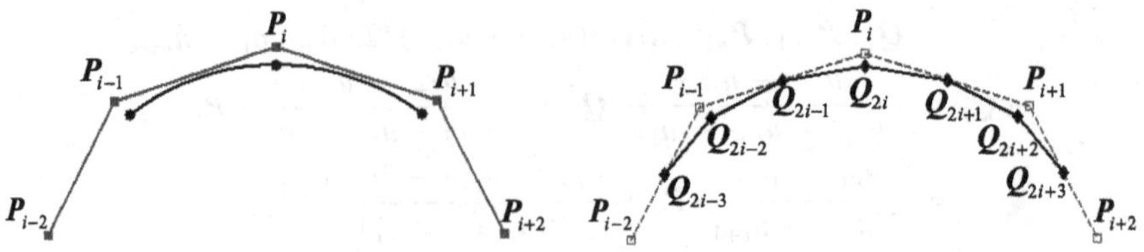

(a) Cubic B-spline Curves (b) Subdivision once

Fig. 2.18 Cubic B-spline curves and their subdivisions

After doubling the knots, we have:

$$
\begin{aligned}
\boldsymbol{Q}_{2i+1} &= \frac{(d_i + 2d_{i+1})\boldsymbol{P}_i + (d_i + 2d_{i-1})\boldsymbol{P}_{i+1}}{2(d_{i-1} + d_i + d_{i+1})} \\
\boldsymbol{Q}_{2i} &= \frac{d_i \boldsymbol{Q}_{2i-1} + (d_{i-1} + d_i)\boldsymbol{P}_i + d_{i-1}\boldsymbol{Q}_{2i+1}}{2(d_{i-1} + d_i)}.
\end{aligned}
\tag{2.11}
$$

Edges of the new control polygon have parameters:

$$\boldsymbol{Q}_{2i}\boldsymbol{Q}_{2i+1} \to d_i/2, \quad \boldsymbol{Q}_{2i+1}\boldsymbol{Q}_{2i+2} \to d_i/2.$$

2.3.2 *Subdivision of B-Spline Surface*

Firstly, we investigate bi-quadric B-spline surfaces:

$$\boldsymbol{p}(u, v) = \sum_{i=-\infty}^{+\infty} \sum_{j=-\infty}^{+\infty} \boldsymbol{P}_{i,j} N_{i,2}(u) N_{j,2}(v),$$

where $\boldsymbol{U} = [\ldots, u_{-1}, u_0, u_1, \ldots]$, $\boldsymbol{V} = [\ldots, v_{-1}, v_0, v_1, \ldots]$.

Since B-spline surfaces are tensor surfaces, we can double their knots in two parameter directions, as shown in Fig. 2.19. According to the subdivision Formula (2.9) of quadric B-spline curves, we can refine every column control vertices in u parameter direction:

$$
\begin{aligned}
\boldsymbol{Q}_{i,2j} &= \frac{(d_j + 2d_{j+1})P_{i,j} + d_j \boldsymbol{P}_{i,j+1}}{2(d_j + d_{j+1})} \\
\boldsymbol{Q}_{i+1,2j} &= \frac{(d_j + 2d_{j+1})\boldsymbol{P}_{i+1,j} + d_j \boldsymbol{P}_{i+1,j+1}}{2(d_j + d_{j+1})},
\end{aligned}
$$

where $\boldsymbol{P}_{i,j} \to d_j$.

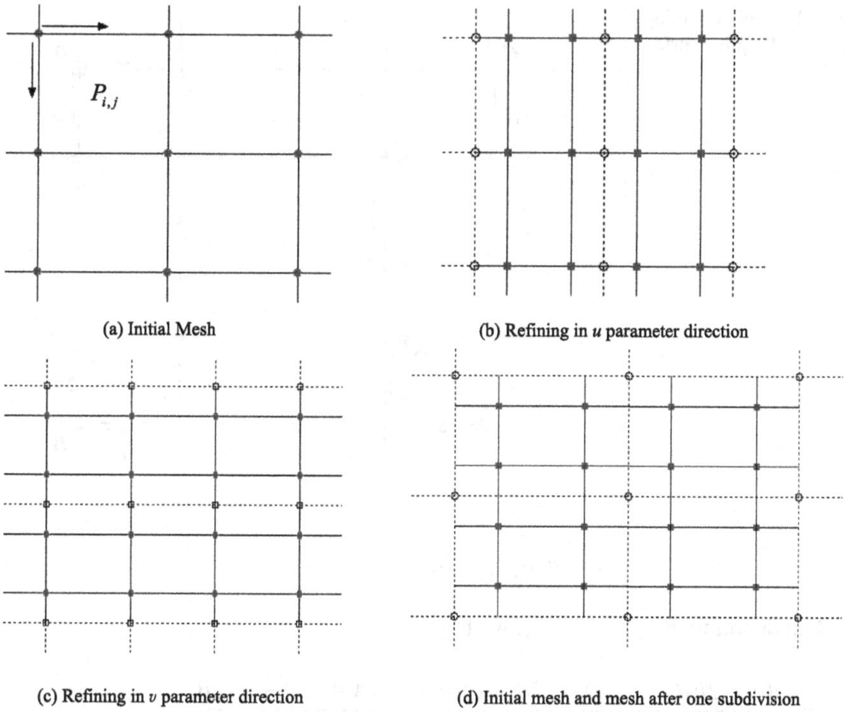
$P_{i,j}$
(a) Initial Mesh
(b) Refining in u parameter direction
(c) Refining in v parameter direction
(d) Initial mesh and mesh after one subdivision

Fig. 2.19 Subdivision of bi-quadric B-spline surface

Consequently, we can obtain a new mesh M_Q. We now refine the control vertices in each row in v parameter direction:

$$\begin{aligned}
\boldsymbol{P}'_{2i,2j} &= \frac{(e_i + 2e_{i+1})\boldsymbol{Q}_{i,2j} + e_i\boldsymbol{Q}_{i+1,2j}}{2(e_i + e_{i+1})} \\
&= ((e_i + 2e_{i+1})(d_j + 2d_{j+1})\boldsymbol{P}_{i,j} + (e_i + 2e_{i+1})d_j\boldsymbol{P}_{i,j+1} \\
&\quad + e_i(d_j + 2d_{j+1})\boldsymbol{P}_{i+1,j} + e_i d_j \boldsymbol{P}_{i+1,j+1})/ \\
&\quad (4(e_i + e_{i+1})(d_j + d_{j+1})) \\
\boldsymbol{P}'_{2i+1,2j} &= \frac{e_{i+1}\boldsymbol{Q}_{i,2j} + (2e_i + e_{i+1})\boldsymbol{Q}_{i+1,2j}}{2(e_i + e_{i+1})} \\
&= (e_{i+1}(d_j + 2d_{j+1})\boldsymbol{P}_{i,j} + e_{i+1}d_j\boldsymbol{P}_{i,j+1} + (2e_i + e_{i+1})(d_j + 2d_{j+1}) \\
&\quad \boldsymbol{P}_{i+1,j} + (2e_i + e_{i+1})d_j\boldsymbol{P}_{i+1,j+1})/ \\
&\quad (4(e_i + e_{i+1})(d_j + d_{j+1})),
\end{aligned}$$

where $\boldsymbol{Q}_{i,j} \to e_j$.

Assume that $\boldsymbol{P}_{i,j} = \boldsymbol{C}$, $\boldsymbol{P}_{i,j+1} = \boldsymbol{D}$, $\boldsymbol{P}_{i+1,j} = \boldsymbol{A}$, $\boldsymbol{P}_{i+1,j+1} = \boldsymbol{B}$,

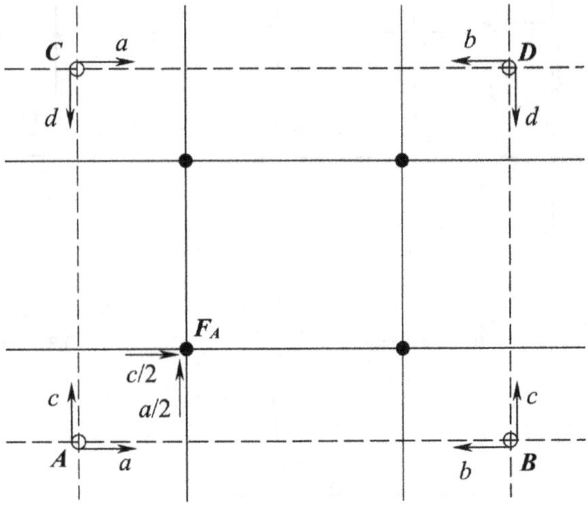

Fig. 2.20 Knot doubling of quadratic B-spline surface

$$d_j = a, d_{j+1} = b, e_i = d, e_{i+1} = c,$$

And denoting $\boldsymbol{P}'_{2i+1,2j}$ by $\boldsymbol{F}_A$, we have

$$\begin{aligned}\boldsymbol{F}_A &= \frac{c(a+2b)\boldsymbol{C} + ac\boldsymbol{D} + (2d+c)(a+2b)\boldsymbol{A} + (2d+c)a\boldsymbol{B}}{4(a+b)(c+d)} \\ &= \frac{(ac+2bc+2ad+4bd)\boldsymbol{A} + (2ad+ac)\boldsymbol{B} + (2bc+ac)\boldsymbol{C} + ac\boldsymbol{D}}{4(a+b)(c+d)}. \quad (2.12)\end{aligned}$$

Figure 2.20 gives other labels for Formula (2.12). Formula (2.12) is straightforward if we generalize these subdivision rules to any mesh, which can be found in the literature [25].

Now, we research bi-cubic B-spline surfaces and insert midpoint to every knot interval successively in u and v parameter directions, as shown in Fig. 2.21. From Fig. 2.21c, we can know that the process to obtain the mesh M^1 from the mesh M^0 uses the following rules:

- Every face in M^0 has a corresponding new vertex in M^1, and the new vertex is called the new face point;
- Every edge in M^0 has a corresponding new vertex in M^1, and the new vertex is called the new edge point; and
- Every vertex in M^0 has a corresponding new vertex in M^1, and the new vertex is called the new vertex point.

We now deduce the calculating formulas for these three types of new points, which can be obtained from the knot-inserting algorithm of cubic B-spline curves:

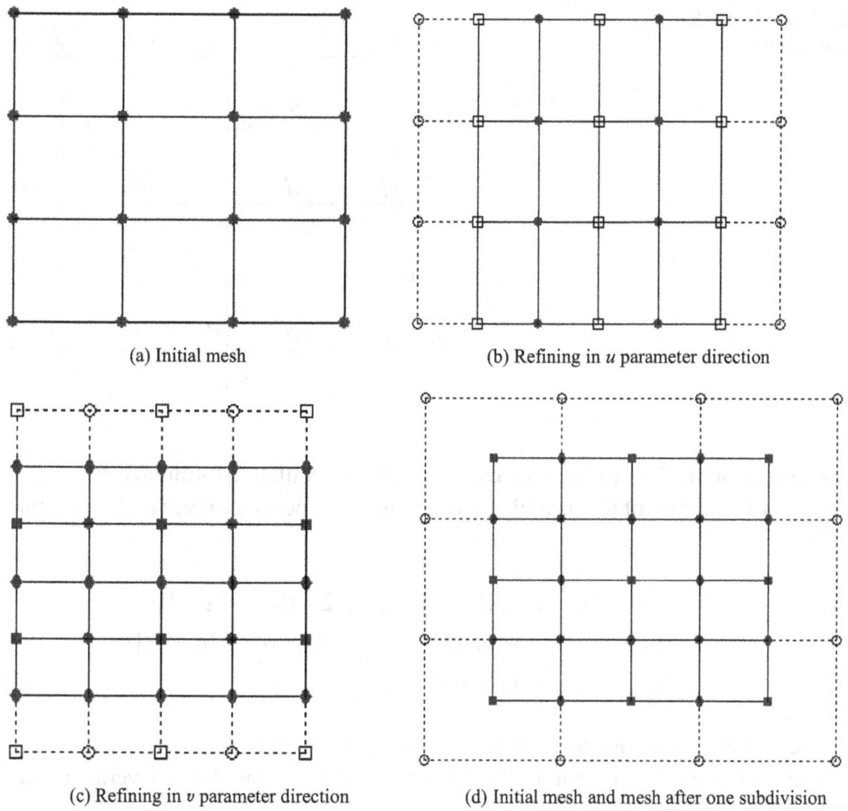
(a) Initial mesh
(b) Refining in u parameter direction
(c) Refining in v parameter direction
(d) Initial mesh and mesh after one subdivision

Fig. 2.21 Subdivision of bi-cubic B-spline surface

$$\boldsymbol{Q}_{i,2j+1} = \frac{(d_j + 2d_{j+1})\boldsymbol{P}_{i,j} + (d_j + 2d_{j-1})\boldsymbol{P}_{i,j+1}}{2(d_{j-1} + d_j + d_{j+1})}$$

$$\boldsymbol{Q}_{i,2j} = \frac{d_j \boldsymbol{Q}_{2j-1} + (d_{j-1} + d_j)\boldsymbol{P}_j + d_{j-1}\boldsymbol{Q}_{2j+1}}{2(d_{j-1} + d_j)},$$

where $\boldsymbol{P}_{i,j}\boldsymbol{P}_{i,j+1} \to d_j$.

While we double knots in v parameter direction, we get

$$\boldsymbol{P}^1_{2i+1,2j+1} = \frac{(e_i + 2e_{i+1})\boldsymbol{Q}_{i,2j+1} + (e_i + 2e_{i-1})\boldsymbol{Q}_{i+1,2j+1}}{2(e_{i-1} + e_i + e_{i+1})}$$
$$= \frac{a\boldsymbol{P}_{i,j} + b\boldsymbol{P}_{i,j+1} + c\boldsymbol{P}_{i+1,j} + d\boldsymbol{P}_{i+1,j+1}}{4(e_{i-1} + e_i + e_{i+1})(d_{j-1} + d_j + d_{j+1})},$$

where $a = (e_i + 2e_{i+1})(d_j + 2d_{j+1})$, $b = (e_i + 2e_{i+1})(d_j + 2d_{j-1})$, $c = (e_i + 2e_{i-1})(d_j + 2d_{j+1})$, $d = (e_i + 2e_{i-1})(d_j + 2d_{j-1})$, and $\boldsymbol{Q}_{i,j}\boldsymbol{Q}_{i+1,j} \to e_j$.

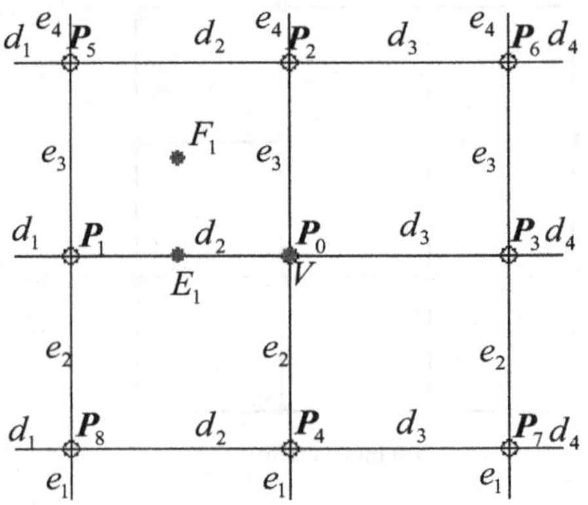

Fig. 2.22 Face, edge, and vertex points

After replacing the variables, we can find that the calculating formula of $P^1_{2i+1,2j+1}$ is in accordance with of the formula for calculating the new face point in the literature [25]:

$$\begin{aligned}F_1 = {} & [(e_3+2e_4)(d_2+d_1)P_0+(e_3+2e_4)(d_2+2d_3)P_1+\\ & (e_3+2e_2)(d_2+2d_3)P_5+(e_3+2e_2)(d_2+2d_1)P_2]/\\ & 4(e_2+e_3+e_4)(d_1+d_2+d_3).\end{aligned} \tag{2.13}$$

The variables in above formula are shown in Fig. 2.22.

Similarly, we can get formulas for the new edge point and the new vertex point:

$$E_1 = \frac{e_2F_1+e_3F_4+(e_2+e_3)B_1}{2(e_2+e_3)}, \tag{2.14}$$

where

$$B_1 = \frac{(2d_1+d_2)P_0+(d_2+2d_3)P_1}{2(d_1+d_2+d_3)}$$

$$\begin{aligned}V = {} & \frac{P_0}{4}+\frac{d_3e_2F_1+d_2e_2F_2+d_2e_3F_3+d_3e_3F_4}{4(d_2+d_3)(e_2+e_3)}+[d_3(e_2+e_3)B_1+\\ & e_2(d_2+d_3)B_2+d_2(e_2+e_3)B_3+e_3(d_2+d_3)B_4]/[4(d_2+d_3)(e_2+e_3)].\end{aligned} \tag{2.15}$$

Consequently, subdivision of B-spline surface can be realized by subdivision of B-spline curves in two parameter directions based on tensor product property of B-spline surfaces.

2.4 Uniform Subdivision of B-Spline Curves and Surfaces

These subdivision formulas obtained with knot-inserting algorithm are complex. A majority of existing subdivision rules, such as Doo–Sabin subdivision, Catmull–Clark subdivision, Loop subdivision, do not involve in parameters of vertices and edges of meshes. Since knot distribution of uniform B-spline curves and surfaces is uniform, those parameters of vertices and edges are identical. Consequently, we need not consider parameters when we research subdivision of uniform B-spline curves and surfaces. So, simpler subdivision formulas can be obtained.

2.4.1 Subdivision of Uniform B-Spline Curves

Uniform B-spline has a definition equivalent to the recursive Formula (2.1)—the convolution definition. It is easy to deduce subdivision rules of uniform B-spline curves and surfaces with the convolution definition [8, 86].

Definition 2.2 The convolution of two continuous functions $f(u)$ and $g(u)$ can be defined as:

$$(f \otimes g)(u) = \int_s f(s)g(u-s)ds. \tag{2.16}$$

Theorem 2.1 *The convolution has properties as follows:*
Linearity: $f(u) \otimes (g(u) + h(u)) = f(u) \otimes g(u) + f(u) \otimes h(u)$,
Time shift: $f(u-i) \otimes g(u-j) = p(u-i-j)$,
Time scaling: $f(2u) \otimes g(2u) = \frac{1}{2}p(2u)$,
where $p(u) = f(u) \otimes g(u)$.

Proof We only deduce the time scaling property. The deduction of other two properties is similar to the following deduction.

$$\int_s f(2s)g(2u-2s)ds = \frac{1}{2}\int_s f(2s)g[2(u-s)]d(2s)$$

Let $2s = s'$, $2u = u'$, we have

$$\frac{1}{2}\int_s f(2s)g(2u-2s)d(2s) = \frac{1}{2}\int_{s'} f(s')g(u'-s')ds' = \frac{1}{2}p(u') = \frac{1}{2}p(2u) \quad \#$$

Theorem 2.2 $f(2u-i) \otimes g(2u-j) = \frac{1}{2}p(2u-i-j)$, *where* $p(u) = f(u) \otimes g(u)$

Proof $f(2u-i) \otimes g(2u-j) = \int_s f(2s-i)g(2u-j-2s)ds$

Assume $s' = 2s - i, u' = 2u$, we have

$$f(2u-i) \otimes g(2u-j) = \frac{1}{2}\int_{s'} f(s')g(u'-j-i-s')ds' = \frac{1}{2}p(2u-j-i)$$

Definition 2.3 Let zero-degree B-spline basic function be:

$$N_0(u) = \begin{cases} 1, & 0 \leqslant u \leqslant 1 \\ 0, & other. \end{cases} \tag{2.17}$$

k-degree B-spline basic function can be defined as:

$$N_k(u) = \int_s N_{k-1}(s)N_0(u-s)ds. \tag{2.18}$$

Based on Definition 2.3, when $k = 1$,

$$N_1(u) = \int_0^u N_0(s)N_0(u-s)ds = u, u \in [0, 1].$$

If $u \in [1, 2]$,

$$N_1(u) = \int_{u-1}^1 N_0(s)N_0(u-s)ds = 2-u,$$

$$\therefore \quad N_1(u) = \begin{cases} u & u \in [0, 1] \\ 2-u & u \in [1, 2]. \end{cases}$$

In the case of $k = 2$, we have
If $u \in [0, 1]$,

$$N_2(u) = \int_0^u N_1(s)N_0(u-s)ds = \int_0^s sds = \frac{1}{2}u^2.$$

If $u \in [1, 2]$,

$$\begin{aligned} N_2(u) = \int_{u-1}^u N_1(s)N_0(u-s)ds &= \int_{u-1}^1 N_1(s)N_0(u-s)ds + \\ &\int_1^u N_1(s)N_0(u-s)ds \\ &= \int_{u-1}^1 sds + \int_1^u (2-s)ds = -u^2 + 3u - \frac{3}{2}. \end{aligned}$$

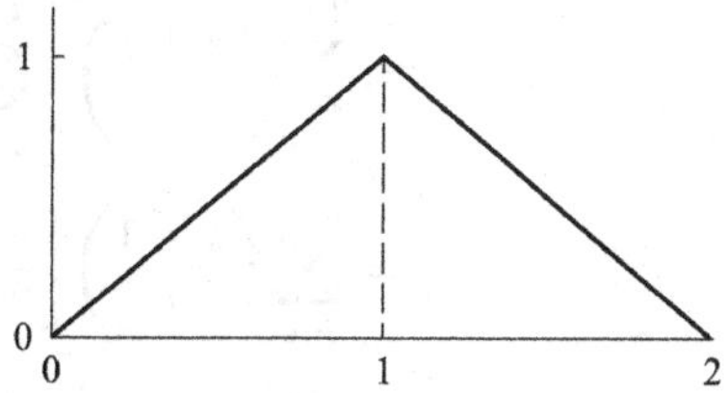

Fig. 2.23 Shape of $N_1(u)$

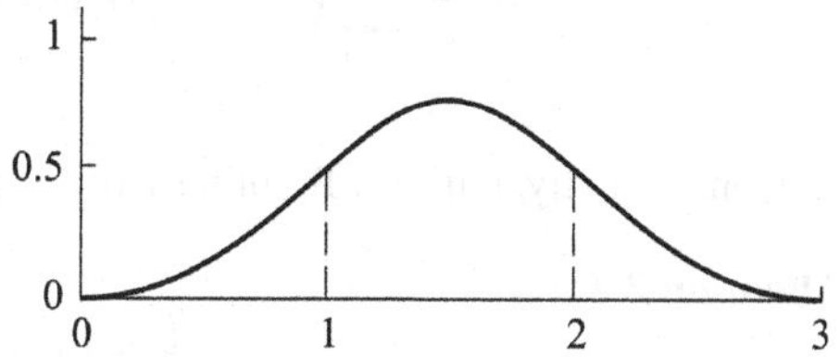

Fig. 2.24 Shape of $N_2(u)$

If $u \in [2, 3]$,

$$N_2(u) = \int_{u-1}^{2} N_1(s)N_0(u-s)ds = \int_{u-1}^{2} (2-s)ds = \frac{1}{2}u^2 - 3u + \frac{9}{2}$$

$$\therefore \quad N_2(u) = \begin{cases} u^2/2, & u \in [0, 1] \\ -u^2 + 3u - 3/2, & u \in [1, 2] \\ u^2/2 - 3u + 9/2, & u \in [2, 3] \\ 0, & other. \end{cases}$$

The curves of $N_1(u)$ and $N_2(u)$ are shown in Figs. 2.23 and 2.24.

Let $\boldsymbol{U} = [\ldots, -2, -1, 0, 1, 2, \ldots]$, it can be obtained from the above deduction:

$$\begin{cases} N_k(u) = N_{0,k}(u) \\ N_k(u-i) = N_{i,k}(u). \end{cases}$$

We now double knots of $\boldsymbol{U}$:

$$\boldsymbol{U}^1 = [\ldots, -1, -1/2, 0, 1/2, 1, \ldots].$$

Obviously, $\boldsymbol{N}_0(2u)$ is the function defined on $\boldsymbol{U}^1$ and it satisfies the formula:

$$N_0(u) = N_0(2u) + N_0(2u-1). \tag{2.19}$$

Based on Definitions 2.2, 2.3, and Theorem 2.1, we can know

$$N_k(u) = \bigotimes_{i=0}^{k} N_0(u) = \bigotimes_{i=0}^{k} [N_0(2u) + N_0(2u-1)]$$

$$= \sum_{i=0}^{k+1} \binom{k+1}{i} \left[\bigotimes_{j=0}^{k-i} N_0(2u) \right] \otimes \left[\bigotimes_{j=0}^{i-1} N_0(2u-1) \right]$$

$$= \sum_{i=0}^{k+1} \binom{k+1}{i} \left[\frac{1}{2^{k-i}} N_{k-i}(2u) \right] \otimes \left[\frac{1}{2^{i-1}} N_i(2u-i) \right]$$

$$= \frac{1}{2^k} \sum_{i=0}^{k+1} \binom{k+1}{i} N_k(2u-i).$$

Consequently, uniform B-spline satisfies the following theory,

Theorem 2.3

$$N_k(u) = \frac{1}{2^k} \sum_{i=0}^{k+1} \binom{k+1}{i} N_k(2u-i), \tag{2.20}$$

where the knot vector of $N_k(u)$ is $\boldsymbol{U}$.

Based on Theorem 2.3, we can know that when $k = 2$, the quadratic uniform B-spline basic function is:

$$\begin{aligned} N_2(u) &= \frac{1}{4} \sum_{i=0}^{3} \binom{3}{i} N_2(2u-i) \\ &= \frac{1}{4} N_2(2u) + \frac{3}{4} N_2(2u-1) + \frac{3}{4} N_2(2u-2) + \frac{1}{4} N_2(2u-3) \\ &= \frac{1}{4} N_2(2u) + \frac{3}{4} N_2 \left[2\left(u - \frac{1}{2}\right) \right] + \\ &\quad \frac{3}{4} N_2 [2(u-1)] + \frac{1}{4} N_2 \left[2\left(u - \frac{3}{2}\right) \right] \end{aligned}$$

$$\therefore \quad N_{i,2}(u) = \frac{1}{4} N^1_{2i-2}(u) + \frac{3}{4} N^1_{2i-1}(u) + \frac{3}{4} N^1_{2i}(u) + \frac{1}{4} N^1_{2i+1}(u). \tag{2.21}$$

Positions of $N_{0,2}(u)$ and $N^1_{j,2}(u)$, $j = 0, 1/2, 1, 3/2$ are shown in Fig. 2.25.

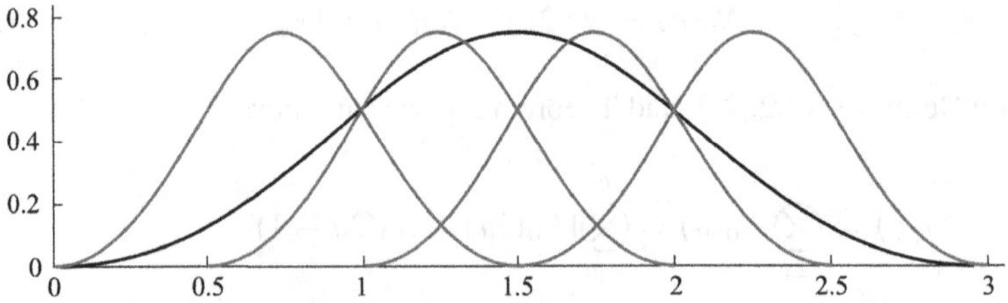

Fig. 2.25 Positions of $N_{0,2}(u)$ and $N^1_{j,2}(u) (j = 0, 1/2, 1, 3/2)$

With Formula (2.21), the quadratic uniform B-spline curve can be represented as:

$$
\begin{aligned}
\boldsymbol{p}(u) &= \sum_{i=-\infty}^{+\infty} \boldsymbol{P}_i N_{i,2}(u) \\
&= \cdots + \boldsymbol{P}_{i-1}\left[\frac{1}{4}N^1_{2i-4}(u) + \frac{3}{4}N^1_{2i-3}(u) + \frac{3}{4}N^1_{2i-2}(u) + \frac{1}{4}N^1_{2i-1}(u)\right] + \\
&\quad \boldsymbol{P}_i\left[\frac{1}{4}N^1_{2i-2}(u) + \frac{3}{4}N^1_{2i-1}(u) + \frac{3}{4}N^1_{2i}(u) + \frac{1}{4}N^1_{2i+1}(u)\right] + \\
&\quad \boldsymbol{P}_{i+1}\left[\frac{1}{4}N^1_{2i}(u) + \frac{3}{4}N^1_{2i+1}(u) + \frac{3}{4}N^1_{2i+2}(u) + \frac{1}{4}N^1_{2i+3}(u)\right] + \cdots
\end{aligned}
$$

Let

$$\boldsymbol{P}^1_{2i} = \frac{3}{4}\boldsymbol{P}_i + \frac{1}{4}\boldsymbol{P}_{i+1},\ \boldsymbol{P}^1_{2i+1} = \frac{1}{4}\boldsymbol{P}_i + \frac{3}{4}\boldsymbol{P}_{i+1}. \tag{2.22}$$

We have,

$$\boldsymbol{p}(u) = \boldsymbol{p}^1(u) = \sum_{-\infty}^{+\infty} \boldsymbol{P}^1_i N^1_{i,2}.$$

It is easy to know that Formula (2.22) is an especial example of Formula (2.9) while all parameters are equal. When we continuously subdivide the control polygon P by Formula (2.22), we can get the series of polygons:

$$\boldsymbol{P}^0, \boldsymbol{P}^1, \ldots, \boldsymbol{P}^i, \ldots$$

And,

$$\boldsymbol{P}^\infty = \boldsymbol{p}(u).$$

This is Chaikin algorithm which renders curves by cutting corners of polygons. The process can be represented by Fig. 2.15.

Similarly, the cubic uniform B-spline basic function can be written as:

$$N_{i,3}(u) = \frac{1}{8}N^1_{2i-2,3}(u) + \frac{4}{8}N^1_{2i-1,3}(u) + \frac{6}{8}N^1_{2i,3}(u) + \frac{4}{8}N^1_{2i+1,3}(u) + \frac{1}{8}.N^1_{2i+2,3}(u) \tag{2.23}$$

Based on Formula (2.23), we can know that

$$
\begin{aligned}
\boldsymbol{p}(u) &= \sum_{i=-\infty}^{+\infty} \boldsymbol{P}_i N_{i,3}(u) \\
&= \cdots + \boldsymbol{P}_{i-1}\Big[\frac{1}{8}N^1_{2i-4,3}(u) + \frac{4}{8}N^1_{2i-3,3}(u) + \frac{6}{8}N^1_{2i-2,3}(u) + \\
&\quad \frac{4}{8}N^1_{2i-1,3}(u) + \frac{1}{8}N^1_{2i,3}(u)\Big] +
\end{aligned}
$$

$$
\begin{aligned}
&\boldsymbol{P}_i\Big[\frac{1}{8}N^1_{2i-2,3}(u)+\frac{4}{8}N^1_{2i-1,3}(u)+\frac{6}{8}N^1_{2i,3}(u)+\\
&\frac{4}{8}N^1_{2i+1,3}(u)+\frac{1}{8}N^1_{2i+2,3}(u)\Big]+\\
&\boldsymbol{P}_{i+1}\Big[\frac{1}{8}N^1_{2i,3}(u)+\frac{4}{8}N^1_{2i+1,3}(u)+\frac{6}{8}N^1_{2i+2,3}(u)+\\
&\frac{4}{8}N^1_{2i+3,3}(u)+\frac{1}{8}N^1_{2i+4,3}(u)\Big]+\cdots\\
&=\cdots+\frac{1}{2}(\boldsymbol{P}_{i-1}+\boldsymbol{P}_i)N^1_{2i-1,3}+\left(\frac{1}{8}\boldsymbol{P}_{i-1}+\frac{4}{8}\boldsymbol{P}_i+\frac{1}{8}\boldsymbol{P}_{i+1}\right)N^1_{2i,3}+\\
&\frac{1}{2}(\boldsymbol{P}_i+\boldsymbol{P}_{i+1})N^1_{2i+1,3}+\cdots
\end{aligned}
$$

Let

$$
\boldsymbol{P}^1_{2i}=\frac{1}{8}\boldsymbol{P}_{i-1}+\frac{6}{8}\boldsymbol{P}_i+\frac{1}{8}\boldsymbol{P}_{i+1},\ \boldsymbol{P}^1_{2i+1}=\frac{1}{2}\boldsymbol{P}_i+\frac{1}{2}\boldsymbol{P}_{i+1}. \tag{2.24}
$$

We can get

$$
\boldsymbol{p}(u)=\boldsymbol{p}^1(u)=\sum_{-\infty}^{+\infty}\boldsymbol{P}^1_iN^1_{i,3}.
$$

Formula (2.24) is a special case of Formula (2.11) while all parameters are equal.

2.4.2 Subdivision of Uniform B-Spline Surfaces

Assume that $\boldsymbol{U}=\boldsymbol{V}=[\ldots,-2,-1,0,1,2,\ldots]$ are knot vectors in two parameter directions. We study the change of control meshes of B-spline surfaces while we double knots of $\boldsymbol{U}$ and $\boldsymbol{V}$. We firstly investigate the uniform bi-quadric B-spline surface:

$$
\boldsymbol{p}(u,v)=\sum_{-\infty}^{+\infty}\boldsymbol{P}_{i,j}N_{i,2}(u)N_{j,2}(v),\ (u,v)\in(-\infty,+\infty).
$$

When each knot interval is bisected, we can know from Formula (2.21):

$$
N_{i,2}(u)N_{j,2}(v)=\frac{1}{16}\sum_{s=0}^{3}\binom{3}{s}N^1_{2i+s-2,2}(u)\sum_{t=0}^{3}\binom{3}{t}N^1_{2i+t-2,2}(v)
$$

$$
\therefore\ N_{i,2}(u)N_{j,2}(v)=\frac{1}{16}\sum_{s=0}^{3}\sum_{t=0}^{3}\binom{3}{s}\binom{3}{t}N^1_{2i+s-2,2}(u)N^1_{2i+t-2,2}(v). \tag{2.25}
$$

Let $N^1_{2i+s-2}N^1_{2j+t-2} = N^1_{2i+s-2,2j+t-2} = (2i+s-2, 2j+t-2)$, basic functions and coefficients in every term of (2.25) have the following correspondence relations:

$$\begin{bmatrix} (2i-2,2j-2) & (2i-2,2j-1) & (2i-2,2j) & (2i-2,2j+1) \\ (2i-1,2j-2) & (2i-1,2j-1) & (2i-1,2j) & (2i-1,2j+1) \\ (2i,2j-2) & (2i,2j-1) & (2i,2j) & (2i,2j+1) \\ (2i+1,2j-2) & (2i+1,2j-1) & (2i+1,2j) & (2i+1,2j+1) \end{bmatrix} \to$$

$$\frac{1}{16}\begin{bmatrix} 1 & 3 & 3 & 1 \\ 3 & 9 & 9 & 3 \\ 3 & 9 & 9 & 3 \\ 1 & 3 & 3 & 1 \end{bmatrix}. \tag{2.26}$$

The matrix on the left of Formula (2.26) is called the subdivision coefficient matrix of $\boldsymbol{P}_{i,j}$. For any term $N^1_{2i+s-1,2j+t-1}$ in the subdivision coefficient matrix of $\boldsymbol{P}_{i,j}$, there are always subdivision coefficient matrixes of four vertices in which the $N^1_{2i+s-1,2j+t-1}$ is. Consequently,

$$\begin{aligned} \boldsymbol{S}(u,v) &= \frac{1}{16}\sum_{-\infty}^{+\infty}\sum_{-\infty}^{+\infty}\boldsymbol{P}_{i,j}\left[\sum_{s=0}^{3}\sum_{t=0}^{3}\binom{3}{s}\binom{3}{t}N^1_{2i+s-2}N^1_{2j+t-2}\right] \\ &= \cdots + \left(\frac{9}{16}\boldsymbol{P}_{i-1,j-1} + \frac{3}{16}\boldsymbol{P}_{i-1,j} + \frac{3}{16}\boldsymbol{P}_{i,j-1} + \frac{1}{16}\boldsymbol{P}_{i,j}\right)N^1_{2i-2,2j-2} + \\ &\quad \left(\frac{3}{16}\boldsymbol{P}_{i-1,j-1} + \frac{1}{16}\boldsymbol{P}_{i-1,j} + \frac{9}{16}\boldsymbol{P}_{i,j-1} + \frac{3}{16}\boldsymbol{P}_{i,j}\right)N^1_{2i-2,2j-1} + \\ &\quad \left(\frac{1}{16}\boldsymbol{P}_{i-1,j-1} + \frac{3}{16}\boldsymbol{P}_{i-1,j} + \frac{9}{16}\boldsymbol{P}_{i,j-1} + \frac{3}{16}\boldsymbol{P}_{i,j}\right)N^1_{2i-1,2j-2} + \\ &\quad \left(\frac{1}{16}\boldsymbol{P}_{i-1,j-1} + \frac{3}{16}\boldsymbol{P}_{i-1,j} + \frac{3}{16}\boldsymbol{P}_{i,j-1} + \frac{9}{16}\boldsymbol{P}_{i,j}\right)N^1_{2i-1,2j-1}. \end{aligned}$$

Let

$$\begin{aligned} \boldsymbol{P}^1_{2i-2,2j-2} &= \frac{9}{16}\boldsymbol{P}_{i-1,j-1} + \frac{3}{16}\boldsymbol{P}_{i-1,j} + \frac{3}{16}\boldsymbol{P}_{i,j-1} + \frac{1}{16}\boldsymbol{P}_{i,j}, \\ \boldsymbol{P}^1_{2i-2,2j-1} &= \frac{3}{16}\boldsymbol{P}_{i-1,j-1} + \frac{1}{16}\boldsymbol{P}_{i-1,j} + \frac{9}{16}\boldsymbol{P}_{i,j-1} + \frac{3}{16}\boldsymbol{P}_{i,j} \\ \boldsymbol{P}^1_{2i-1,2j-2} &= \frac{1}{16}\boldsymbol{P}_{i-1,j-1} + \frac{3}{16}\boldsymbol{P}_{i-1,j} + \frac{9}{16}\boldsymbol{P}_{i,j-1} + \frac{3}{16}\boldsymbol{P}_{i,j} \\ \boldsymbol{P}^1_{2i-1,2j-1} &= \frac{1}{16}\boldsymbol{P}_{i-1,j-1} + \frac{3}{16}\boldsymbol{P}_{i-1,j} + \frac{3}{16}\boldsymbol{P}_{i,j-1} + \frac{9}{16}\boldsymbol{P}_{i,j}. \end{aligned}$$

So we can get the subdivision masks of uniform biquadratic B-spline surfaces as shown in Fig. 2.26. A subdivision mask is a mesh formed by old vertices connected with a calculated new vertex. There is the coefficient of the old vertex on the position

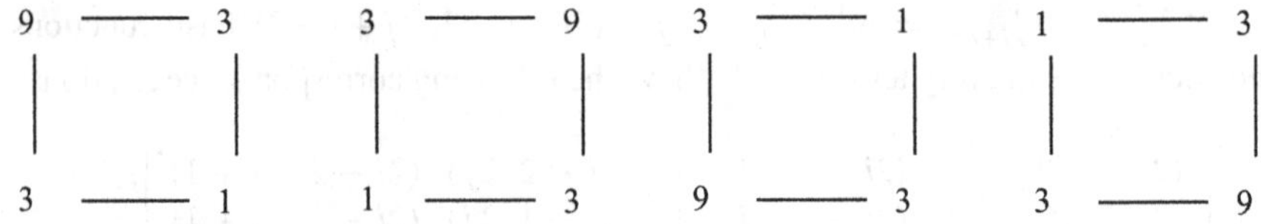

Fig. 2.26 Subdivision masks of uniform biquadratic B-spline surfaces

of the old vertex on the mask. For a mask, all vertices and their coefficients form a linear expression whose value is the geometric position of the corresponding new vertex. A set of subdivision rules is probably described by several masks. The new mesh topology structure is the topology structure shown in Fig. 2.21d.

For the uniform bi-cubic B-spline surface:

$$\boldsymbol{p}(u, v) = \sum_{-\infty}^{+\infty} \boldsymbol{P}_{i,j} N_{i,3}(u) N_{j,3}(v),\ (u, v) \in (-\infty, +\infty).$$

When every knot interval is bisected, we can know through Formula (2.23):

$$N_{i,3}(u)N_{j,3}(v) = \frac{1}{64}\sum_{s=0}^{4}\binom{4}{s}N^1_{2i+s-2,3}(u)\sum_{t=0}^{4}\binom{4}{t}N^1_{2i+t-2,3}(v)$$

$$\therefore\ N_{i,3}(u)N_{j,3}(v) = \frac{1}{64}\sum_{s=0}^{4}\sum_{t=0}^{4}\binom{4}{s}\binom{4}{t}N^1_{2i+s-2,3}(u)N^1_{2i+t-2,3}(v). \tag{2.27}$$

Let $N^1_{2i+s-2}N^1_{2j+t-2} = N^1_{2i+s-2,2j+t-2} = (2i+s-2, 2j+t-2)$, basic functions and coefficients in each term of (2.22) have the following correspondence relation:

$$\begin{bmatrix} (2i-2,2j-2) & (2i-2,2j-1) & (2i-2,2j) & (2i-2,2j+1) & (2i-2,2j+2) \\ (2i-1,2j-2) & (2i-1,2j-1) & (2i-1,2j) & (2i-1,2j+1) & (2i-1,2j+2) \\ (2i,2j-2) & (2i,2j-1) & (2i,2j) & (2i,2j+1) & (2i,2j+2) \\ (2i+1,2j-2) & (2i+1,2j-1) & (2i+1,2j) & (2i+1,2j+1) & (2i+2,2j+1) \\ (2i+2,2j-2) & (2i+2,2j-1) & (2i+2,2j) & (2i+2,2j+1) & (2i+2,2j+2) \end{bmatrix} \rightarrow$$

$$\frac{1}{64}\begin{bmatrix} 1 & 4 & 6 & 4 & 1 \\ 4 & 16 & 24 & 16 & 4 \\ 6 & 24 & 36 & 24 & 6 \\ 4 & 16 & 24 & 16 & 4 \\ 1 & 4 & 6 & 4 & 1 \end{bmatrix}. \tag{2.28}$$

Through the correspondence relation (2.28), we have known that there are subdivision matrixes of nine vertices which contain the term $N^1_{2i-2,2j-2}$: $\boldsymbol{P}_{i-2,j-2}$, $\boldsymbol{P}_{i-1,j-2}$,

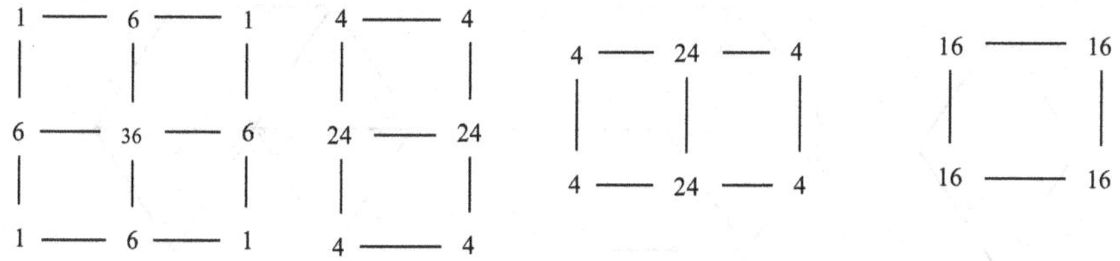

(a) Mask of new vertex points (b) Mask of new edge point (c) Mask of new edge point (d) Mask of new face points

Fig. 2.27 Subdivision masks of uniform bi-cubic B-spline surfaces

$\boldsymbol{P}_{i,j-2}$, $\boldsymbol{P}_{i-2,j-1}$, $\boldsymbol{P}_{i-1,j-1}$, $\boldsymbol{P}_{i,j-1}$, $\boldsymbol{P}_{i-2,j}$, $\boldsymbol{P}_{i-1,j}$, $\boldsymbol{P}_{i,j}$, and coefficient of $N^1_{2i-2,2j-2}$ in each subdivision matrix is shown in Fig. 2.27a:

As for $N^1_{2i-2,2j-1}$, there are subdivision matrixes of six vertices which contain the term: $\boldsymbol{P}_{i-2,j-1}$, $\boldsymbol{P}_{i-1,j-1}$, $\boldsymbol{P}_{i,j-1}$, $\boldsymbol{P}_{i-2,j}$, $\boldsymbol{P}_{i-1,j}$, $\boldsymbol{P}_{i,j}$. Coefficient of $N^1_{2i-2,2j-1}$ in each subdivision matrix is shown in Fig. 2.27b. Similarly, the subdivision mask of $N^1_{2i-1,2j-2}$ is shown in Fig. 2.27c. The subdivision mask of $N^1_{2i-1,2j-1}$ is shown in Fig. 2.27d. Consequently, the subdivision rules of uniform bi-cubic B-spline surface are found.

In the above discussion, we always assume that the control meshes are infinite. The case does not exist at all in engineer application. However, we can apply subdivision rules on finite control meshes, as shown in Figs. 2.19d and 2.21d. So, we can obtain a mesh series which continuously shrinks, and its limitation is a B-spline surface defined by the control mesh. The subdivision of finite polygons is an analogy of infinite meshes.

It is a basic idea of this section to obtain subdivision rules of spline surfaces from subdivision rules of basic functions. The deduction of subdivision rules of box spline surfaces also follows the idea.

2.5 Box Spline

The spline on triangular grids is one of the simplest forms of box splines. It is found by Doo–Sabin and evolved into current box splines due to the researches of de-Boor et al. The uniform B-spline can be regarded as a special case of box splines. Subdivision algorithms on box splines can be founded in the literature [91–93]. These subdivision algorithms can refine control meshes of any box spline surface. This section mainly introduces integral induction definition of bi-variant box splines.

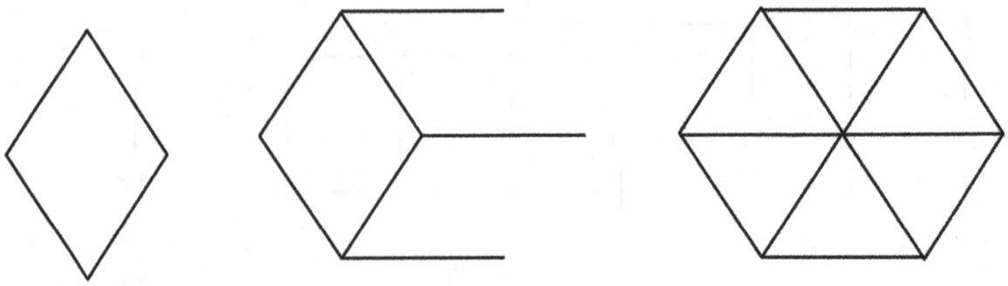

(a) Grid $G_{D^{k-1}}$ (b) Obtain all new vertices (c) Form grid G_{D^3}

Fig. 2.28 Process from G_D^2 to G_D^3

Fig. 2.29 Triangle grid

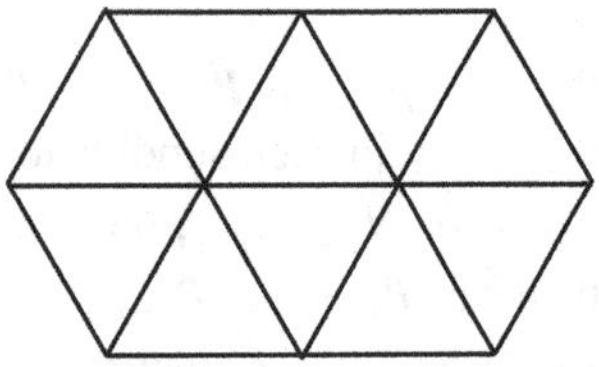

2.5.1 Vector Group and Support Mesh

Definition 2.4 Let $\boldsymbol{s}_0 = [1, 0]^{\mathrm{T}}, \boldsymbol{s}_1 = [0, 1]^{\mathrm{T}}, \boldsymbol{s}_2 = \boldsymbol{s}_0 + \boldsymbol{s}_1, \boldsymbol{s}_3 = \boldsymbol{s}_0 - \boldsymbol{s}_1$, and

$$D^2 = \{\boldsymbol{s}_i, \boldsymbol{s}_j\}, (i \neq j, i, j = 0, 1, 2, 3)$$
$$D^k = D^{k-1} \cup \{\boldsymbol{s}_i\}, (i = 0, 1, 2, 3, k \geqslant 2).$$

We call that $D^k (k \geqslant 2)$ is a vector group. Without loss of generality, we let $D^2 = \{\boldsymbol{s}_0, \boldsymbol{s}_1\}$ in the latter discussion. The grid generated by D^k is denoted by G_D^k and is called the support grid. Obviously, G_D^2 is a rhombus shown in Fig. 2.28a. To obtain G_D^k from G_D^{k-1}, the following process can be used:
Step 1: Make a unit length extend in direction $\boldsymbol{u}_i$ on boundary vertices of G_D^{k-1}.If a vertex obtained by extend does not superpose any vertex of G_D^{k-1}, a new vertex is obtained.
Step 2: All new vertices and vertices of G_D^{k-1} form vertices of G_D^k. Link vertices of G_D^k with vectors in D^k, and then the grid G_D^k is formed.

Figure 2.28 gives the process from G_D^2 to G_D^3, where $D^3 = \{\boldsymbol{s}_0, \boldsymbol{s}_1, \boldsymbol{s}_2\}$.

By translating G_D^k to each vertex of the existing grid, we can obtain an infinite mesh covering the whole plane. As for a vector group D^k,
If $\{\boldsymbol{s}_0, \boldsymbol{s}_1\} \subseteq D^k$ and $\{\boldsymbol{s}_0, \boldsymbol{s}_1, \boldsymbol{s}_2\} \not\subset D^k$ and $\{\boldsymbol{s}_0, \boldsymbol{s}_1, \boldsymbol{s}_2, \boldsymbol{s}_3\} \not\subset D^k$, an infinite quadrilateral grid can be formed by D^k.
If $\{\boldsymbol{s}_0, \boldsymbol{s}_1, \boldsymbol{s}_2\} \subseteq D^k$ and $\{\boldsymbol{s}_0, \boldsymbol{s}_1, \boldsymbol{s}_2, \boldsymbol{s}_3\} \not\subset D^k$, an infinite triangle grid can be formed by D^k. The triangle grid usually becomes the shape as shown in Fig. 2.29 by the affine transform.
If $\{\boldsymbol{s}_0, \boldsymbol{s}_1, \boldsymbol{s}_2, \boldsymbol{s}_3\} \subseteq D^k$, an infinite tri-quadrangle mesh can be formed by D^k. It is shown in Fig. 2.30.

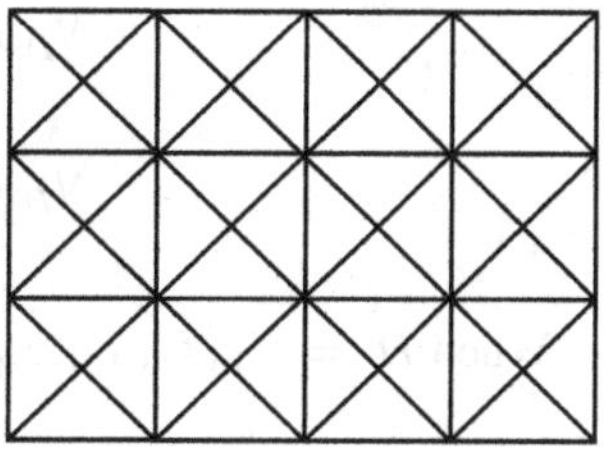

Fig. 2.30 Tri-quadrangle grid

It is easy to know that any vertex of a grid can be denoted as:

$$\boldsymbol{i} = i_0 \boldsymbol{s}_0 + i_1 \boldsymbol{s}_1 = (i_1, i_2),$$

where $\{\boldsymbol{e}_1, \boldsymbol{e}_2\}$ forms an affine reference frame. $\boldsymbol{i}$ is called the knot of the grid. In this case, any point on the plane can be denoted as:

$$\boldsymbol{x} = u\boldsymbol{s}_0 + v\boldsymbol{s}_1 = (u, v).$$

Definition 2.5 The support region of a vector group D^k is the region that is enclosed by the boundary of the grid G_D^k and can be denoted as:

$$\text{supp}(D^k) = [\boldsymbol{t}_0, \boldsymbol{t}_1, \ldots, \boldsymbol{t}_{k-1}][0, 1)^k,$$

where $D^k = \{\boldsymbol{t}_0, \boldsymbol{t}_1, \ldots, \boldsymbol{t}_{k-1}\}, \boldsymbol{t}_i \in \{\boldsymbol{s}_0, \boldsymbol{s}_1, \boldsymbol{s}_2, \boldsymbol{s}_3\}$. In the latter discussion, elements in D^k are denoted by $\boldsymbol{s}_0, \boldsymbol{s}_1, \ldots, \boldsymbol{s}_{k-1}$ successively, unless otherwise specified. Obviously, if $\boldsymbol{x} \in \text{supp}(D^k)$,

$$\boldsymbol{x} = \alpha_0 \boldsymbol{s}_0 + \cdots + \alpha_{k-1}\boldsymbol{s}_{k-1}, \alpha_i \in [0, 1].$$

Definition 2.6 Translate G_D^k to a knot of the plane grid produced by L_D^k and obtain a new grid $G_{i,D}^k$, and the region that is enclosed by $G_{i,D}^k$ is also called the support region of the vector group D^k and denoted as:

$$\text{supp}(D_i^k) = \boldsymbol{i} + \text{supp}(D^k).$$

When $D^k = \{\boldsymbol{s}, \ldots, \boldsymbol{s}\}$ (where $\boldsymbol{s} \in \{\boldsymbol{s}_0, \boldsymbol{s}_1, \boldsymbol{s}_2, \boldsymbol{s}_3\}$), the discussions in this section and latter sections are also applicable. In this case, uniform B-spline is obtained.

2.5.2 *Inductive Definition of Box Splines*

Definition 2.7 Let $D^k (k \geqslant 2)$ be vector groups, the inductive definition of box splines can be given as [94]:

$$N_{D^2}(\boldsymbol{x}) = \begin{cases} 1, & \boldsymbol{x} \in \text{supp } D^2 \\ 0, & \text{other} \end{cases} \tag{2.29}$$

$$N_{D^k}(\boldsymbol{x}) = \int_0^1 N_{D^{k-1}}(\boldsymbol{x} - t\boldsymbol{t}_k)dt \tag{2.30}$$

When $D^3 = \{\boldsymbol{s}_0, \boldsymbol{s}_1, \boldsymbol{s}_2\}$, L_D^3 is shown in Fig. 2.29a. Let $\boldsymbol{x} = (u, v)$, we get

$$\boldsymbol{x} - t\boldsymbol{s}_2 = (u - t, v - t).$$

Due to Definition 2.7, it easy to know that when

$$\max\{u - 1, v - 1, 0\} \leqslant t \leqslant \min\{u, v, 1\},$$

$N_{D^3}(\boldsymbol{x}) \neq 0$. Consequently, when $\boldsymbol{x}$ is in the region enclosed by (0, 0), (1, 0), (1, 1),

$$v \leqslant u \leqslant 1, 0 \leqslant v \leqslant 1$$
$$\therefore 0 \leqslant t \leqslant v$$
$$\therefore N_{D^3}(\boldsymbol{x}) = \int_0^v dt = v.$$

When $\boldsymbol{x}$ is in the region enclosed by (1, 1), (2, 2), (2, 1),

$$v \leqslant u \leqslant 2, 1 \leqslant v \leqslant 2$$
$$\therefore N_{D^3}(\boldsymbol{x}) = \int_{u-1}^1 dt = 2 - u,$$

When $\boldsymbol{x}$ is in the region enclosed by (1, 0), (1, 1), (2, 1),

$$1 \leqslant u \leqslant 1 + v, 0 \leqslant v \leqslant 1$$
$$N_{D^3}(\boldsymbol{x}) = \int_{u-1}^v dt = 1 - u + v,$$

Similarly, we can deduce those expressions of $N_{D^3}(\boldsymbol{x})$ when $\boldsymbol{x}$ is in other regions of supp (D^3). Those expressions are shown in Fig. 2.31a. Figure 2.31b gives out the shape of the function $N_{D^3}(\boldsymbol{x})$. When $(u, v) \notin \text{supp}(D^3)$,

$$\boldsymbol{x} - t\boldsymbol{s}_0 = (u - t, v - t) \notin \text{supp}(D^3)$$
$$\therefore N_{D^3}(\boldsymbol{x}) = \int_0^1 0dt = 0.$$

We now investigate the box spline defined by $D^4 = \{\boldsymbol{s}_0, \boldsymbol{s}_1, \boldsymbol{s}_2, \boldsymbol{s}_3\}$. G_D^4 is shown in Fig. 2.32a. Let $\boldsymbol{x} = (u, v)$, we have

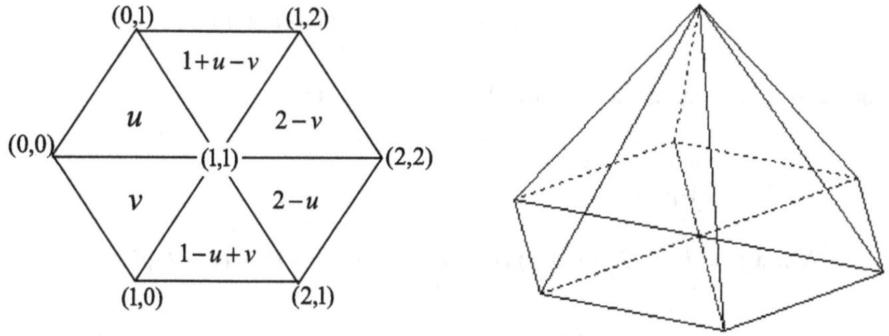

(a) Expressions of $N_{D^3}(\boldsymbol{x})$ (b) Shape of $N_{D^3}(\boldsymbol{x})$

Fig. 2.31 Box spline defined by $D^3 = \{s_0, s_1, s_2\}$

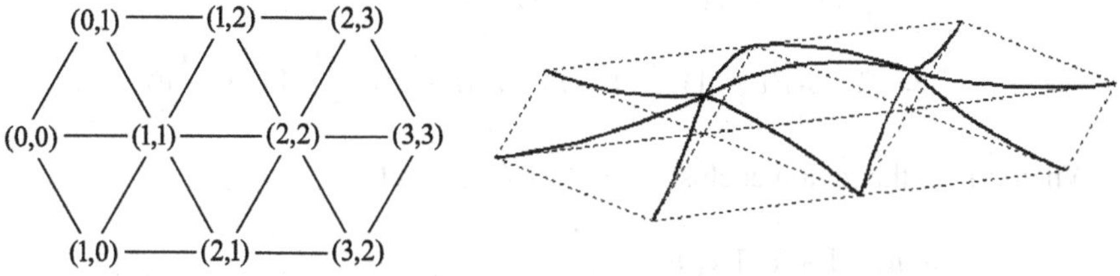

(a) Support region of D^k (b) Shape of $N_{D^4}(\boldsymbol{x})$

Fig. 2.32 Box spline defined by $D^k = \{s_0, s_1, s_2, s_3\}$

$$\boldsymbol{x} - ts_2 = (u - t, v - t)$$

$\because \boldsymbol{x} \in [\boldsymbol{v}_1, \boldsymbol{v}_2, \boldsymbol{v}_3, \boldsymbol{v}_4][0, 1]^3$ where $N_{D^4}(\boldsymbol{x}) \neq 0$
$\therefore \max\{u - 2, v - 2, 0\} \leqslant t \leqslant \min\{u, v, 1\}$.
When $\boldsymbol{x}$ is in the region enclosed by (0, 0), (1, 0), (1, 1),

$$v \leqslant u \leqslant 1, 0 \leqslant v \leqslant 1$$
$$\therefore 0 \leqslant t \leqslant v$$
$$\therefore N_{D^4}(\boldsymbol{x}) = \int_0^v (v - t)dt = \frac{1}{2}v^2.$$

When $\boldsymbol{x}$ is in the region enclosed by (1, 0), (1, 1), (2, 1),

$$1 \leqslant u \leqslant 1 + v, 0 \leqslant v \leqslant 1$$
$$\therefore 0 \leqslant t \leqslant v$$
$$N_{D^4}(\boldsymbol{x}) = \int_0^{u-1} (1 - u + v)dt + \int_{u-1}^{v} (v - t)dt$$

$$= (1 - u + v)(u - 1) + \frac{1}{2}(1 - u + v)^2.$$

When $\boldsymbol{x}$ is in the region enclosed by (1, 1), (2, 2), (2, 1),

$$v \leqslant u \leqslant 2, 1 \leqslant v \leqslant 2$$
$$\therefore 0 \leqslant t \leqslant 1$$
$$N_{D^4}(\boldsymbol{x}) = \int_0^{v-1} (2 - u + t)dt + \int_{v-1}^{u-1} [1 - (u - t) + (v - t)]dt + \int_{u-1}^{1} (v - t)dt$$
$$= (2 - u)(v - 1) + \frac{1}{2}(v - 1)^2 + (1 - u + v)(u - v) + \frac{1}{2}(1 - u + v)^2 - \frac{1}{2}(v - 1)^2$$
$$= (2 - u)(v - 1) + (1 - u + v)(u - v) + \frac{1}{2}(1 - u + v)^2.$$

When $\boldsymbol{x}$ is in the region enclosed by (2,2), (3,2), (2,1),

$$2 \leqslant u \leqslant 1 + v, 1 \leqslant v \leqslant 2$$
$$\therefore u - 2 \leqslant t \leqslant 1$$
$$N_{D^4}(\boldsymbol{x}) = \int_{u-2}^{v-1} (2 - u + t)dt + \int_{v-1}^{1} (1 - u + v)dt$$
$$= (2 - u)(v - u + 1) + \frac{1}{2}(v - 1)^2 - \frac{1}{2}(u - 2)^2 + (1 - u + v)(2 - v).$$

When $\boldsymbol{x}$ is in the region enclosed by (2,2), (3,3), (3,2),

$$v \leqslant u \leqslant 3, 2 \leqslant v \leqslant 3$$
$$\therefore u - 2 \leqslant t \leqslant 1$$
$$N_{D^4}(\boldsymbol{x}) = \int_{u-2}^{1} (2 - u + t)dt = (2 - u)(3 - u) + \frac{1}{2} - \frac{1}{2}(u - 2)^2.$$

Similarly, we can obtain expressions of $N_{D^4}(\boldsymbol{x})$ in other regions of supp (D^4). The shape of $N_{D^4}(\boldsymbol{x})$ is shown in Fig. 2.32b.

Based on Formula (2.30), it is easy to know that it only makes a convolution for $N_{D^{k-1}}(\boldsymbol{x})$ and $h(t\boldsymbol{t}_k)$ in the $\boldsymbol{t}_k$ direction to obtain $N_{D^k}(\boldsymbol{x})$. The process is shown in Fig. 2.33, where

$$h(t\boldsymbol{t}_k) = \begin{cases} 1 & t \in [0, 1) \\ 0 & t \in (-\infty, 0) \cup [1, +\infty). \end{cases}$$

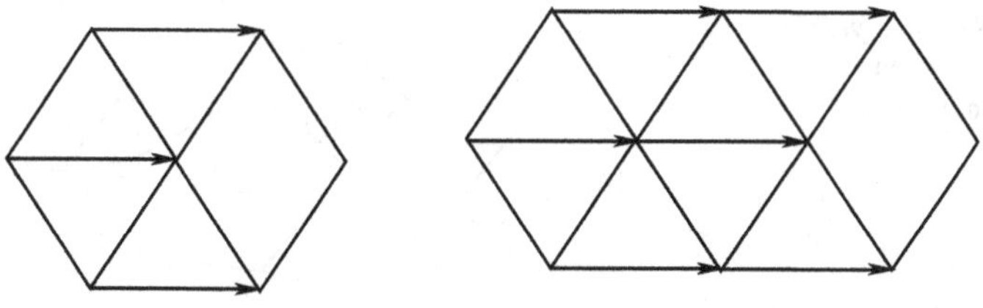

(a) From $N_D{}^2(x)$to $N_D{}^3(x)$ (b) From $N_D{}^3(x)$to $N_D{}^4(x)$

Fig. 2.33 Process of convolution

Actually, when we evaluate a box spline $N_{D^4}(\boldsymbol{x})$, the following recursive formula is applicable [95]:

$$(k-2)N_{D^k}(\boldsymbol{x}) = \sum_{i=1}^{k}[t_i N_{\{D^k\backslash \boldsymbol{t}_i\}}(\boldsymbol{x}) + (1-t_i)N_{\{D^k\backslash \boldsymbol{t}_i\}}(\boldsymbol{x}-\boldsymbol{t}_i)],$$

where $\{D^k \setminus \boldsymbol{t}_i\}$ denotes the vector group obtained by deleting $\boldsymbol{t}_i$ for D^k. If we only want to show the shape of a box spline $N_{D^k}(\boldsymbol{x})$, subdivision can be used. How to make a subdivision will be discussed in later sections.

2.5.3 Basic Properties of Box Splines

From Definition 2.7 and the discussion of Sect. 2.5.2, box splines have the following properties [94]:

(1) It does not depend on the ordering of vectors $\boldsymbol{t}_i (i = 1, \ldots, k)$;

(2) It is non-negative over the support region supp (D^k):

$$N_{D^k}(\boldsymbol{x}) \begin{cases} > 0 & O(\text{supp}(D^k)) \\ = 0 & \boldsymbol{x} \notin O(\text{supp}(D^k)), \end{cases}$$

where $O(\text{supp}(D^k))$ is an open set which is the inner region of supp (D^k).

(3) It is symmetric with respect to the center of its support region.

(4) It is $k-2$ degree polynomial over each tile of this partition.

Further, let $N_{D^k}(\boldsymbol{y}) := N_{D^k}(\boldsymbol{x}+y\boldsymbol{t}_r)$. If $\boldsymbol{t}_r \notin \text{span}\{D^k \backslash \boldsymbol{t}_r\}$, $N_{D^k}(\boldsymbol{y})$ is piece constant in $\boldsymbol{t}_r$ direction. It is easy to be found in the case that $D^4 = \{\boldsymbol{t}_0, \boldsymbol{t}_0, \boldsymbol{t}_0, \boldsymbol{t}_1\}$. In the $\boldsymbol{t}_1$ direction, $N_{D^k}(\boldsymbol{y})$ is piece constant which is shown in Fig. 2.34.

If $\boldsymbol{t}_r \in \text{span}\{D^k \setminus \boldsymbol{t}_r\}$, then $N_{D^k}(\boldsymbol{y})$ is continuous since it can be obtained by a convolution from $N_{D^k\backslash \boldsymbol{v}_r}(\boldsymbol{y}) = N_{D^k\backslash \boldsymbol{v}_r}(\boldsymbol{x} + y\boldsymbol{t}_r)$,

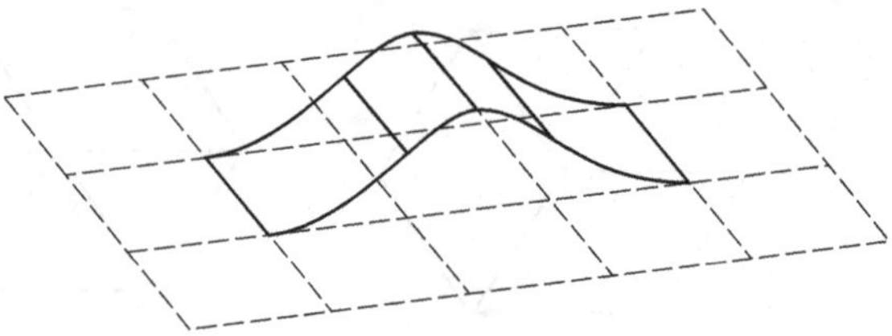

Fig. 2.34 $N_{D^k}(\boldsymbol{x} + y\boldsymbol{t}_r)$ is piece constant when $D^4 = \{\boldsymbol{s}_0, \boldsymbol{s}_0, \boldsymbol{s}_0, \boldsymbol{s}_1\}$

$$N_{D^k}(y) = \int_0^1 N_{D^k\backslash t_r}(y-t)dt = \int_{y-1}^{y} N_{D^k\backslash t_r}(t)dt$$
$$= \int_{-\infty}^{y} [N_{D^k\backslash t_r}(t) - N_{D^k\backslash t_r}(t-1)]dt.$$

Further, the directional derivative with respect to $\boldsymbol{v}_r$ is given by

$$\frac{N_{D^k}(\boldsymbol{x})}{\partial \boldsymbol{t}_r} = \left.\frac{N_{D^k}(y)}{\partial y}\right|_{y=0} = N_{D^k\backslash t_r}(\boldsymbol{x}) - N_{D^k\backslash t_r}(\boldsymbol{x} - \boldsymbol{v}_r).$$

If $D^k \setminus \boldsymbol{t}_r$ spans space R^2 (i.e., plane), then $N_{D^k}(\boldsymbol{x})$ is continuous and its directional derivatives can be written as a translating linear combination of box splines $N_{D^k\backslash \boldsymbol{v}_r}(\boldsymbol{x})$. Applying this argument repeatedly, we can see that

(5) $N_{D^k}(\boldsymbol{x})$ is $s-1$ times continuously differentiable in $\boldsymbol{t}_r$ direction if there are $s\boldsymbol{t}_r$ vectors in the vector group D^k.

(6) $N_{D^k}(\boldsymbol{x})$ is normative:

$$\int_{R^s} N_{D^k}(\boldsymbol{x})d\boldsymbol{x} = 1 \quad \text{or} \quad \sum_{\boldsymbol{i}} N_{D^k}(\boldsymbol{x} - \boldsymbol{i}) = 1,$$

where $R^s = \text{span}(D^k)$. Actually,

$$\int_{R^s} N_{D^k}(\boldsymbol{x})d\boldsymbol{x} = \int_{R^s}\int_0^1 D^{k-1}(\boldsymbol{x} - t\boldsymbol{v}_k)dtd\boldsymbol{x}$$
$$= \int_0^1 \left[\int_{R^s} N_{D^{k-1}}(\boldsymbol{x} - t\boldsymbol{v}_k)d\boldsymbol{x}\right]dt = \int_0^1 dt = 1.$$

2.6 Box Spline Surfaces

2.6.1 Definition and Properties

Definition 2.8 Let Γ_{D^k} be the plane grid formed by gradually translating G_{D^k} to each knot. $\boldsymbol{i}$ denotes the knot of Γ_{D^k}. We call

$$p(x) = \sum_i P_i N_{D^k}(x - i) \tag{2.31}$$

the box spline surface, where $P_i = (P_{ix}, P_{iy}, P_{iz})$ is the point in a three-dimensional space. It is called control vertices of $S(x)$. The mesh M formed by $\{P_i\}$ is topology isomorphic to Γ_{D^k} and is the control mesh of $S(x)$. In the case of not raising confusion, Formula (2.30) is usually denoted as:

$$p(x) = \sum_i P_i N_i(x) \tag{2.31$'$}$$

Based on basic properties of $N_{D^k}(x)$, we can know that:

• $p(x)$ is a polynomial over each tile of the given plane partition

• $p(x)$ is $s - 1$ times continuously differentiable in t_r direction if there are s t_r vectors in the vector group D^k.

• $p(x)$ is in convexity formed by the mesh M.

• Pulling a control vertex P_i, only a local region of $p(x)$ changes its shape.

Obviously, uniform B-spline surfaces are unique cases of box spline surfaces.

2.6.2 Subdivision of Box Spline Surfaces

Let $D = \{t_1, t_2, \cdots, t_k\}$ and the basic function defined by the vector group be $N^1(x)$. Its support mesh is G_D, and the grid Γ^1 is obtained after translating G_D to each knot. The box spline surface defined by $N^1(x)$ is:

$$p(x) = \sum_i P_i^1 N_i^1(x) = \sum_i P_i^1 N^1(x - i).$$

We now make a halving refinement, i.e., add a midpoint for each edge of the grid Γ^1, and then link new vertices and old vertices to form a grid Γ^2 topology isomorphic to Γ^1. So, we can calculate a new control vertex set $\{P_j^2 | j \in Z^2/2\}$ and make

$$p(x) = \sum_j P_j^2 N^1[2(x - j)], \quad j \in Z^2/2.$$

Rules according to which we calculate new P_j^2 are as follows: [94]

$$\begin{aligned} d_j^0 &= \begin{cases} 0 & j \notin Z^2 \\ P_j^1 & j \in Z^2 \end{cases} \\ d_j^r &= d_j^{r-1} + d_{j-t_r/2}^{r-1}, r = 1, \ldots, k \\ P_j^2 &= 2^{-(k-2)} d_j^k. \end{aligned} \tag{2.32}$$

Based on the above recursive refinement formula, we can find the relationship between basic functions $N_D(\boldsymbol{x}-\boldsymbol{i})$ and $N_D(2(\boldsymbol{x}-\boldsymbol{j}))$, where $\boldsymbol{i} \in \mathbf{Z}^2$, $\boldsymbol{j} \in \mathbf{Z}^2/2$. As for a box spline basic function $N^1(\boldsymbol{x})$, we have

$$N^1(\boldsymbol{x}) = \sum_i \boldsymbol{P}_i^1 N_i^1(\boldsymbol{x}) = \sum_i \boldsymbol{P}_i^1 N^1(\boldsymbol{x}-\boldsymbol{i}),$$

where $\boldsymbol{P}_i^1 = \begin{cases} 1 & \boldsymbol{i} = (0,0) \\ 0 & other \end{cases}$

By the recursive refinement formula (2.32), we have

$$N_D(\boldsymbol{x}) = \sum_j c_j N_D[2(\boldsymbol{x}-\boldsymbol{j})], \tag{2.33}$$

where $\boldsymbol{j} \in \text{knot}(G_D)/2$, $\text{knot}(G_D)$ is the knot set of the mesh G_D. When $D = \{\boldsymbol{s}_0, \boldsymbol{s}_1, \boldsymbol{s}_2, \boldsymbol{s}_3\}$, the recursive process is shown in Fig. 2.35.

Based on Fig. 2.35, we know that the recursive process can be executed directly on the initial grid L_D. Figure 2.36 gives the recursive process when $D = \{\boldsymbol{s}_0, \boldsymbol{s}_0, \boldsymbol{s}_1, \boldsymbol{s}_1, \boldsymbol{s}_2, \boldsymbol{s}_2\}$. It should be known that the result is not connected with the direction order adopted. For example, to obtain the result in Fig. 2.36c, we can also execute in the following steps: once in $\boldsymbol{s}_2$ direction, twice in $\boldsymbol{s}_0$ direction, twice in $\boldsymbol{s}_1$ direction, once in $\boldsymbol{s}_2$ direction.

In Figs. 2.35 and 2.36, those coefficients should lastly multiply $2^{-(k-2)}$, i.e., the third in Formula (2.32) should be executed.

2.6.3 *Generating Function of Box Spline*

To obtain those coefficients c_j in Formula (2.33), we can use those recursive rules in Formula (2.32) and we can also use the generating function [95] introduced in this section. Assume vector group $D = \{\boldsymbol{s}_0, \boldsymbol{s}_1\}$. Obviously, $\boldsymbol{s}_0, \boldsymbol{s}_1$ are linear independent. Similar to subdivision Formula (2.21) of the univariate basic function, we can get subdivision formula of bivariate box spline:

$$\begin{aligned} N_D(\boldsymbol{x}) &= \sum_{i_1=0}^{1}\sum_{i_2=0}^{1} N_D(2\boldsymbol{x}-\boldsymbol{i}) \\ &= N_D(2\boldsymbol{x}) + N_D(2\boldsymbol{x}-\boldsymbol{s}_0) + N_D(2\boldsymbol{x}-\boldsymbol{s}_1) + N_D(2\boldsymbol{x}-\boldsymbol{s}_2). \end{aligned} \tag{2.34}$$

So, we can define a generating function for $N_D(\boldsymbol{x})$:

$$C(z) = \prod_{j=0}^{1}(1+z_j) = (1+z_0)(1+z_1) = 1 + z_0 + z_1 + z_0 z_1.$$

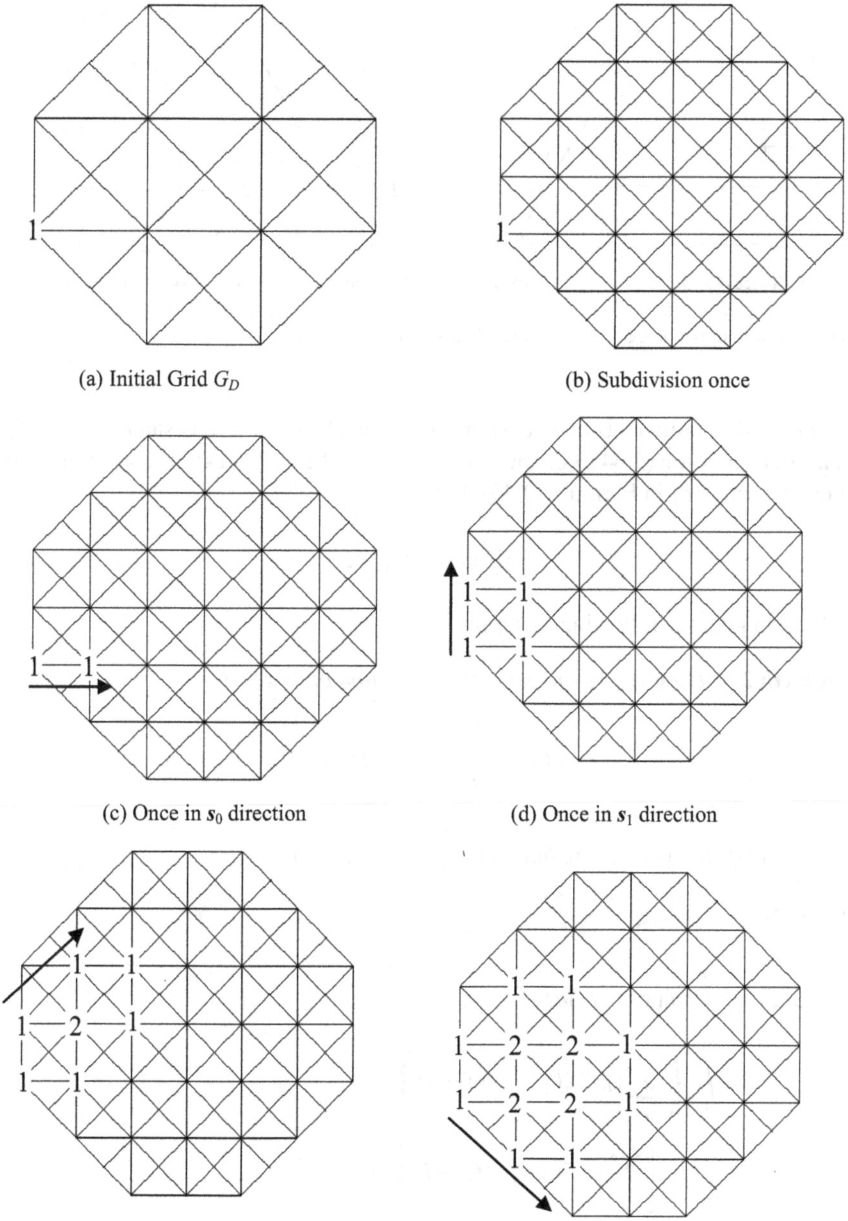

(a) Initial Grid G_D

(b) Subdivision once

(c) Once in s_0 direction

(d) Once in s_1 direction

(e) Once in s_2 direction

(f) Once in s_3 direction

Fig. 2.35 Recursive process defined by Formula (2.32)

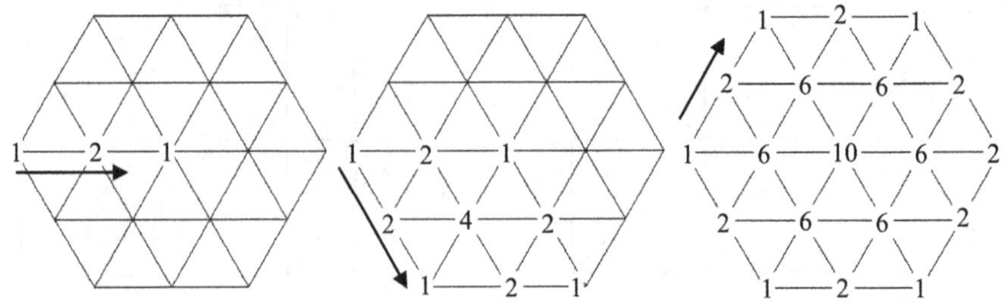

(a) Twice in s_2 direction (b) Twice in s_0 direction (c) Twice in s_1 direction

Fig. 2.36 Recursive process on G_D, $D = \{s_0, s_0, s_1, s_1, s_2, s_2\}$

Obviously, coefficients of extension expression of $C(z)$ are consistent with coefficients of expression (2.34), i.e., members z_0 and z_1 of z are grid knots and coefficient of every term is c_j in Formula (2.33). Let

$$z^{t_k} = z_0^{t_k[0]} z_1^{t_k[1]}.$$

We can give the following theorems:

Theorem 2.4 *Let $C(z)$ be the generating function of $f(x)$. If*

$$g(x) = \int_0^1 f(x - tt_k)dt,$$

then $g(x)$ has the generating function: $\frac{1}{2}C(z)(1 + z^{v_k})$.

Proof By hypothesis,

$$\begin{aligned} g(x) &= \int_0^1 f(x - tt_k)dt \\ &= \int_0^1 \left(\sum_i c_i f[2(x - tt_k) - i] \right) dt \\ &= \frac{1}{2}\int_0^2 \left[\sum_i c_i f(2x - tt_k - i) \right] dt \\ &= \frac{1}{2}\int_0^1 \left[\sum_i c_i f(2x - tt_k - i) + \sum_i c_i f(2x - tt_k - i - t_k) \right] dt \end{aligned}$$

$$= \frac{1}{2}\int_0^1 \left[\sum_i (c_i + c_{i-t_k}) f(2\boldsymbol{x} - t\boldsymbol{t}_k - \boldsymbol{i})\right] dt$$

$$= \frac{1}{2}\left[\sum_i (c_i + c_{i-t_k}) \int_0^1 f(2\boldsymbol{x} - t\boldsymbol{t}_k - \boldsymbol{i}) dt\right]$$

$$= \frac{1}{2}\sum_i [(c_i + c_{i-t_k}) g(2\boldsymbol{x} - \boldsymbol{i})]$$

So, the generating function of $g(\boldsymbol{x})$ is exactly $\frac{1}{2}C(z)(1 + z^{\boldsymbol{v}_k})$. #

Based on Theorem 2.4, we know that the generating function associated with the subdivision Formula (2.33) is:

$$C_D(z) = 4\prod_{i=0}^{k-1} \frac{1}{2}(1 + z^{t_k}).$$

When $D = \{\boldsymbol{s}_0, \boldsymbol{s}_1, \boldsymbol{s}_2\}$,

$$\begin{aligned} C_D(z) &= \frac{1}{2}(1 + z_0)(1 + z_1)(1 + z_0 z_1) \\ &= \frac{1}{2} + \frac{1}{2}z_0 + \frac{1}{2}z_1 + z_0 z_1 + \frac{1}{2}z_0^2 z_1 + \frac{1}{2}z_0 z_1^2 + \frac{1}{2}z_0^2 z_1^2. \end{aligned}$$

2.7 Subdivision Mask of Box Spline Surface

Based on the subdivision Formula (2.33), we can deduce subdivision mask of box spline surfaces. We firstly investigate quadric three-directional box spline surfaces [1] whose basic function is $N_D(\boldsymbol{x})$ with the vector group $D = \{\boldsymbol{e}_1, \boldsymbol{e}_1, \boldsymbol{e}_2, \boldsymbol{e}_2, \boldsymbol{e}_3, \boldsymbol{e}_3\}$. Coefficients in subdivision Formula (2.33) are shown in Fig. 2.34. We regard the grid start vertex of $N_D(\boldsymbol{x})$ as(0,0). After translating $N_D(\boldsymbol{x})$ and let its grid start vertex be (i, j), we denote $N_D(\boldsymbol{x})$ by $N_{i,j}(\boldsymbol{x})$. Obviously,

$$N_D(\boldsymbol{x}) = N_{0,0}(\boldsymbol{x}).$$

In this case, box spline surface (2.31′) can be written as:

$$\boldsymbol{p}(\boldsymbol{x}) = \sum_i \sum_j \boldsymbol{P}_{i,j} N_{i,j}(\boldsymbol{x}). \tag{2.35}$$

Now, we refine the plane grid formed by G_D and knots of the grid of being refined are still numbered by integers. On the new grid, the box spline whose start vertex is

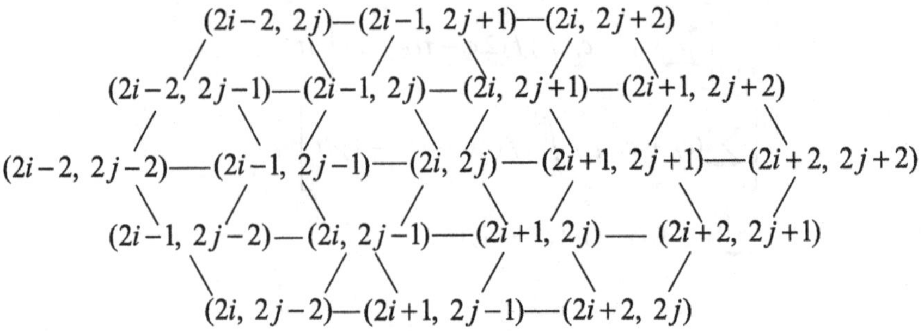

Fig. 2.37 Subdivision coefficient matrix of quadric three-directional box spline surface

(i, j) is denoted by $n^1_{i,j}(\boldsymbol{x})$. So, Formula (2.33) can be rewritten as:

$$N_{i,j}(\boldsymbol{x}) = \sum_{s=0}^{2}\sum_{t=0}^{2}\sum_{l=0}^{2} \beta_{s+l,t+l} N^1_{2i-2+s+l,2j-2+t+l}(\boldsymbol{x}). \tag{2.36}$$

It is easy to know that, when

$$\boldsymbol{x} = \alpha_0 \boldsymbol{t}_0 + \cdots + \alpha_{k-1}\boldsymbol{t}_{k-1}, \alpha_i \in \{0, 1, 2\},$$

different vectors $[\alpha_0, \cdots, \alpha_{k-1}]$ can correspond to the same $\boldsymbol{x}$. Based on this reason, in the extended expression of (2.36), if the frequency of $\beta_{s+l,t+l}N^1_{s+l,t+l}(\boldsymbol{x})$ is a larger one, we only count one for $\beta_{s+l,t+l}N^1_{s+l,t+l}(\boldsymbol{x})$. Consequently, there are only 19 terms in the extended expression of (2.36). Let

$$N^1_{2i-2+s+l,2j-2+t+l} \to (2i-2+s+l, 2j-2+t+l).$$

Consequently, we can get the subdivision coefficient matrix of the control point $\boldsymbol{P}_{i,j}$ of the quadric three-directional box spline surfaces (2.34) and the matrix is shown in Fig. 2.37. The coefficient of every basic function in Fig. 2.37 is the value on the corresponding position of Fig. 2.36c.

Based on Fig. 2.37, we can know that there are seven control vertices whose subdivision coefficient matrixes include the basic function $N^1_{2i-2,2j-2}$:

$$\boldsymbol{P}_{i,j}, \boldsymbol{P}_{i-1,j-1}, \boldsymbol{P}_{i-2,j-2}, \boldsymbol{P}_{i,j-1}, \boldsymbol{P}_{i,j-2}, \boldsymbol{P}_{i-1,j}, \boldsymbol{P}_{i-2,j-1}.$$

Consequently, the new mesh vertex $\boldsymbol{P}^1_{2i-2,2j-2}$ is a linear combination of the above seven control vertices and coefficient of every vertex is shown in Fig. 2.38a. Similarly, $\boldsymbol{P}^1_{2i-1,2j-2}$ is a linear combination of four control vertices: $\boldsymbol{P}_{i,j}$, $\boldsymbol{P}_{i-1,j-1}$, $\boldsymbol{P}_{i,j-1}$, $\boldsymbol{P}_{i-1,j-2}$ and their coefficients are shown in Fig. 2.38b.

It is easy to know that the new vertex corresponding to a basic function in Fig. 2.37 has a subdivision mask in Fig. 2.38. Each vertex in the old mesh corresponds to a

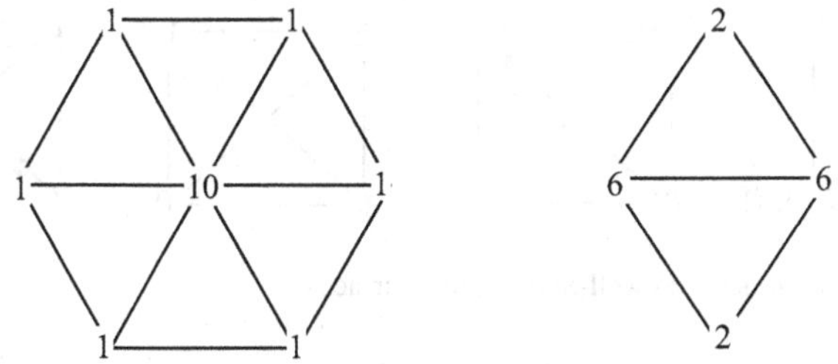

(a) Subdivision mask of new vertex point (b) Subdivision mask of new edge point

Fig. 2.38 Masks of quadric three-directional box spline surfaces

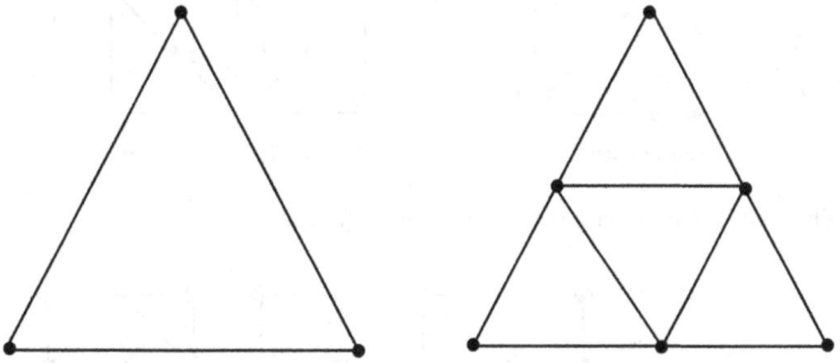

(a) Initial mesh (b) Mesh after subdivision

Fig. 2.39 Mesh topology of subdivision about quadric three-directional box spline surface

new vertex point, and each edge in the old mesh corresponds to a new edge point. The topology relation between the new mesh and the old mesh is shown in Fig. 2.39.

We now research the Powell-Sabin spline surface [96]. The vector group of its basic function is $D = \{\boldsymbol{e}_1, \boldsymbol{e}_2, \boldsymbol{e}_3, \boldsymbol{e}_4\}$. In this case, Formula (2.33) can be rewritten as:

$$N_{i,j} = \sum_{s=0}^{1}\sum_{t=0}^{1}\sum_{m=0}^{1}\sum_{n=0}^{1} \beta_{s+m+n,t+m-n} N^1_{2i-2+s+m+n,2j-2+s+m-n}. \tag{2.37}$$

In the extended expression of (2.31), if the frequency of $\beta_{s+m+n,t+m-n}$ $N^1_{2i-2+s+m+n,2j-2+s+m-n}$ is a larger one, we only count one for $\beta_{s+m+n,t+m-n}$ $N^1_{2i-2+s+m+n,2j-2+s+m-n}$. Consequently, there are only 12 terms in the extended expression of (2.37). Through the definition of the subdivision coefficient matrix, we can know that the new mesh vertex $\boldsymbol{P}^1_{2i-2,2j-2}$ is a linear combination of these three control vertices: $\boldsymbol{P}_{i,j}$, $\boldsymbol{P}_{i-1,j-1}$, $\boldsymbol{P}_{i-1,j}$. Their coefficients are shown in Fig. 2.40. It is easy to know that the new vertex corresponding to a basic function in right expression of Eq. (2.37) has a subdivision mask in Fig. 2.41.

Only concerned with topology structure, subdivision of the Powell-Sabin spline surface is similar to subdivision of the uniform bi-quadric B-spline surface. The process is shown in Fig. 2.42a, c. Since there are four directions, we can transform

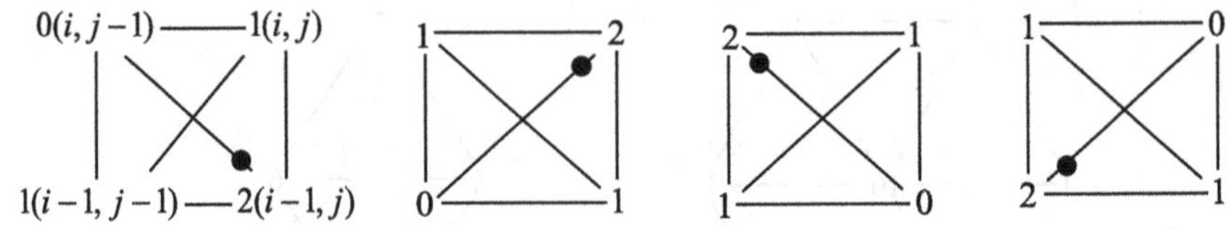

Fig. 2.40 Subdivision masks of Powell-Sabin spline surfaces

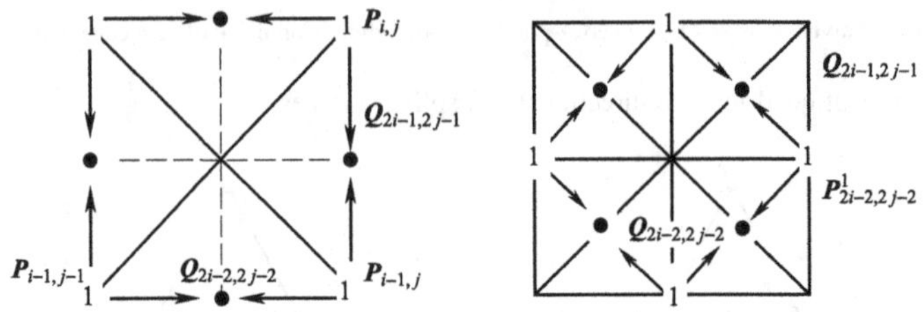

(a) Mask of step one (b) Mask of step two

Fig. 2.41 Split of subdivision mask of Powell-Sabin spline surface

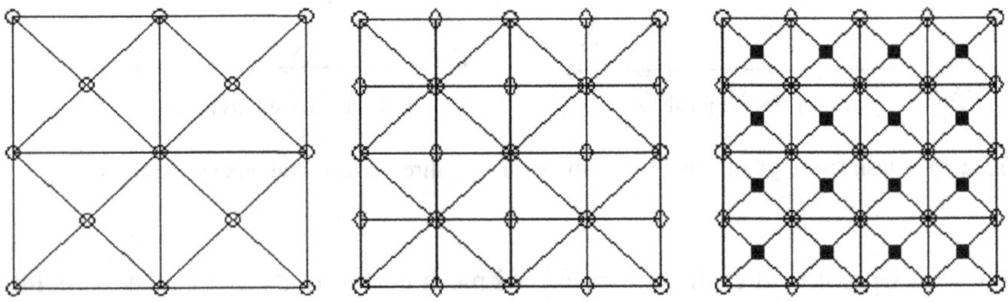

(a) Initial mesh (b) First subdivision (c) Second subdivision

Fig. 2.42 Subdivision by split masks of Powell-Sabin spline surface

the mesh shown in Fig. 2.42a to the mesh shown in Fig. 2.42c by two steps, i.e., a subdivision shown in Fig. 2.40 can be split into two steps shown in Fig. 2.41. Figure 2.42 gives the full process of subdivision by split masks. It should be noticed that every coefficient in Fig. 2.40 should be multiplied by 1/4 and every coefficient in Fig. 2.31 should be multiplied by 1/2.

You are probably thinking about why these two subdivision masks in Fig. 2.42 can be united to the subdivision masks in Fig. 2.41. For the new vertex $\boldsymbol{P}^1_{2i-2,2j-2}$ (see Fig. 2.41),

$$\boldsymbol{Q}_{2i-2,2j-2} = \frac{1}{2}\boldsymbol{P}_{i-1,j-1} + \frac{1}{2}\boldsymbol{P}_{i-1,j}, \quad \boldsymbol{Q}_{2i-1,2j-1} = \frac{1}{2}\boldsymbol{P}_{i-1,j} + \frac{1}{2}\boldsymbol{P}_{i,j}$$

$$\therefore \boldsymbol{p}^1_{2i-2,2j-2} = \frac{1}{4}\boldsymbol{P}_{i-1,j} + \frac{1}{2}\boldsymbol{P}_{i-1,j} + \frac{1}{4}\boldsymbol{P}_{i,j}.$$

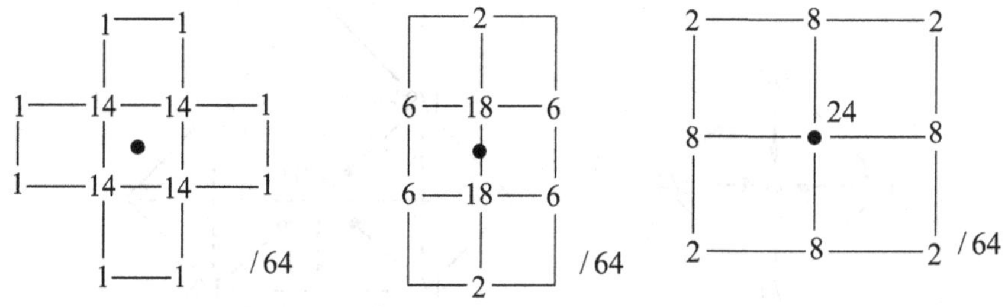

Fig. 2.43 Face, edge, and vertex masks of C^4 box spline surface

Consequently, the two subdivision masks in Fig. 2.41 can indeed be united to the subdivision masks in Fig. 2.40. Masks given in Fig. 2.41 are smaller than masks in Fig. 2.40.

We now research the quadric four-directional box spline surface [97]. The vector group of its basic function is $D = \{s_0, s_0, s_1, s_1, s_{2,} s_2, s_3, s_3\}$. By the subdivision coefficient matrix, we can get its subdivision masks shown in Fig. 2.43.

Only concerned with topology structure, subdivision of C^4 box spline surface is similar to subdivision of the uniform bi-cubic B-spline surface. However, the large support of the masks makes the implementation of the subdivision of C^4 box spline surface difficult. We decompose the C^4 smoothing operator into two masks as shown in Fig. 2.44. The face mask calculates positions of new added vertices with valance 4 in the new mesh. The vertex mask calculates positions of vertices with valance 8 in the new mesh. In the old mesh, valances of these vertices are 4 or 8. Figure 2.45 gives meshes on three consecutive subdivision levels, $j-1$, j, and $j+1$. The subdivision is executed by masks in Fig. 2.44.

Let M be a mesh. We obtain a mesh M^1 after we subdivide M once with the masks in Fig. 2.43.

We obtain another mesh $(M^1)'$ after we subdivide M twice with the masks in Fig. 2.44. Why is $(M^1)'$ the same as M^1? In order to ask the question, we label those vertices of the mesh in the subdivision level j in Fig. 2.43. These labels are shown in Fig. 2.46.

At the level $j+1$, the value of vertex u is computed using the vertex mask in Fig. 2.44:

$$\boldsymbol{u}^{j+1} = \frac{4}{8}\boldsymbol{u}^j + \frac{1}{8}(\boldsymbol{f}^j + g^j + \boldsymbol{k}^j + \boldsymbol{j}^j).$$

The value of $\boldsymbol{u}^j$ is calculated using the face mask in Fig. 2.44:

$$\boldsymbol{u}^j = \frac{1}{4}(\boldsymbol{f}^{j-1} + g^{j-1} + \boldsymbol{k}^{j-1} + \boldsymbol{j}^{j-1}).$$

The value of $\boldsymbol{f}^j$, g^j, $\boldsymbol{k}^j$, and $\boldsymbol{j}^j$ is calculated using the vertex mask in Fig. 2.44. For $\boldsymbol{f}^j$, we have

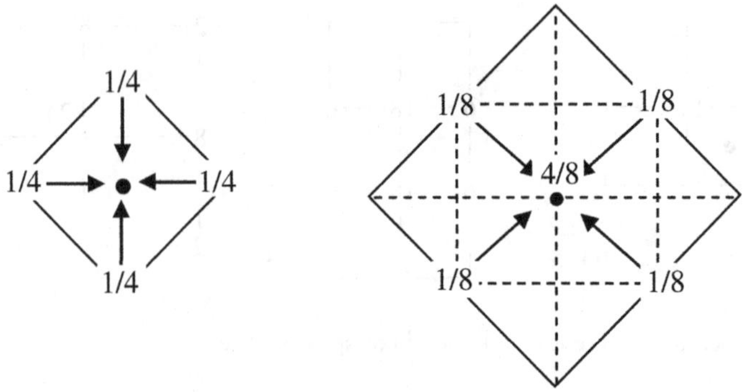

Fig. 2.44 Factorized face and vertex mask of C^4 box spline surface

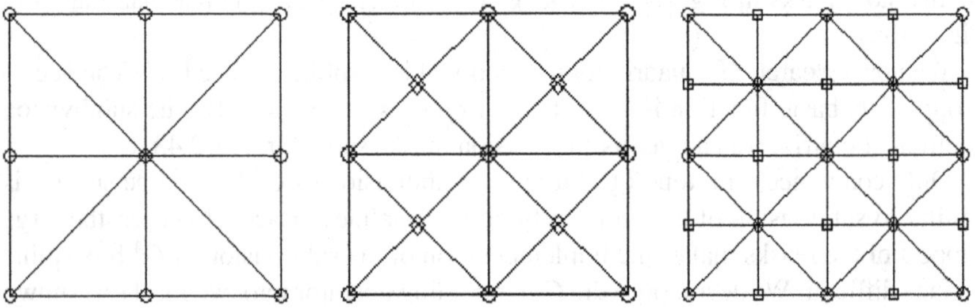

Fig. 2.45 Three consecutive subdivision levels, $j-1$, j, and $j+1$

Fig. 2.46 Labels of vertices of the mesh in the subdivision level j

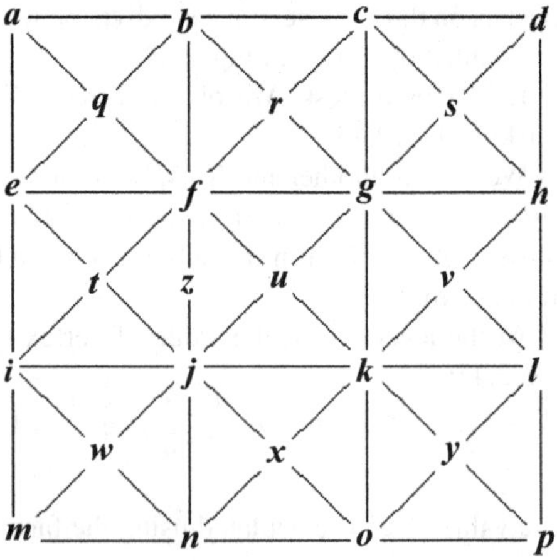

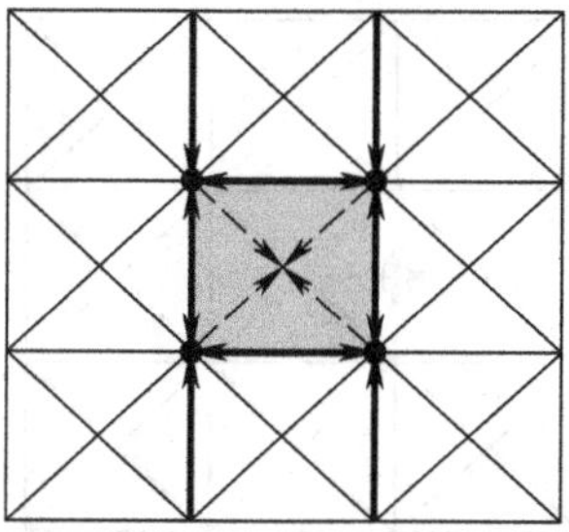

(a) New face point appear (diamond) in the first step

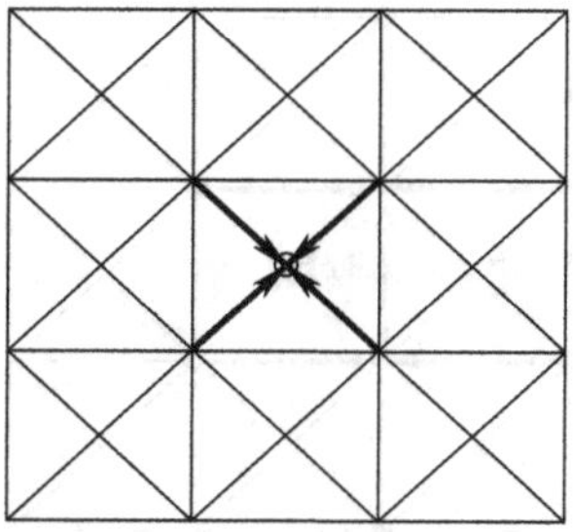

(b) The new face point is renewed using a vertex

Fig. 2.47 Decomposition of the face mask of C^4 box spline surface

$$f^j = \frac{4}{8} f^{j-1} + \frac{1}{8}(b^{j-1} + g^{j-1} + j^{j-1} + e^{j-1}).$$

Substituting the formulas of vertices u^j, f^j, g^j, k^j, and j^j into the equation for u^{j+1}, we obtain the face mask in Fig. 2.43, which gives the value of u^{j+1} in terms of the values of the vertices of the mesh at level $j-1$. Consequently, the face mask is decomposed into an application of face and vertex masks at level $j-1$, followed by an application of vertex mask at level j. Figure 2.47 shows the flow of computations used for updating vertex u.

At level $j+1$, the value of vertex z is computed using the face mask in Fig. 2.44:

$$z^{j+1} = \frac{1}{4}(f^j + u^j + j^j + t^j).$$

The values of f^j and j^j are calculated using the vertex mask, and the values of u^j and t^j are calculated using the face mask. The calculation for j^j is similar to the calculation of f^j, while the calculation for t^j is similar to the calculation of u^j. Substituting the formulas of f^j, u^j, j^j, and t^j into equation for z^{j+1}, we obtain the edge mask. Consequently, the edge mask is decomposed into an application of face and vertex masks at level $j-1$, followed by an application of face mask at level j. Figure 2.48 shows the flow of computations used for updating vertex z.

At level $j+1$, the value of vertex f is calculated using the vertex mask:

$$f^{j+1} = \frac{4}{8} f^j + \frac{1}{8}(r^j + u^j + t^j + q^j).$$

The values of f^j, r^j, u^j, t^j, and q^j are calculated using the vertex mask for calculating f^j and the face mask for calculating r^j, u^j, t^j, and q^j. The calculation for r^j, t^j, and q^j is similar to the calculation of u^j. Again, substituting the formulas of f^j, r^j, u^j, t^j, and q^j into the equation for f^{j+1}, we obtain the vertex mask. Consequently, the vertex mask is decomposed into an application of face and vertex

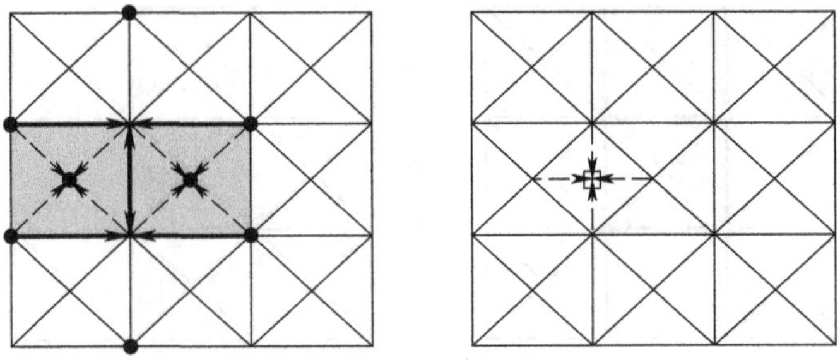

(a) New face point appears (diamond) in the first step

(b) The new edge point appears using a face

Fig. 2.48 Decomposition of the edge mask of C^4 box spline surface

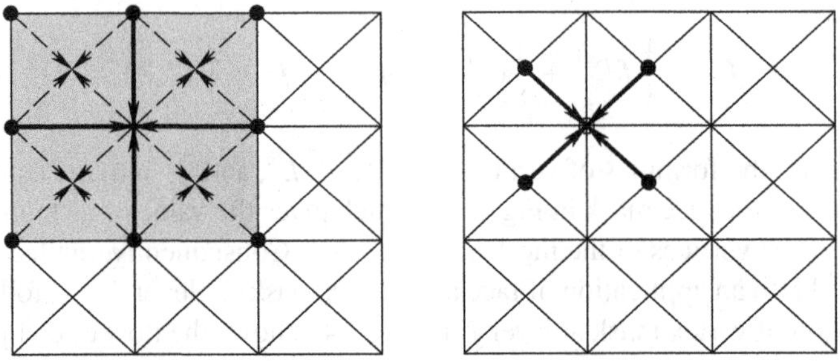

(a) New face point appears (diamond) in the first step

(b) The vertex point is renewed using a vertex mask (circle)

Fig. 2.49 Decomposition of the edge mask of C^4 box spline surface

masks at level j-1, followed by an application of vertex mask at level j. Figure 2.49 shows the flow of computations used for updating vertex $\boldsymbol{f}$.

Remarks

This chapter gives refinement rules for control meshes of spline surfaces. The refinement has two meanings: (1) double knot intervals and obtain a new spline function space; (2) densify vertices of old control meshes and obtain new control meshes. The refinement of control meshes of spline surfaces is also called subdivision of spline surfaces. We firstly obtain refinement rules of basic functions, and then deduce subdivision rules of spline surfaces by refinement rules of basic functions. For four-directional box spline surfaces, we decompose subdivision masks to obtain smaller subdivision masks because of subdivision masks directly derived from refinement rules of basic functions are large. There are several methods to deduce refinement rules of basic functions: the de-Boor algorithm and the convolution definition for B-spline basic functions, the recursive method, and the generating function for box spline basic functions. It should be noticed that B-spline is a unique case of box splines.

Exercises

1. For the knot vector $U = [\ldots, -3, -2, -1, 0, 1, 2, 3, \ldots]$, please write program codes to compute the B-splines: $N_{0,2}(u)$, $N_{0,3}(u)$, $N_{0,4}(u)$, $N_{0,5}(u)$ using the formula (2.1) and render the curves of these spline functions.
2. Assume that $\boldsymbol{P}_0 = [0, 0]$, $\boldsymbol{P}_1 = [1, 1]$, $\boldsymbol{P}_3 = [3, 0]$. The cubic spline curve is $\boldsymbol{p}(u) = \sum_{i=0}^{3} \boldsymbol{P}_i N_{i,3}(u)$, where $U = [0\,0\,0\,0\,1\,1\,1\,1]$. Compute its point using the de-Boor algorithm (see Formula (2.7)) and render the curve.
3. For the curve in the above item, double the knot interval [01] as [0 0.51] and then obtain the knot vector $U^1 = [0\,0\,0\,0\,0.5\,1\,1\,1\,1]$. Please compute the new control vertices for the curve.
4. For the curve in the item 2, double knot intervals with nonzero length k times and then the knot vector $U^k = [0, 0, 0, 0, u_1, u_2, \ldots, u_n, 1, 1, 1, 1]$. Please compute the new control vertices after every doubling step.
5. Using Formula (2.11), write program codes to subdivide the control polygon of a non-uniform cubic B-spline curve.
6. Using Formula (2.13)~(2.15), write program codes to subdivide the control mesh of a non-uniform bi-cubic B-spline surface.
7. For uniform cubic B-spline, when double knots, please deduce the formula:

$$N_{i,3}(u) = \frac{1}{8}N^1_{2i-2,3}(u) + \frac{4}{8}N^1_{2i-1,3}(u) + \frac{6}{8}N^1_{2i,3}(u) + \frac{4}{8}N^1_{2i+1,3}(u) + \frac{1}{8}N^1_{2i+2,3}(u).$$

8. See Definition 2.4. Write program codes to construct the grid G^3_D according to vectors in D^3.

Chapter 3
Meshes and Subdivision

By generalizing refinement rules of control meshes of spline surfaces, we can obtain many subdivision schemes of meshes with arbitrary topology. Most subdivision schemes only fit for two-manifold meshes though there are also non-manifold subdivision schemes [98] and volume subdivision schemes [16]. This chapter gives a description for two-manifold meshes and these regular two-manifold meshes are classified. In light of the mesh classification, some typical subdivision schemes are introduced. For most subdivision schemes introduced in this chapter, they make subdivision meshes convergent to spline surfaces if the subdivided initial meshes are regular. That is, most subdivision schemes are generalizations of refinement rules of control meshes of spline surfaces.

3.1 Topological Structure of Meshes for Subdivision

We have motioned meshes several times—control meshes, polygon meshes, and mesh surfaces. *Mesh* is one of the most common concepts in subdivision modeling. What is *mesh* on earth? Does it have a strict definition? Here we give a description for mesh. *Mesh* is a geometric shape formed by *vertices*, *edges*, and *faces* in a space. Compared with *graphs*, it has an excessive element—face. Subdivision schemes obtained by generalizing refinement rules of control meshes of spline surfaces request that subdivided meshes are two-manifolds. In this book, we also mainly refer to two-manifold meshes. However, what is a two-manifold? If we lose sight of geometric positions of vertices and only consider topology relations of vertices, we can obtain a several mesh called simplicial complex. Referring to the definition of simplicial complex [19], Li [99] gives the definition of generalized simplicial complex as following:

W. Liao et al., *Subdivision Surface Modeling Technology*,
DOI 10.1007/978-981-10-3515-9_3

Definition 3.1 A generalized simplicial complex is a triple $K = (V, E, F)$, where V is the set of vertices and $V \subset \mathbf{Z}$ ($\mathbf{Z}$ denotes the integer set), E is the set of edges and $E \subset \{(i, j) \in V \otimes V\}$, F is the set of faces and

$$F \subset \bigcup_{k=3}^{|V|} \overbrace{(V \otimes V \otimes \cdots \otimes V)}^{k}.$$

In the above expression, $|V|$ denotes number of vertices. Let those arrays with the same cycle order denote the same element, such as, $(i, j) = (j, i)$, $(i_0, i_1, \cdots, i_k) = (i_1, \ldots, i_k, i_0)$, *etc.*. A generalized simplicial complex should satisfy the following conditions:

(1) Any edge of any face is in the set E: $\forall (i_0, i_1, \cdots, i_k) \in F$, $(i_s, i_{(s+1)\%(k+1)}) \in E$ $(0 \leqslant s \leqslant k)$.

(2) Any edge in the set E is an edge of a certain face: $\forall (i, j) \in E, \exists (\cdots, i, j, \cdots) \in F$.

(3) Any vertices in the set V is a endpoint of a certain edge: $\forall i \in V, \exists j$ so that $(i, j) \in E$.

(4) No more than two faces share an edge.

(5) For any two edges e_1 and e_2 whose common endpoint is $i \in V$, there exists a series of faces with the common vertex i : $f_0, f_1, \ldots, f_k$, so that e_1 and e_2 are, respectively, edges of f_0 and f_k. For every pair face, f_s and $f_{s+1} (s = 0, \ldots, k-1)$, have a shared edge.

(6) Any two faces at most have a shared edge.

This book calls this generalized simplicial complex as the *simplicial complex*. Conditions in Definition 3.1 ensure that meshes defined by simplicial complexes are two-manifold meshes. There are also non-manifold examples in Fig. 3.1 which respectively do not satisfy the conditions (2), (3), (4), (5), (5), (6) from left to right.

If an edge in a simplicial complex is shared by two faces, the edge is called an *internal edge*. Otherwise, it is called a *boundary edge*. If a vertex is an endpoint of a boundary edge, then the vertex is called a *boundary vertex*. If a face has a boundary vertex, the face is called a *boundary face*. Non-boundary vertex and face are, respectively, *internal vertex* and *internal face*. In Fig. 3.2, e_1 is a boundary edge and e_2 and e_3 are internal edges. v_1 is a boundary vertex and v_2 is an internal vertex. f_1 and f_2 are boundary faces, and f_3 is an internal face. A simplicial complex is

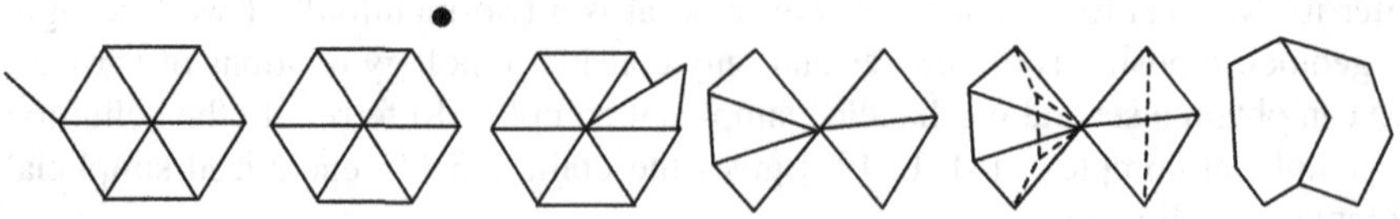

Fig. 3.1 Examples of non-manifold

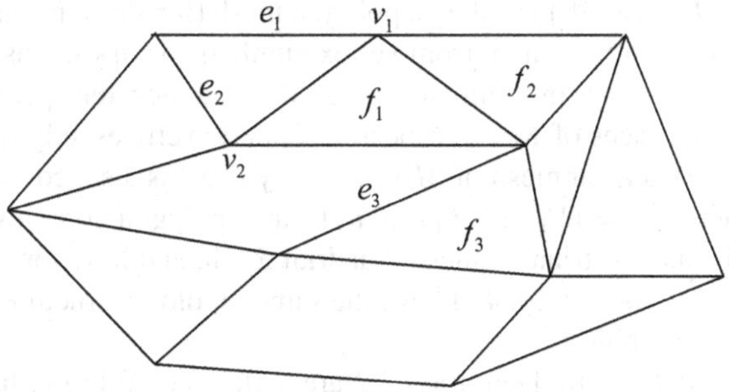

Fig. 3.2 Simplicial complex

called an open simplicial complex if it has boundaries. Otherwise it is called a close simplicial complex.

A subcomplex $K' \subseteq (V', E', F')$ of a complex K is a complex that satisfies $V' \subseteq V, E' \subseteq E$, and $F' \subseteq F$. It is easy to know that all vertices of a closed complex are internal vertices. If all faces in a complex have the same vertex number n, the complex is called an n-polygon complex. Two simplicial complexes K_1 and K_2 are *isomorphic* if there is a bijective simplicial map $K_1 \rightarrow K_2$.

For a vertex $i \in V$, if $\exists j \in V$ such that $e = (i, j) \in E$, e is called a neighbor edge of i. j is called a neighbor vertex of i. The number of neighbor edges of a vertex is called *valance* of the vertex. If there are vertices $i_1, \ldots, i_k \in V$ such that $f = (i, i_1, \ldots, i_k) \in F$, f is called a neighbor face of the vertex.

A 1-neighborhood $Ne_1(W)$ of a set of vertex $W \subset V$ is the subcomplex of K consisting of all polygons with at least one vertex in W, and their edges and vertices. An m-neighborhood $Ne_m(W)$ is defined recursively as a 1-neighborhood of the $(m-1)$-neighborhood. We also use notation $Ne_m(W, K)$ when we want to emphasize the complex in which we find the neighborhood. Figure 3.3 gives examples of 1-neighborhood and 2-neighborhood of the vertex v. In this case, $\{v\} = W$.

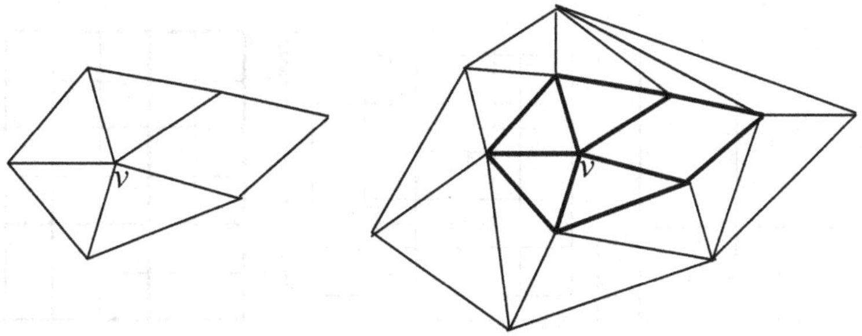

(a) 1-neighbourhood (b) 2-neighbourhood

Fig. 3.3 Neighborhood of v

Definition 3.2 $M = (K, \Phi)$ is called a polygon mesh (for short, mesh), where K is a complex, $\Phi : V \to R^n$ is a map from vertex numbers to a n-dimension space.

We only discuss meshes in 2-dimension space or 3-dimension space in this book. Vertices, edges, and faces of K are respectively called vertices, edges, and faces of M. $M_1 = (K_1, \Phi_1)$ is a submesh of M if and only if K_1 is a subcomplex of K and $\Phi|_{V_1} = \Phi_1$, where $K_1 = (V_1, E_1, F_1)$. If K is an n-polygon complex, M is called n-polygon mesh, such as triangle mesh, quadrial mesh, and hexagon mesh. We call $M_1 = (K_1, \Phi_1)$ and $M_2 = (K_2, \Phi_2)$ have the same topology structure if and only if K_1 and K_2 are isomorphic.

These meshes defined by Definition 3.2 are called manifold mesh. Most subdivision rules are given on two-manifold mesh. If there is no special explanation, all meshes in this book are manifold meshes. Non-manifold meshes are those meshes whose topology are not simplicial complexes, Ying [98] and Zorin [16] discusses the subdivision of non-manifold meshes and the subdivision of volume meshes.

3.2 Regular Mesh

What is a regular mesh? In order to have a clear explanation, we start from the control meshes of spline surfaces, which are regular meshes. Those control meshes are topologically equivalent to regular tilings of the plane composed of regular triangles, quadrilaterals, or hexagons. There are only three types of plane tilings formed by tiles that are congruent to a regular polygon. They correspond to uniform tessellations generated by equilateral triangles, squares, and hexagons. The above tilings have the desired property: It is possible to refine the tiles obtaining a new tiling made by similar elements of smaller size, as shown in Fig. 3.4. The most common ones are the triangle and quadrilateral tessellations.

A larger class of refinable tilings is the monohedral tilings with regular vertices, also known as Laves tilings [96]. In a monohedral tiling, if every tile is congruent to one fixed tile, it is called the prototile [97]. This means that all faces in the

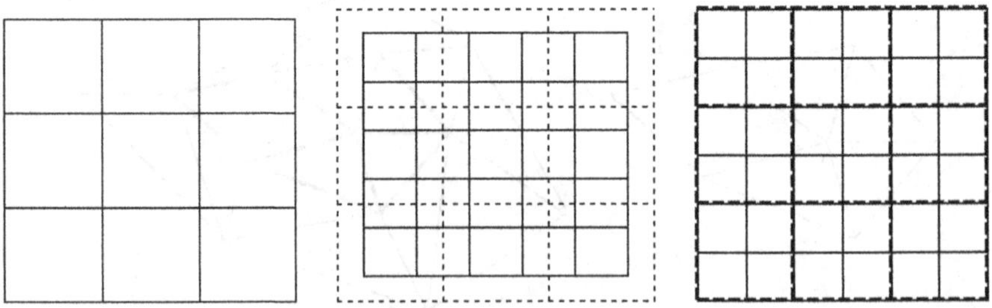

(a) Original tiles (b) Subdivision method I (c) Subdivision method II

Fig. 3.4 Subdivide quadrilateral tiles

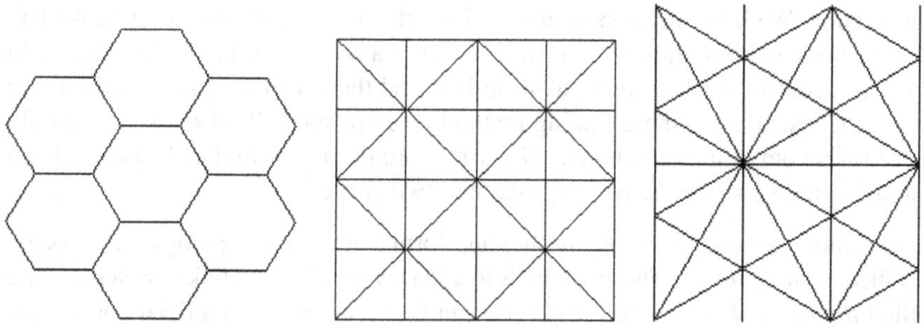

(a) Hexagon tiling (b) [4,8,8] Lave tiling (c) [4,6,12] Lave tiling

Fig. 3.5 Several tilings

tessellation have the same shape and size. A vertex $\boldsymbol{v}$ of a tiling is called regular if the angle between each consecutive pair of edges that are adjacent to $\boldsymbol{v}$ and is equal to $2\pi/n$, where n is the valence of $\boldsymbol{v}$. There are eleven tilings satisfying these two conditions. One is those whose tiles are regular polygons and the other is that whose vertices are regular. Regular vertex means that any angle around the vertex is $2\pi/n$ if the valence of the vertex is n. Figures 3.4 and 3.5 respectively gives 3 types of tilings.

We classify these tilings by listing the valence of the vertices of their prototiles in cyclic order. Prototiles can be regarded as faces of meshes. Thus, they are named using the notation $[n_0, n_1, \ldots, n_k]$, where n_i is the valence of vertex $\boldsymbol{v}_i$ (we also use superscripts to indicate repetition of symbols). As expected, regular triangle, quadrilateral, and hexagonal tilings belong to the Lave tiling class. They are, respectively, the Lave tilings of types $[6^3]$, $[4^4]$, and $[3^6]$. It is possible to extend all refinement concepts to the Lave tilings. For example, the control mesh of the quadric four-directional box spline surface has the structure of Lave tiling [4, 8, 8].

If a mesh has the topology structure of one of the Lave tilings, we call that the mesh is a *regular mesh*. Assume that a mesh has a similar topology structure with one of the Lave tilings. However, there are some vertices whose valance $|\boldsymbol{V}| \notin [n_0, n_1, \cdots, n_k]$ or some faces do not have the shape of one of the Lave tiles of the Lave tiling. We call that these vertices are *extraordinary vertices* and these faces are *extraordinary faces*. When a mesh is finite, *i.e.*, the mesh has a boundary, we have to consider boundary and distinguish extraordinary boundary vertices and regular boundary vertices in actual situations.

3.3 Subdivision Scheme

A subdivision scheme is a set of subdivision rules by which we densify vertices of meshes to obtain meshes with the same topology structure of original meshes but with smaller faces. If there is no confusion, we also call a subdivision scheme as a subdivision. We have mentioned subdivision schemes because we always calculate new vertices and link those new vertices to form a new control mesh in Chap. 2. In this case, control meshes are regular meshes, and these vertex linking methods are simple. Consequently, these linking methods are not specially discussed, and only these rules computing coordinates of vertices are discussed in details. Generally, a subdivision scheme can be decomposed into two parts:

(1) Add new vertices for old meshes and form a new mesh topology. The process is called *Mesh Splitting*. Those methods to add new vertices and link new vertices are called *topological rules*. Mesh Splitting can be executed on simplicial complexes. We can find the Mesh Splitting in Fig. 3.4.

(2) Calculate geometric positions of all new vertices. The process is called averaging, and corresponding average methods are called *geometric rules*. Geometric rules are usually given by subdivision masks such as those in Figs. 2.26 and 2.27.

We should notice that a mesh sequence appears when we continuously apply a subdivision scheme on a mesh. Topology rules decide topology structures of these meshes, and geometric rules are related to the shapes of limit surfaces. When a mesh is regular, Mesh Splitting is the refinement of Lave tiles. We can construct a new subdivision scheme by altering topology rules or geometric rules for a mesh with a certain Lave tiling topology.

We now denote a subdivision scheme by a subdivision operator T, and let M^0 denote a initial mesh and

$$M^k = T(M^{k-1}).$$

If there is

$$\boldsymbol{S} = \lim_{k\to\infty} M^k.$$

We call $\boldsymbol{S}$ a subdivision surface. Subdivision surfaces are sometimes called limit surfaces. $M^k(k \geqslant 0)$ are control meshes of the subdivision surface. M^0 is called the initial mesh, and $M^k(k > 0)$ are subdivision meshes or semi-regular meshes because only those vertices or those faces corresponding to extraordinary vertices or extraordinary faces of the initial mesh are extraordinary vertices and extraordinary faces of a new mesh. That is, new extraordinary vertices or extraordinary faces are not added in subdivision meshes after one subdivision step. We can find the rule in Figs. 3.6 and 3.7. If a subdivision scheme can produce semi-regular meshes whose faces are quadrangles except for several extraordinary faces, we call the subdivision scheme is a quadrangle subdivision scheme, for short, quadrangle subdivision. Similarly, we can define triangle subdivision and hexagonal subdivision.

Most subdivision schemes are constructed on the basis of refinement rules of regular meshes and generalize those rules so that they can cover extraordinary vertices and faces. We can find the idea in Fig. 3.7. Those subdivision rules used in Fig. 3.7 come from the subdivision of bi-cubic B-spline surfaces. Generally, there are several logical steps in the definition of a subdivision scheme [35]:

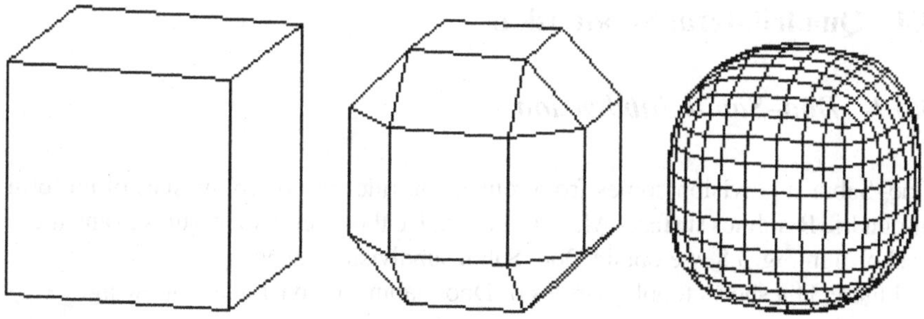

(a) Initial mesh (b) Subdivision once (c) Subdivision thrice

Fig. 3.6 Extraordinary faces are not added after one subdivision step

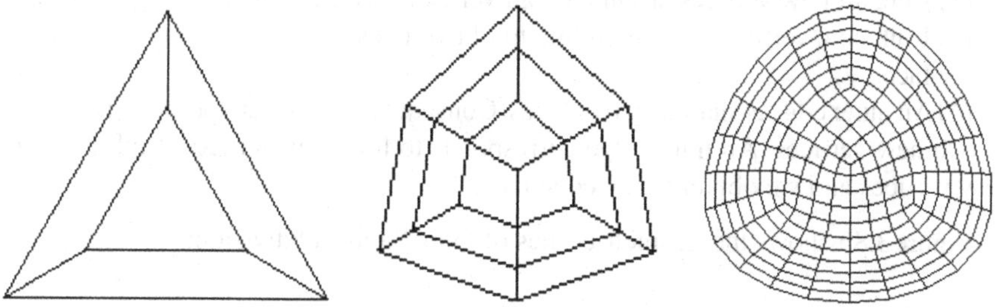

(a) Initial mesh (b) Subdivision once (c) Subdivision thrice

Fig. 3.7 Extraordinary vertices are not added after one subdivision step

(1) The description of the sequence of regular meshes generated by the subdivision process;

(2) The choice of subdivision masks, that is, the selection of a set of existing vertices that will be used to calculate a new vertex in the next step;

(3) The choice of the exact value of those coefficients in masks;

(4) The modification of the rules for the irregular case.

For some subdivision schemes generalized from B-spline and box spline, we have made the steps (1)$\sim$(3) in Chap. 2. In this chapter, we can find that some given subdivision schemes are generalizations of refinement rules of control meshes. In other words, we obtain some subdivision schemes by using step (4) in above. Generally, subdivision surfaces are only G^1 continuous on extraordinary vertices, and their continuity order is larger than G^1 on regular regions of meshes. If a subdivision surface is a generalization of a spline surface, the subdivision surface has the continuity order of the corresponding spline surface on regular regions of meshes. That is, the subdivision surface has the continuity order of the corresponding spline surface almost everywhere except several extraordinary vertices. The continuity analysis method of subdivision surfaces will be given in Chap. 4.

3.4 Quadrilateral Subdivision

3.4.1 Doo–Sabin Subdivision

Doo–Sabin subdivision comes from refinement rules of control meshes of uniform bi-quadric B-splines surface. After we generalize those refinement rules to any mesh as shown in Fig. 3.8, we obtain Doo–Sabin subdivision scheme.

Figure 3.8a shows topology rules of Doo–Sabin subdivision as following:

(1) Every vertex corresponds to a new vertex for every face of the split old mesh. Consequently, for the old mesh, every vertex is split as n new vertices, where n is the number of old faces neighboring to the split old vertex;

(2) Link all new vertices around an old vertex and we obtain a new vertex face;

(3) Link new vertices corresponding to old vertices of an old face and we obtain a new face;

(4) An inner edge is shared by two faces. Consequently, each endpoint corresponds to two new vertices. An inner edge corresponds to four new vertices. Link the four new vertices and we obtain a new edge face.

Figure 3.8b shows the geometric rules of Doo–Sabin subdivision:

$$V_0' = \sum_{i=0}^{n-1} \gamma_i V_i,$$

where, $\gamma_0 = \dfrac{n+5}{4n}$, $\gamma_i = \dfrac{1}{4n}(3+\cos\dfrac{2i\,\pi}{n})(i = 1, \ldots, n-1)$. $V_i'(i = 1, \ldots, n-1)$ have similar computation formulas as that of V_0'. It is easy to know that Doo–Sabin subdivision is the subdivision of bi-quadric B-spline surfaces when n is equal to 4. Fig. 3.6 shows the effect of Doo–Sabin subdivision.

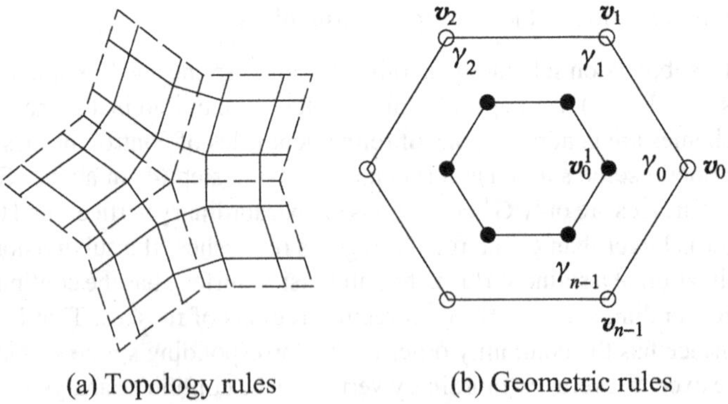

Fig. 3.8 Doo–Sabin subdivision scheme

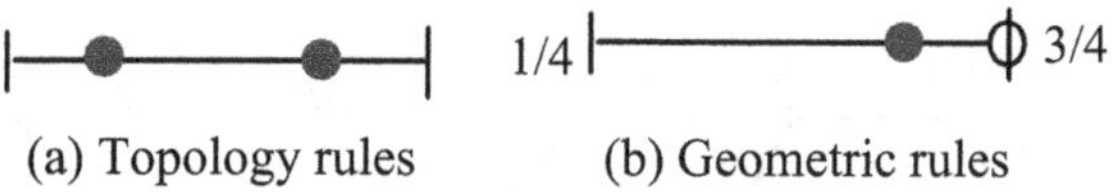

Fig. 3.9 Boundary rules of Doo–Sabin subdivision

It is important to determine the subdivision rules for boundary vertices and boundary edges. There are two methods to subdivide mesh boundaries. One method is that boundary vertices and boundary edges do not produce new faces in every subdivision step. In this case, subdivision surfaces are uniform bi-quadric B-spline surfaces when meshes are regular. The other is that every edge corresponds to two new vertices as shown in Fig. 3.9a. Geometric rules calculating new boundary vertices are shown in Fig. 3.9b. Those boundary rules can make the limit curve of the boundary polygon be a uniform quadric B-spline curve. We link those new vertices with methods in Fig. 3.8a. It is possible to give other boundary subdivision rules when we apply the second method, especially geometric rules to calculate new vertices. We have already known that we can abandon boundaries or reserve boundaries when we subdivision a mesh. In later discussion, we will call the first method as the subdivision without boundaries and the second method as the subdivision with boundaries, respectively. The first method is also called as the skirt-removed approach in [99]. For different subdivision schemes, used rules to abandon boundaries or reserve boundaries are probably different.

3.4.2 *Catmull–Clark Subdivision*

Catmull–Clark subdivision comes from refinement rules of control meshes of uniform bi-cubic B-spline surfaces. Its topology rules are shown Fig. 3.10, and geometric rules are shown in Fig. 3.11.

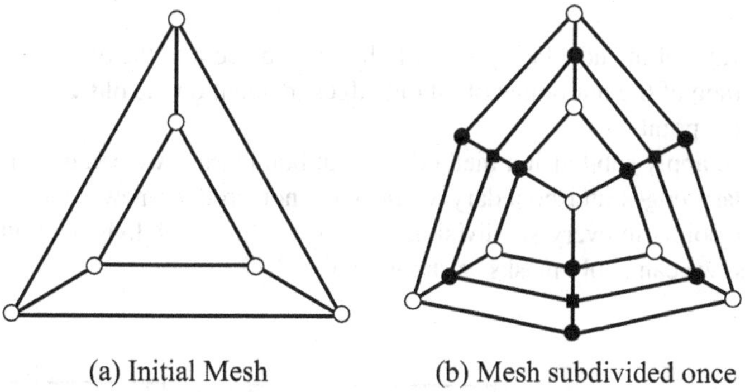

Fig. 3.10 Topology rules of Catmull–Clark subdivision

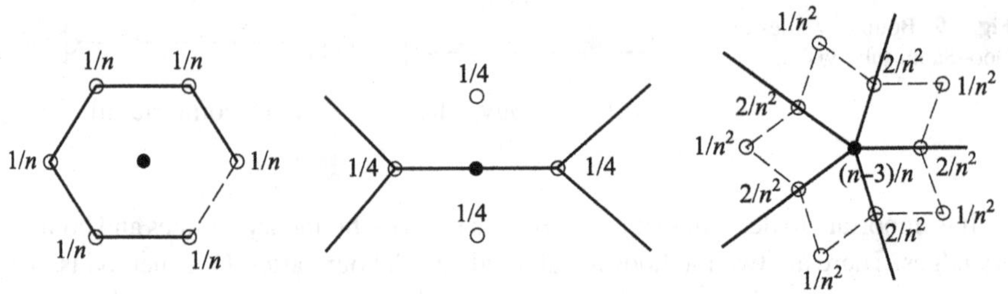

(a) New face point (b) New edge point (c) New vertex point

Fig. 3.11 Masks of Catmull–Clark subdivision

Topology rules shown in Fig. 3.10 can be described as:

(1) Each old face corresponds to a new vertex called new face point; each old edge corresponds to a new vertex called new edge point; each old vertex corresponds to a new vertex called new vertex point;

(2) Connecting each new face point to the new edge points of the edges defining the old face;

(3) Connecting each new vertex point to the new edge points of all old edges incident on the old vertex.

Geometric rules shown in Fig. 3.11 can be described as:

(1) New face point: the average of all old vertices defining the face;

(2) New edge point: the average of the midpoint of the old edge with the average of two new face points of the face sharing the edge;

(3) New vertex point: the average

$$\frac{\boldsymbol{Q}}{n} + \frac{2\boldsymbol{R}}{n} + \frac{\boldsymbol{V}(n-3)}{n}$$

where,

$\boldsymbol{Q}$: the average of the new face points of all faces adjacent to the old vertex point;
$\boldsymbol{R}$: the average of the midpoints of all old edges incident on the old vertex point;
$\boldsymbol{V}$: old vertex point

When we apply subdivision method without boundaries, we can use the method that boundary edges and boundary vertices do not produce new edge points and new vertex points in every subdivision step. If we apply subdivision method with boundaries, we can apply masks as shown in Fig. 3.12.

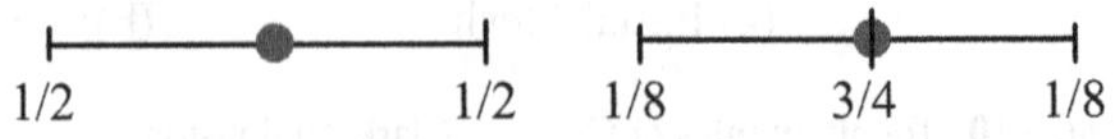

(a) New edge point (b) New vertex point

Fig. 3.12 Masks of Catmull–Clark subdivision on boundaries

From Fig. 3.12, it can be found that every edge corresponds to a new edge point and every vertex corresponds to a new vertex point. The limit curve of a boundary polygon is a uniform cubic B-spline curve. Figure 3.10 gives the effect of C-C subdivision with boundaries.

3.4.3 Non-uniform Subdivision

Sederberg [25] gives the non-uniform Doo–Sabin subdivision scheme and the non-uniform C-C subdivision scheme by generalizing refinement rules of control meshes of non-uniform B-spline surfaces. Just as the non-uniform B-spline surface is a generalization of the uniform B-spline surface, non-uniform subdivision schemes can be considered as generalizations of corresponding uniform subdivision schemes, i.e., the non-uniform Doo–Sabin subdivision scheme is a generalization of the Doo–Sabin subdivision scheme, and the non-uniform C-C subdivision scheme is a generalization of the C-C subdivision scheme. The non-uniform Doo–Sabin subdivision scheme has the same topology rules as the Doo–Sabin subdivision scheme. The non-uniform C-C subdivision scheme has the same topology rules as C-C subdivision scheme. Consequently, this section only discusses geometric rules of non-uniform subdivision schemes. It should be noticed that geometric rules discussed in this section are straightforward generalizations of subdivision rules in Chap. 2.

1. Knot spacing

We firstly define the knot spacing. Knot spacing is a set of parameters. In the non-uniform Doo–Sabin subdivision scheme, each vertex in subdivided meshes is assigned parameters for each edge radiating from it, as shown in Fig. 3.13. Such knot spacing is the same as the knot spacing of the subdivision of non-uniform quadratic B-spline surface when the subdivided mesh is regular. $d^0_{i,j}$ denotes the parameter of edge $\boldsymbol{V}_i\boldsymbol{V}_j$. Rotating $\boldsymbol{V}_i\boldsymbol{V}_j$ counterclockwise around $\boldsymbol{V}_i$, then $d^1_{i,j}$ denotes the parameter of the first edge that is encountered, and $d^2_{i,j}$ denotes the parameter of the second edge, etc. For the non-uniform Catmull–Clark subdivision scheme, each edge is assigned a parameter as shown in Fig. 3.13. Consequently, $d^0_{i,j} = d^0_{j,i}$ in the cubic case.

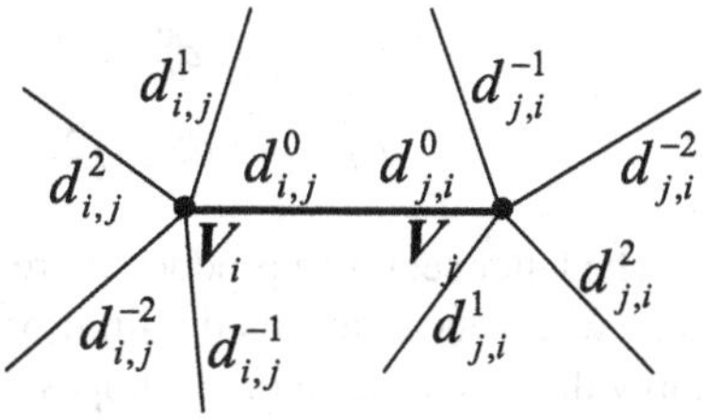

Fig. 3.13 Knot spacing of non-uniform subdivisions

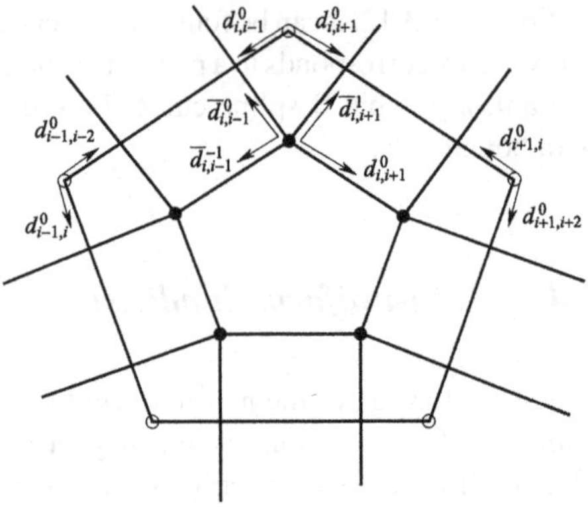

Fig. 3.14 Parameter evaluating rules of non-uniform Doo–Sabin subdivision scheme

2. Non-uniform Doo–Sabin subdivision

Refer to Fig. 3.14 for the non-uniform Doo–Sabin subdivision scheme. The new vertex $\boldsymbol{V}_i'$ is calculated:

$$\boldsymbol{V}_i' = (\boldsymbol{V} + \boldsymbol{V}_i)/2 + d\left(-n\boldsymbol{V}_i + \sum_{j=1}^{n} c_{i,j}\boldsymbol{V}_j\right) \Big/ \left(8\sum_{k=1}^{n} d^0_{k-1,k} d^0_{k+1,k}\right)$$

where, $d = d^0_{i+1,i+2} d^0_{i+3,i+2} + d^0_{i-1,i-2} d^0_{i-3,i-2}$,

$$c_{i,j} = \sum_{j=1}^{n}\left(1 + 2\cos\frac{2\,\pi\,|i-j|}{n}\right)\boldsymbol{V}_j,$$

$$\boldsymbol{V} = \sum_{k=1}^{n} d^0_{k-1,k} d^0_{k+1,k} \boldsymbol{V}_k \Big/ \sum_{k=1}^{n} d^0_{k-1,k} d^0_{k+1,k}.$$

New knot spacing $\overline{d}^k_{i,j}$ can be specified in many ways. Here are two straightforward options:

$$\overline{d}^0_{i,i+1} = \overline{d}^{-1}_{i,i-1} = d^0_{i,i+1}/2, \quad \overline{d}^0_{i,i-1} = \overline{d}^1_{i,i+1} = d^0_{i,i-1}/2$$

or,

$$\overline{d}^0_{i,i+1} = d^0_{i,i+1}/2, \quad \overline{d}^{-1}_{i,i-1} = (d^0_{i,i+1} + d^{-1}_{i,i-1})/4$$
$$\overline{d}^0_{i,i-1} = d^0_{i,i-1}/2, \quad \overline{d}^1_{i,i+1} = (d^0_{i,i-1} + d^1_{i,i+1})/4.$$

The latter seems to produce more satisfactory shapes than the former [25]. However, when one analyzes the limit properties of subdivision surfaces, he or she can apply the former if he or she hopes that analysis become simple.

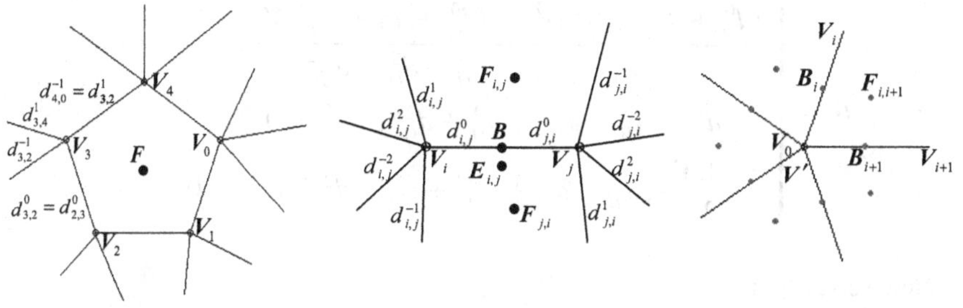

(a) New face point (b) New edge point (c) New vertex point

Fig. 3.15 Masks of non-uniform Catmull–Clark subdivision

3. Non-uniform C-C subdivision

Referring to Fig. 3.15, the new face point, new edge point, and new vertex point can be calculated as following:

New face point: assume that $\boldsymbol{V}_0, \boldsymbol{V}_1, \boldsymbol{V}_2, \ldots, \boldsymbol{V}_{n-1}$ are vertices of the face F.

$$\boldsymbol{F}_F = \begin{cases} \dfrac{\sum\limits_{i=0}^{n-1} \omega_i \boldsymbol{V}_i}{\sum\limits_{i=0}^{n-1} \omega_i}, \sum\limits_{i=0}^{n-1} \omega_i \neq 0 \\ \dfrac{1}{n} \sum\limits_{i=0}^{n-1} \boldsymbol{V}_i, \sum\limits_{i=0}^{n-1} \omega_i = 0 \end{cases}$$

where,

$$\begin{aligned} \omega_i = {} & (d_{i+1,i}^0 + d_{i+1,i}^2 + d_{i+1,i}^{-2} + d_{i-2,i-1}^0 + d_{i-2,i-1}^2 + d_{i-2,i-1}^{-2}) \\ & \times (d_{i-1,i}^0 + d_{i-1,i}^2 + d_{i-1,i}^{-2} + d_{i+2,i+1}^0 + d_{i+2,i+1}^2 + d_{i+2,i+1}^{-2}). \end{aligned}$$

New edge point: assume that $\boldsymbol{V}_i, \boldsymbol{V}_j$ are two endpoints of the internal edge E, $\boldsymbol{F}_{ij}$ and $\boldsymbol{F}_{ji}$ are two new face points of two faces adjacent to the edge E.

$$\boldsymbol{E}_E = (1 - \alpha_{ij} - \alpha_{ji})\boldsymbol{B} + \alpha_{ij}\boldsymbol{F}_{ij} + \alpha_{ji}\boldsymbol{F}_{ji}$$

where,

$$\alpha_{ij} = \begin{cases} \dfrac{d_{ji}^1 + d_{ij}^{-1}}{2(d_{ji}^1 + d_{ji}^{-1} + d_{ij}^1 + d_{ij}^{-1})}, d_{ji}^1 + d_{ji}^{-1} + d_{ij}^1 + d_{ij}^{-1} \neq 0 \\ 0, d_{ji}^1 + d_{ji}^{-1} + d_{ij}^1 + d_{ij}^{-1} = 0 \end{cases}$$

$$\boldsymbol{B} = \begin{cases} \dfrac{(d_{ji}^0 + d_{ji}^2 + d_{ji}^{-2})\boldsymbol{V}_i + (d_{ij}^0 + d_{ij}^2 + d_{ij}^{-2})\boldsymbol{V}_j}{d_{ji}^0 + d_{ji}^2 + d_{ji}^{-2} + d_{ij}^0 + d_{ij}^2 + d_{ij}^{-2}}, \\ d_{ji}^0 + d_{ji}^2 + d_{ji}^{-2} + d_{ij}^0 + d_{ij}^2 + d_{ij}^{-2} \neq 0 \\ \dfrac{\boldsymbol{V}_i + \boldsymbol{V}_j}{2}, d_{ji}^0 + d_{ji}^2 + d_{ji}^{-2} + d_{ij}^0 + d_{ij}^2 + d_{ij}^{-2} = 0 \end{cases} \tag{3.1}$$

New vertex point:

$$\boldsymbol{V}_V = \begin{cases} c\boldsymbol{V}_0 + \dfrac{3\sum\limits_{i=1}^{n}(m_i \boldsymbol{B}_i + f_{i,i+1}\boldsymbol{F}_{i,i+1})}{n\sum\limits_{i=1}^{n}(m_i + f_{i,i+1})}, \sum\limits_{i=1}^{n}(m_i + f_{i,i+1}) \neq 0 \\ \boldsymbol{V}_0, \sum\limits_{i=1}^{n}(m_i + f_{i,i+1}) = 0 \end{cases} \tag{3.2}$$

where $\boldsymbol{V}_i (i = 0, \ldots, n-1)$ is vertex on 1-neighborhood of $\boldsymbol{V}$. $\boldsymbol{B}_i (i = 0, \ldots, n-1)$ is point corresponding to the edge $\boldsymbol{V}_0\boldsymbol{V}_i$ and it is calculated by the Formula (3.1). n is the valence of $\boldsymbol{V}$. $\boldsymbol{F}_{i,j}$ is the new face point corresponding to the face that has the edges $\boldsymbol{V}_0\boldsymbol{V}_i$ and $\boldsymbol{V}_0\boldsymbol{V}_j$. However,

$$m_i = (d_{0,i}^1 + d_{0,i}^{-1})(d_{0,i}^2 + d_{0,i}^{-2})/2$$
$$f_{i,j} = d_{0,i}^1 d_{0,j}^{-1}, c = (n-3)/n.$$

Qin et al. [100] employ some modifications for the Formula (3.2). These modifications seem to make the limit surface has better fairness.

In the subdivision process, each n-polygon face is split into n quadrilateral faces, whose knot spacings are determined in the way as shown in Fig. 3.15. When we subdivide meshes with boundaries, we may use boundary subdivision rules (2.9) for the non-uniform Doo–Sabin subdivision and (2.11) for the non-uniform Catmull–Clark subdivision, respectively (Fig. 3.16).

Compared with uniform subdivision schemes, non-uniform subdivision scheme can express some sharp features by setting knot spacing. Figure 3.17 shows a non-uniform C-C subdivision surfaces. In the control mesh of the subdivision surface, parameters of all edges neighboring to a vertex are set as 0. When all parameters are equal, non-uniform subdivisions become uniform subdivision. From uniform subdivisions to non-uniform subdivisions, the ability of shape expression is advanced because of the knot spacing. However, non-uniform subdivisions are much more trivial than uniform subdivisions also because of the knot spacing. In some literatures, non-uniform subdivision surfaces are shortened as NURSS.

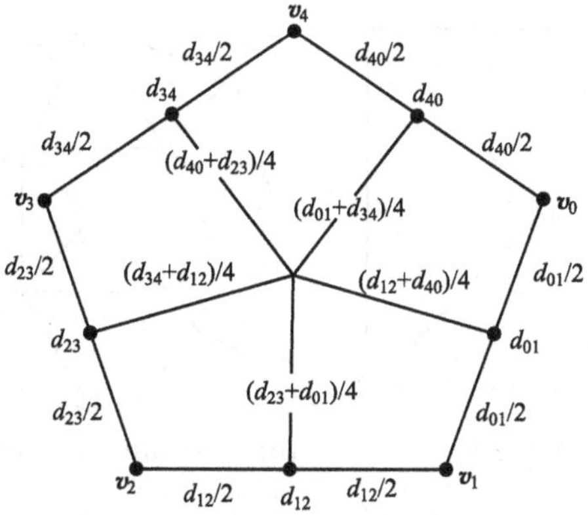

Fig. 3.16 Parameter evaluating rules of non-uniform C-C subdivision scheme

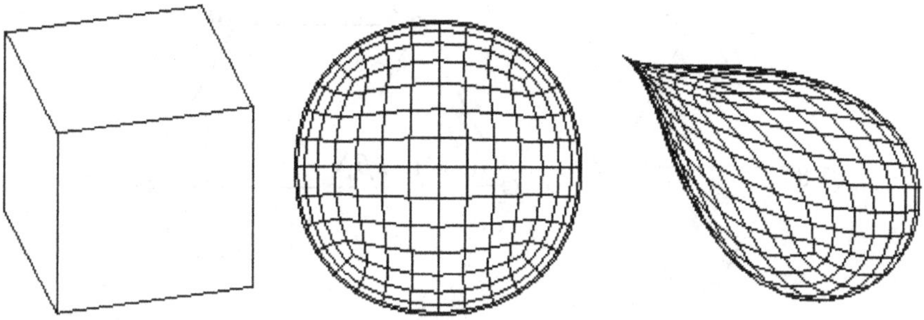

(a) Initial mesh (b) C-C subdivision surface (c) Non-uniform C-C subdivision surface

Fig. 3.17 C-C subdivision surface and non-uniform C-C subdivision surface

3.5 Triangular Subdivision

3.5.1 Loop Subdivision

Loop subdivision comes from refinement rules of control meshes of quadric three-directional box spline surfaces. Its topology rules are shown in Fig. 3.18, and geometric rules are shown in Fig. 3.19.

Topology rules shown in Fig. 3.18 can be described as:

(1) Each old edge corresponds to a new vertex called new edge point; each old vertex corresponds to a new vertex called new vertex point;

(2) Connecting new edge points for each edge of old triangle;

(3) Connecting each new vertex point to the new edge points of all old edges incident to its old vertex.

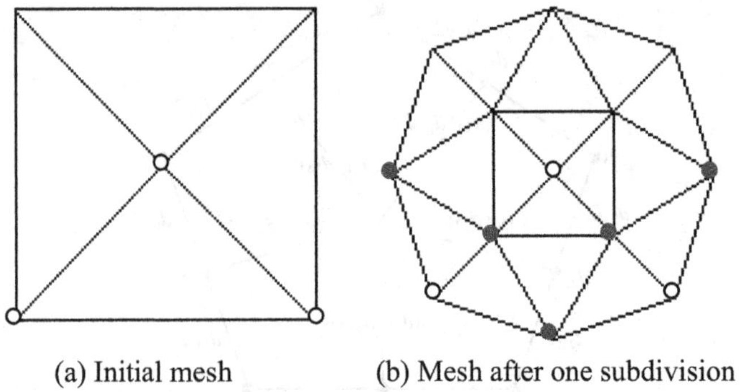
(a) Initial mesh
(b) Mesh after one subdivision

Fig. 3.18 Topology rules of Loop subdivision

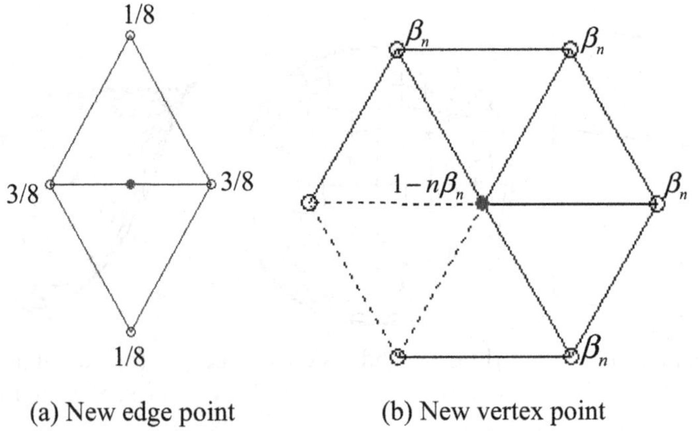
1/8
3/8
3/8
1/8
(a) New edge point
β_n
β_n
$1-n\beta_n$
β_n
β_n
(b) New vertex point

Fig. 3.19 Geometric rules of Loop subdivision

Geometric rules shown in Fig. 3.19 can be described as:
New vertices:

$$\boldsymbol{V}_V = (1 - n\beta_n)\boldsymbol{V} + \beta_n \sum_{i=0}^{n-1} \boldsymbol{V}_i$$

where, $\beta_n = \frac{1}{n}\left(\frac{5}{8} - \left(\frac{3}{8} + \frac{1}{4}\cos\frac{2\pi}{n}\right)^2\right)$, or

$$\beta_n = \begin{cases} 3/16, & n = 3 \\ 3/8n, & n > 3 \end{cases}$$

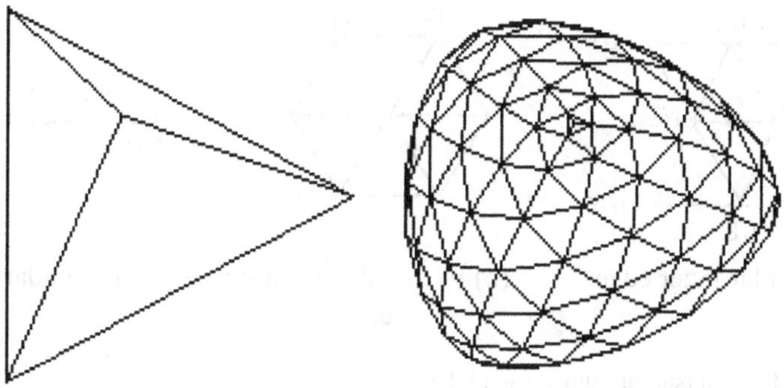

(a) Initial mesh (b) Subdivision mesh

Fig. 3.20 Subdivision effect of Loop subdivision

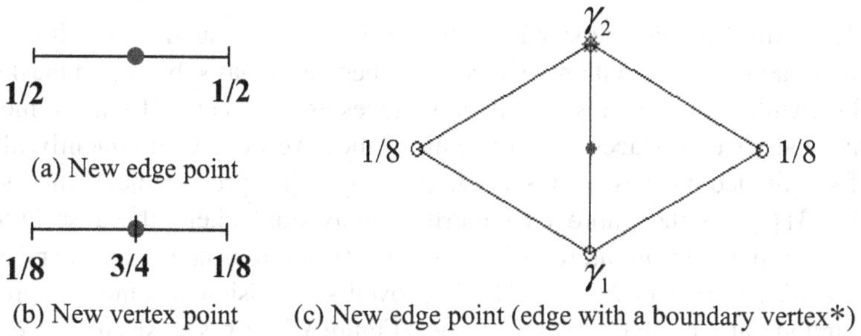

1/2
1/2
(a) New edge point
1/8
3/4
1/8
(b) New vertex point
γ_2
1/8
1/8
γ_1
(c) New edge point (edge with a boundary vertex*)

Fig. 3.21 Masks of Loop subdivision on boundaries

New edge point:

$$\boldsymbol{E}_E = \frac{3}{8}(\boldsymbol{V}_0 + \boldsymbol{V}_1) + \frac{1}{8}(\boldsymbol{V}_2 + \boldsymbol{V}_3).$$

Figure 3.20 gives the subdivision effect of Loop subdivision scheme. It should be noticed that initial meshes should be triangle meshes while valances of vertices can be arbitrary. For Doo–Sabin subdivision and C-C subdivision, initial meshes can be arbitrary two-manifold meshes. When a mesh is an open mesh and we subdivide it by the method with boundaries, boundary rules of Loop subdivision can be described as [55]:

For Fig. 3.21c,

$$\boldsymbol{E}_E = \gamma_0 \boldsymbol{V}_0 + \gamma_1 \boldsymbol{V}_1 + \frac{1}{8}(\boldsymbol{V}_2 + \boldsymbol{V}_3)$$

where $\gamma_0 = \frac{1}{2}, \gamma_1 = \frac{1}{4}$ or $\gamma_0 = \frac{1}{4} + \frac{1}{4}\cos\frac{2\pi}{n-1}, \gamma_1 = \frac{1}{2} - \frac{1}{4}\cos\frac{2\pi}{n-1}$.

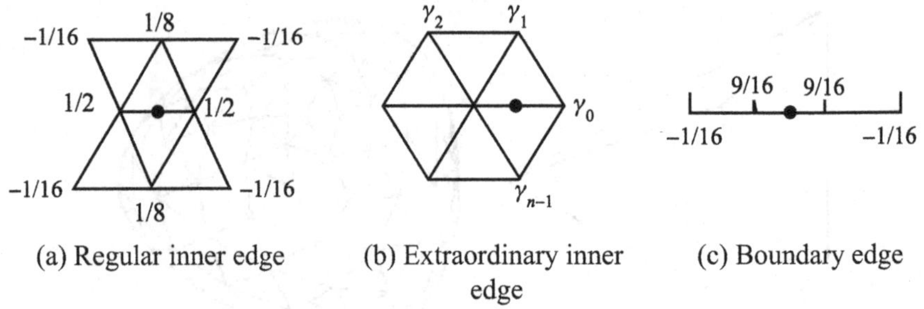

Fig. 3.22 Modified butterfly subdivision scheme

3.5.2 Butterfly Subdivision

Butterfly subdivision has the same topology rules as Loop subdivision. It is firstly given in literature [14], and it obtains the name because it has subdivision masks like butterflies. Unlike the above subdivision surfaces approximating to initial meshes, butterfly subdivision surfaces interpolate initial mesh vertices. Consequently, all vertices of subdivided meshes are reserved, and only edges produce new edge points. Zorin [8, 31] gives the improved butterfly subdivision scheme because butterfly subdivision scheme in literature [14] is only C^0 continuous at extraordinary points whose valance $k = 3$ or $k > 7$. The improved subdivision scheme can produce C^1-continuous surfaces for arbitrary meshes. Figure 3.22 gives masks for three kinds of new edge points for the improved scheme.

Where, the regular inner edge means the inner edge with two regular endpoints, and extraordinary inner edge means the inner edge with a regular endpoint and an extraordinary endpoint. For the Fig. 3.22b:

(1) if $n \geq 5$, $\gamma_i = \dfrac{1}{n}\left(\dfrac{1}{4} + \cos\dfrac{2i\,\pi}{n} + \dfrac{1}{2}\cos\dfrac{4i\,\pi}{n}\right)$;

(2) if $n = 4$, $\gamma_0 = 3/8$, $\gamma_2 = -1/8$, $\gamma_1 = \gamma_3 = 0$;

(3) if $n = 3$, $\gamma_0 = 3/8$, $\gamma_1 = \gamma_2 = -1/12$.

It should be noticed that two extraordinary vertices must not be neighborhoods for the modified butterfly subdivision scheme. Butterfly subdivision surface has worse smoothness than Loop subdivision surfaces because of the reservation of old mesh vertices.

3.5.3 $\sqrt{3}$ Subdivision

Notice that the above subdivision schemes make the number of vertices (and faces) of subdivided meshes increase by the factor 4. That is, if the mesh M^{k-1} has n vertices, then the number of vertices of the mesh M^k is approximately $4n$. When the mesh

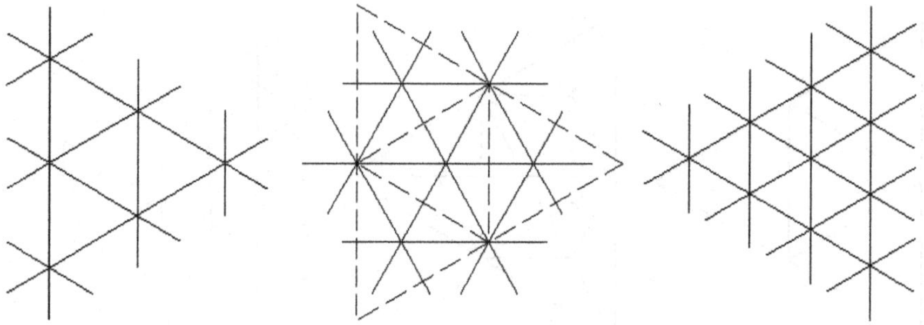

(a) Initial mesh (b) Subdivision once (c) Subdivision twice

Fig. 3.23 $\sqrt{3}$ subdivision scheme

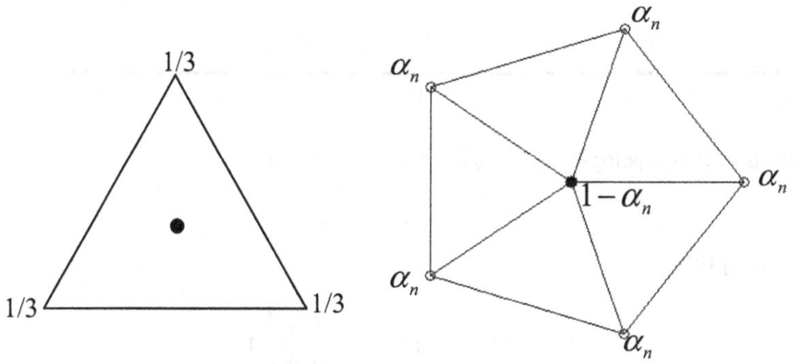

(a) New face point (b) New vertex point

Fig. 3.24 Masks of $\sqrt{3}$ subdivision scheme

M^{k-1} is regular, the number of vertices of the mesh M^k is $4n$. Consequently, those subdivision schemes have fast splitting speed. However, $\sqrt{3}$ subdivision scheme has the increasing factor $\sqrt{3}$, and it has relatively slow splitting speed. Topology rules and geometric rules of the subdivision scheme are shown in Figs. 3.23 and 3.24.

Topology rules shown in Fig. 3.23 can be described as:

(1) Each old face corresponds to a new vertex called new face point; each old vertex corresponds to a new vertex called new vertex point;

(2) Connecting each new face point to new vertex points corresponding to old vertices of the old face;

(3) Connecting two new face points whose two old faces sharing a common edge.

Geometric rules shown in Fig. 3.24 can be described as:

New face point:

$$\boldsymbol{F}' = (\boldsymbol{V}_i + \boldsymbol{V}_j + \boldsymbol{V}_k)/3.$$

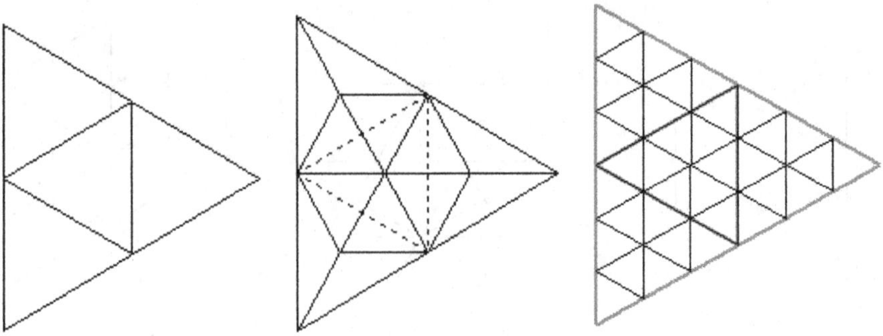

(a) Initial mesh (b) Subdivision once (c) Subdivision twice

Fig. 3.25 $\sqrt{3}$ Subdivision with boundaries

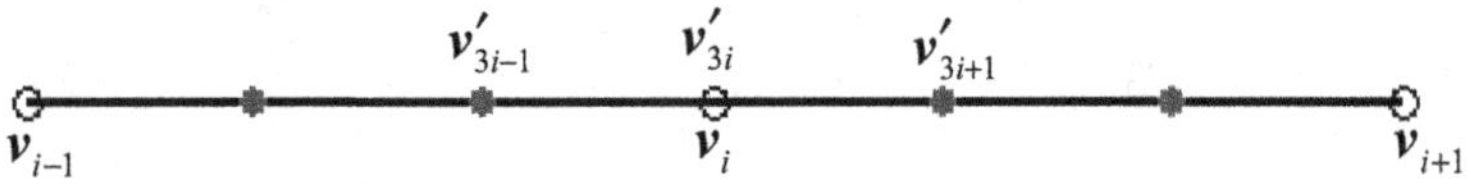

Fig. 3.26 Boundary topology rules for $\sqrt{3}$ subdivision scheme

New vertex point:

$$V' = (1 - \alpha_n)V + \alpha_n \frac{1}{n} \sum_{i=0}^{n-1} V_i$$

where, $\alpha_n = \dfrac{4 - 2\cos(2\,\pi\,/n)}{9}$.

We can know from Fig. 3.23 that an old triangle is replaced by nine new ones by two $\sqrt{3}$ subdivision steps. Consequently, we investigate the subdivision of boundary on two subdivision steps.

We assume that each subdivision process is constituted by two subdivision steps shown in Fig. 3.25b, c. For the first subdivision step, we split all triangular faces 1–3 but flip only the interior edge. Edge flipping at the boundaries is not possible since the opposite triangle-mate is missed. Hence, the boundary polygon is not modified in the first $\sqrt{3}$ subdivision step. When the second subdivision step is executed, we have to apply a univariate trisection rule to the boundary and connect the vertices to the corresponding interior ones so that a uniform 1–9 split is established for each boundary triangle of the mesh before the first subdivision step.

The subdivision mask on boundaries can be described in Fig. 3.26. The geometric rules calculating new vertex positions can be described:

$$V'_{3i-1} = \frac{1}{27}(10V_{i-1} + 16V_i + V_{i+1})$$

$$V'_{3i} = \frac{1}{27}(4V_{i-1} + 19V_i + 4V_{i+1})$$

$$V'_{3i+1} = \frac{1}{27}(V_{i-1} + 16V_i + 10V_{i+1}).$$

A reader is probably very surprised that why these subdivision rules of the scheme are so. These subdivision rules do not come from generalizations of subdivisions of spline surfaces. The topology rules come from the refinement of the equilateral triangle Lave tiling. The geometric rules come from the G^1 continuity constraints, which will be discussed in the Chap. 4. As for the boundary subdivision rules, they come from the trisection mask for the cubic B-spline curve. A reader can deduce the trisection mask by the knot insertion algorithm introduced in Chap. 2.

3.6 Hexagonal Subdivision

Definition 3.3 Let M be a mesh. Link the barycenter of each face according to neighborhood of faces and obtain another mesh M'. M and M' are called dual meshes of each other. Fig. 3.27 shows two meshes dual to each other.

Obviously, the dual mesh of a regular triangle mesh is a regular hexagonal mesh. The dual mesh of a regular quadrangle mesh is a regular quadrangle mesh. If a vertex has the valance n, then it corresponds to an n-polygon in the dual mesh.

Definition 3.4 Let M be a mesh and M' is the dual mesh of M. S and S' are two subdivision operators. If $S(M)$ and $S'(M')$ are two dual meshes of each other, then S and S' are dual subdivision operators or subdivision schemes of each other. If $S''(M')$

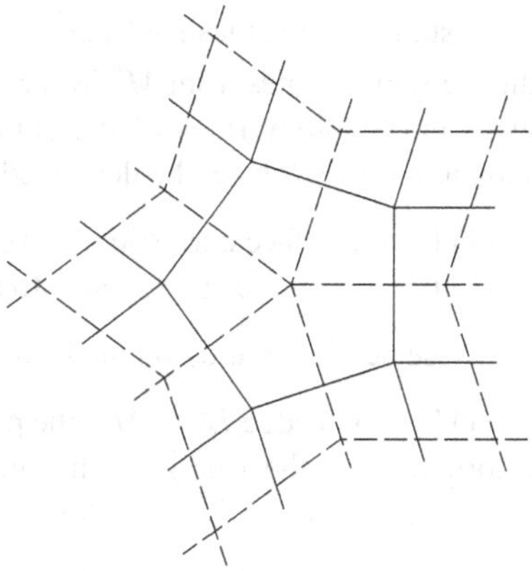

Fig. 3.27 Two meshes dual to each other

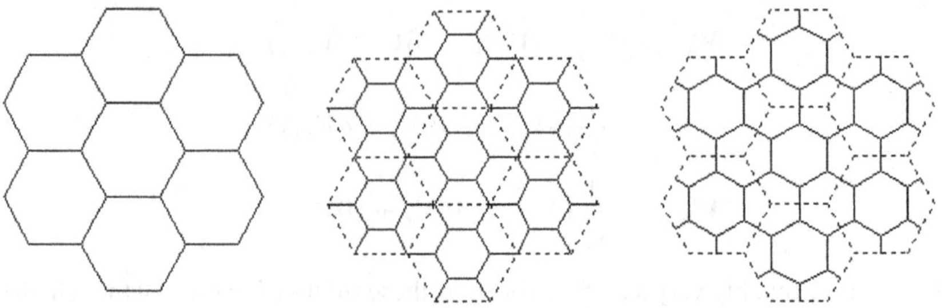

(a) initial mesh (b) Convexity-preserving subdivision (c) Honeycomb subdivision

Fig. 3.28 Topology rules of hexagonal subdivision schemes

and $S'(M')$ have the same topology structure, we call S and S'' dual subdivision operators or dual subdivision schemes in topology.

Generally, hexagonal subdivision schemes can be regarded as dual subdivision schemes of triangle subdivision schemes in topology structure. This section introduces two typical hexagonal subdivision schemes: convexity-preserving subdivision scheme and the honeycomb subdivision scheme. They can respectively be regarded as dual subdivision schemes of Loop subdivision and $\sqrt{3}$ subdivision in topology structure. Their topology rules are shown in Fig. 3.28.

3.6.1 Convexity-Preserving Subdivision Scheme

The convexity-preserving subdivision scheme is given by Dyn N [29]. Though it is called "honeycomb subdivision" in literature [29], we call it as "convexity-preserving subdivision" to distinguish the hexagonal subdivision scheme presented by Zhang and Wang [30] in this book.

Assume that the mesh M^k has vertices $\{V_j^k\}$, edges $\{E_j^k\}$, and faces $\{F_j^k\}$. M^{k+1} is the mesh that comes from M^k by a convexity-preserving subdivision step. Fig. 3.28b gives the topology rules of the convexity-preserving subdivision. Topology rules shown in Fig. 3.28b can be described as:

(1) Each old face corresponds to a new face;
(2) Each old edge corresponds to a new face.

These new faces are obtained by the following steps:

(1) For each edge E_j^k of M^k, the plane T_j^k is assigned and passes through E_j^k with a normal that is the average of the unit normal vectors to the faces forming this edge;

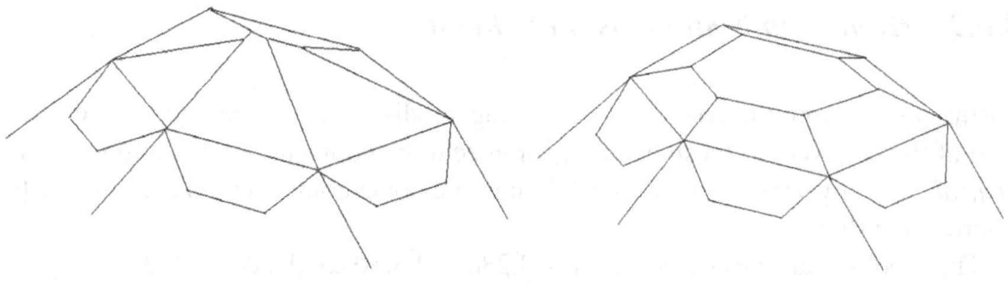

(a) Building polyhedron Δ over edge E_j^k (b) Chopping polyhedron Δ

Fig. 3.29 Form faces by convexity-preserving subdivision

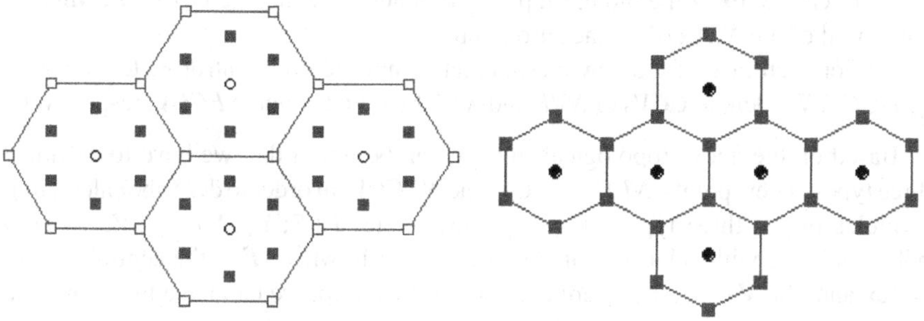

(a) Original mesh; (b) Mesh after subdivision once

Fig. 3.30 Topology rules of honeycomb subdivision. □ Original control vertex; ■ Original control central vertex; ○ New edge vertex (NEV); • New central vertex (NCCV) (including VFCCV and NVCCV)

(2) For each face F_j^k of M^k, the planes $\{T_j^k\}$ that correspond to its edges E_j^k, $j = 1, \ldots, n$, form a closed convex polyhedron outside M^k, as shown in Fig. 3.29a;

(3) Chop the polyhedron at a height h from F_j^k, where h is a half distance from F_j^k to the nearest vertex of that is not on F_j^k, as shown in Fig. 3.30b. So, we obtain a new face that corresponds to the face F_j^k. To each edge of the face F_j^k, there is a 4-sided face on each plane T_j^k, $j = 1, \ldots, n$. The 4-sided faces defined on T_j^k by the two chopped polyhedrons that are pasted to the two faces forming E_j^k constitute a 6-sided face that is assigned to the edge E_j^k.

It should be known that the initial mesh should be convex when we use the convexity-preserving subdivision scheme. We also can find that vertices of any face of a convexity-preserving subdivision mesh are coplanar. The limit surface interpolates the initial mesh. The subdivision scheme is an interpolating subdivision scheme.

3.6.2 Honeycomb Subdivision Scheme

Distinguished from the convexity-preserving subdivision scheme, the vertices of a face of the honeycomb mesh cannot guarantee to be coplanar. So, we introduce the *central control vertex*, denoted by *CCV*, into the honeycomb subdivision scheme to improve the shape control.

The topological rules shown in Figs. 3.28c and 3.30 can be described as:

(1) Every edge of an old face produces a new edge vertex, denoted by *NEV*.

(2) Every old face produces a new face, denoted by *NFF*, by connecting the corresponding *NEV*s of its edges in turn.

(3) Every vertex of the old mesh produces a new face, denoted by *NVF*, which is composed of the *NEV*s of all face around it;

(4) For each face of the new mesh, create a new central control vertex, denoted by *NCCV*. We denote *CCV*s of *NVF* and *NFF* by *NVCCV* and *NFCCV*, respectively.

Based on the above topological rules, it can be found that we have to calculate three types of new points: *NEV*, *NVCCV*, and *NFCCV*. In order to describe calculating formulas of the three types of new points, we let $F[\boldsymbol{F}; \boldsymbol{V}_0, \boldsymbol{V}_1, \ldots, \boldsymbol{V}_{n-1}]$ be a polygonal face with valence n on the original mesh, where $\boldsymbol{F}$ is the central control vertex and $\boldsymbol{V}_0, \boldsymbol{V}_1, \ldots, \boldsymbol{V}_{n-1}$ are the vertices of F. So, we can have the geometric rules:

(1) The new edge vertex *NEV* of F:

$$\boldsymbol{V}_i = (1 - b_n)\boldsymbol{F} + b_n \frac{\boldsymbol{V}_i + \boldsymbol{V}_{i+1}}{2}.$$

(2) The new face central control vertex *NFCCV* of F:

$$\boldsymbol{F}' = (1 - a_n)\boldsymbol{F} + a_n \boldsymbol{V}_b$$

where, $a_n, b_n \in [0, 1]$, $\boldsymbol{V}_b$ is the barycenter of F. To obtain a symmetric mask, we pick $b_n = 2/3$ and the new edge vertex will be the barycenter of the triangle composed by two old edge vertices and the central control vertex. To be similar to the $\sqrt{3}$ subdivision, we pick

$$a_n = (4 - \cos\varphi_n)/9, \quad \varphi_n = 2\pi/n$$

(3) The new vertex central control vertex *NVCCV* of $\boldsymbol{v}$:

$$\boldsymbol{F}' = (1 - c_n)\boldsymbol{V} + c_n \left(\sum_{i=1}^{n} \alpha_i \boldsymbol{V}_i + \sum_{i=1}^{n} \beta_i \boldsymbol{F}_i \right)$$

with $c_n, \alpha_n, \beta_n \in [0, 1]$ and $\sum_{i=1}^{n}(\alpha_i + \beta_i) = 1$. $\boldsymbol{V}_i$ is the i-th vertex connected to $\boldsymbol{V}$, and $\boldsymbol{F}_i$ is the *CCV* of F_i that contains $\boldsymbol{V}$. For α_i, β_i, they can be picked as:

$$\alpha_i = \beta_i = 1/2n.$$

c_n can be considered as a shape factor of the limit surface.

If you generalize hexagonal subdivisions on open meshes, probably it is a good idea to abandon boundaries after every subdivision step, as shown in Fig. 3.30.

3.7 4–8 Subdivision

3.7.1 4–8 Meshes

The 4–8 subdivision scheme is generalized from the subdivision method of the quadric four-directional box spline surface [27]. These control meshes are [4, 8] meshes, from which the name 4–8 subdivision scheme comes. In Chap. 2, we have obtained the 4–8 subdivision rules on regular meshes and we now generalize those to "*arbitrary 4–8 meshes*". An arbitrary 4–8 mesh is such a mesh that valances of its most vertices are 4 or 8, and it is composed of pairs of triangles shown in Fig. 3.31. A pair of triangles is called a *block*. The shared edge is called the *interior edge* of the block, and other edges of the block are *exterior edges*. Arbitrary 4–8 meshes and regular 4–8 meshes can be all called 4–8 meshes. We also call such meshes as triangulated quadrilateral meshes or triquad meshes.

A problem is how we obtain the initial 4–8 mesh before we execute the 4–8 subdivision. A simple way to convert an arbitrary polygonal mesh into a triquad

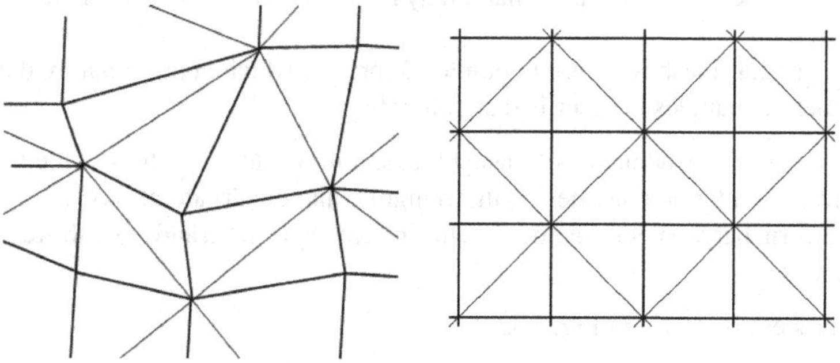

(a) Arbitrary 4-8 mesh (b) Regular 4-8 mesh and its blocks

Fig. 3.31 4–8 Mesh

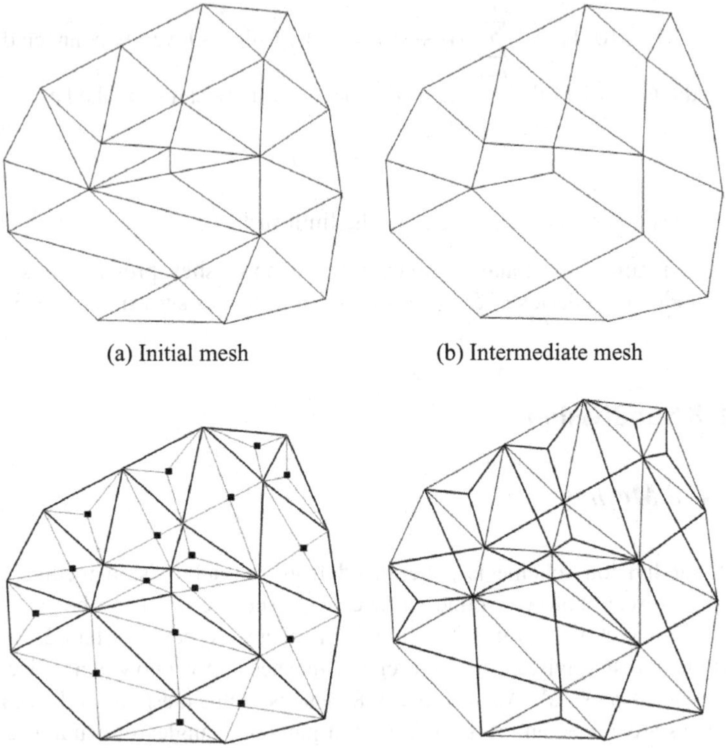

(a) Initial mesh (b) Intermediate mesh

(c) Barycenter subdivision (d) Basic blocks

Fig. 3.32 Obtain basic blocks from *triangle* meshes

mesh is to use a single step of C-C refinement to obtain a quadrilateral mesh and then split each quad into two triangles. The disadvantage of this method, for arbitrary triangular meshes in particular, is that it may result in sixfold increase in the number of triangles.

For triangular meshes, a more complex approach which approximately doubles the number of triangles is described as following:

(1) Find an independent set of basic blocks, remove interior edges of the blocks; the result is an intermediate mesh with triangular and quadrilateral faces.

(2) Perform barycenter refinement on the intermediate mesh and mark the resulting basic blocks.

The process is shown in Fig. 3.32.

3.7.2 Subdivision Rules

Topology rules of 4–8 subdivision schemes are simple: Each interior edge is added with a midpoint and each triangle is bisected, as shown in Fig. 3.33. By a subdivision step, all exterior edges become interior edges, and an interior edge becomes two exterior edges.

Geometric rules of 4–8 schemes are the direct generalization of the subdivision method of the quadric four-directional box spline surface. These subdivision masks are shown in Fig. 3.34. They can be described as following:

(1) Face rules: Each new vertex inserted as a result of bisection refinement of basic block is computed as the barycenter of that block;

(2) Vertex rules: The new position of an existing vertex $\boldsymbol{V}$ is computed as the average of the old position and barycenter of the vertices sharing an exterior block edge with $\boldsymbol{V}$.

When a mesh is open, there are two cases for boundary edges: exterior edges of blocks and interior edges of blocks. If a boundary edge is an exterior edge of a block, it is not split. If a boundary edge is not an exterior edge of a block, the edge is called

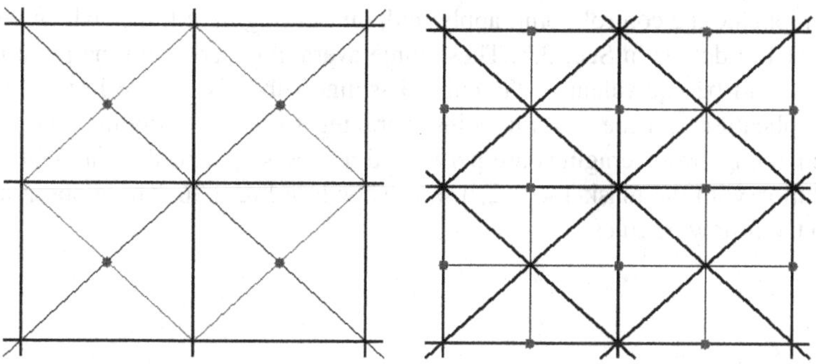

(a) The first subdivision step (b) The second subdivision step

Fig. 3.33 Triangles are bisected by 4–8 subdivision

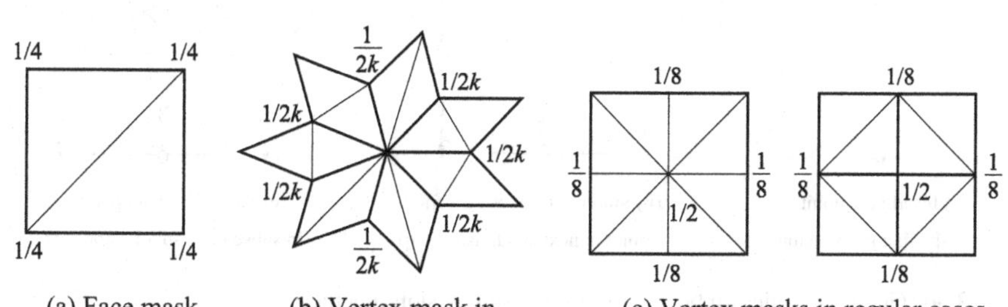

(a) Face mask (b) Vertex mask in extraordinary case (c) Vertex masks in regular cases

Fig. 3.34 Masks of the 4–8 subdivision scheme

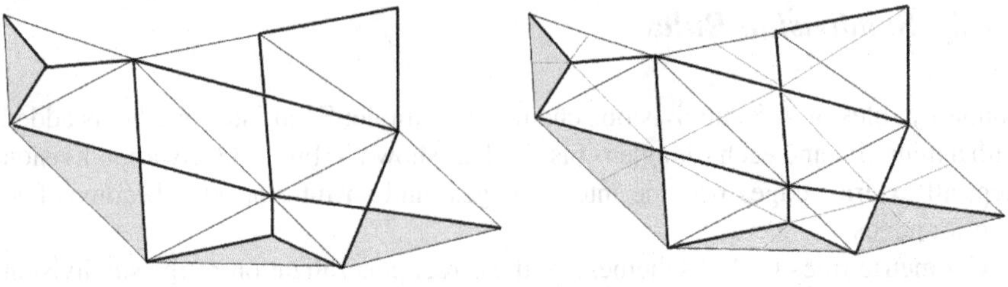

(a) Interior edges on boundary (b) Bisection of boundary basic blocks

Fig. 3.35 Interior edges and their bisections on boundary

an interior edge, and the triangle with the edge is a boundary basic block. In this case, we still apply bisection to the boundary basic block by inserting the split vertex on the boundary edge. Consequently, the boundary edge becomes two exterior edges on the next refinement level. Interior boundary edges and their bisection are shown in Fig. 3.35.

Geometric rules calculating these boundary points are shown in Fig. 3.36. These boundary subdivision rules are obtained by Lane–Riesenfeld algorithm [101]. A degree n B-spline curve can be computed by recursively applying the two-step algorithm: replicate the control point; apply midpoint averaging n times, which will be introduced in details in Sect. 3.8. These three averaging steps with the point replication should be equivalent to the cubic B-spline subdivision rule in (2.11). The point replication and the first midpoint averaging form the midpoint subdivision. The remaining two averagings are performed as one step, and they amount to the convolution with the mask 1/4, 1/2, 1/4, as shown in Fig. 3.36. These observations lead to the following rules:

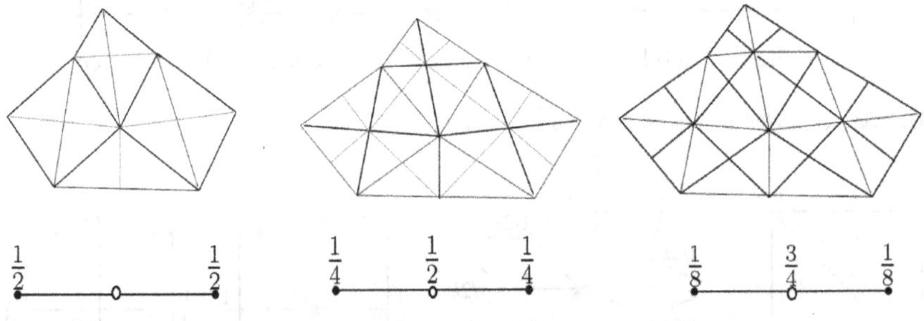

(a) Insert new point in the first subdivision (b) Smooth the newly inserted point in next subdivision step (c) Smooth the old point in subsequent subdivision step

Fig. 3.36 Geometric rules on boundaries of 4–8 subdivision scheme

(1) To insert a new vertex on the boundary, use midpoint subdivision.

(2) If for a boundary vertex the incident boundary edges are exterior block edges, apply the smoothing mask. If these edges are edges of single-triangle boundary blocks, apply cubic B-spline vertex mask.

A shortcoming of these rules is that they lead to surfaces which are not C^1 continuous near extraordinary vertices on the boundary. A solution is given to the problem in [27]. A reader may refer to it if he or she wants to make his/her surface with C^1 continuity.

Velho [96] also gives the semi-regular 4–8 subdivision which is a generalization of the subdivision of **Zwart–Powell** spline surfaces. He points out that the generalization to irregular meshes is immediate since the factorization of mask (shown in Fig. 2.41) can also be applied to semi-regular 4–8 mesh. Except for mask shown in Fig. 2.41, we should still update vertices of old meshes. From Fig. 2.40, we can find that there are some vertices that have not been used after two refinement steps. For this reason, at every double step, all old vertices are updated to be the average of new mesh vertices surrounding it.

3.8 Classification of Subdivision Schemes

The classification of subdivision schemes is helpful to understand characters of subdivision schemes. Generally, classification rules can be given according to geometric rules and topology rules. Geometric rules are connected with the convergence of mesh sequences, the continuity of limit surfaces, and the shape of limit surfaces. Two subdivision schemes with the same topology rules and different geometric rules probably have different properties. Consequently, subdivision scheme can be classified on the basis of geometric rules or topology rules. The following classifications can be regarded as classifications based on geometric rules:

(1) Approximating versus interpolatory. If the limit surface of a subdivision scheme does not go through control vertices of the initial meshes, the subdivision scheme is called as an approximating subdivision scheme. Otherwise, the scheme is an interpolatory subdivision scheme. The butterfly scheme and the convexity-preserving subdivision scheme are interpolatory subdivision schemes. One can alter geometric rules of classic approximating subdivision schemes to obtain interpolatory subdivision schemes [47], which will be discussed in Chap. 5. However, surfaces obtained by approximating subdivision schemes have better smoothness than those surfaces obtained by interpolatory subdivision schemes, which is due to the reason that people research how to construct interpolatory surfaces using approximating subdivision schemes.

(2) Uniform versus non-uniform subdivision. Most of the existing subdivision schemes are uniform subdivision schemes by which an existing mesh is refined uniformly through midpoint knot insertion over the entire surface for all levels of subdivision. Otherwise, it is called a non-uniform subdivision scheme. The NURSS subdivision scheme [25] is a non-uniform subdivision scheme, while C-C subdivision scheme and Doo–Sabin subdivision scheme are uniform subdivision schemes.

(3) Stationary versus non-stationary subdivision. If the subdivision rules do not change during the subdivision process, the scheme is called a stationary subdivision scheme and, otherwise, a non-stationary subdivision scheme. Most of the existing subdivision schemes are stationary subdivision schemes. To produce some specific classes of shapes, such as a perfect circle, a non-stationary subdivision scheme may have to be used [47, 82].

The following classification rules are based on topology structures:

(1) Triangle, quadrilateral, hexagonal, or 4–8 subdivision, etc.. The classification is based on types of regular tilings corresponding to subdivision meshes. In other words, we classify these subdivision schemes based the mesh type.

(2) 1–4 split, 1–3 split, or 1–2 split subdivision. The classification is given based on the split speed of vertices or faces of meshes in the process of subdivisions. Such as, Loop subdivision, Catmull–Clark subdivision and Doo–Sabin subdivision are 1-4 split subdivisions, $\sqrt{3}$ subdivision is a 1-3 split subdivision and 4–8 subdivision is a 1-2 split subdivision.

(3) Primary subdivision versus dual subdivision. These two kinds of topology rules are respectively given in Fig. 3.4b, c. The primary subdivision is also called the face splitting subdivision. The face splitting is to insert new vertices for every face and every edge. The new vertices are respectively called new edge points and new face points. For some subdivision schemes, such as Loop subdivision, new face points probably do not need to be inserted. The dual subdivision is also called the vertex splitting subdivision. Referring to Fig. 3.37, the vertex splitting method splits a vertex i to $|i|_F$ new vertices, and every new vertices corresponds to a neighboring face of the vertex i. $|i|_F$ is the number of neighboring face of the vertex i. From Fig. 3.37, it can also be found that every vertex corresponds to a new vertex for any old faces. Consequently, faces in the new mesh can be divided into three types: $V-$face, E-face, and F-face. In Fig. 3.37, these new faces are labeled as V, E, and F, respectively.

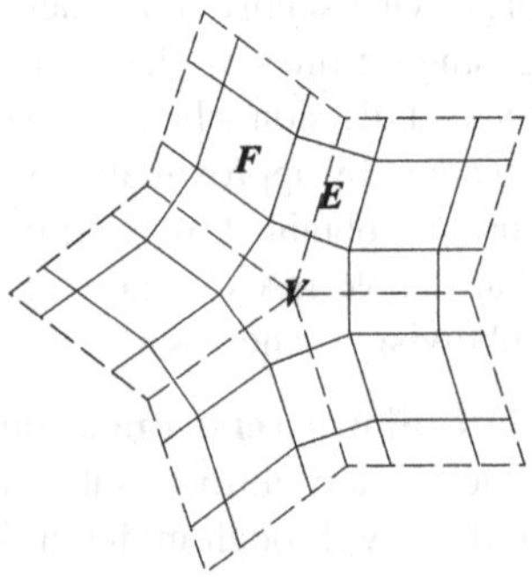

Fig. 3.37 Dual subdivision and categories of new faces

The classification about the primary subdivision and the dual subdivision is important because it can be regarded as a guide to generalize existing subdivision schemes and present new subdivision schemes.

Using the idea classifying subdivision schemes as primary subdivision and dual subdivision, Stam [37] and Zorin et al. [38] respectively give a beautiful solution for the problem generalizing uniform B-spline surfaces of arbitrary degree to arbitrary two-manifold meshes. They all increase the degree of subdivision surfaces by calculating the dual mesh (See Fig. 3.26) of the current mesh again and again. For example, we firstly make a linear subdivision and then calculate dual meshes $k - 1$ times. Just as what Zorin et al.[38] have pointed out, we can obtain a midpoint subdivision when only by executing a linear subdivision step. The Doo–Sabin subdivision is obtained if we calculate the dual only once, two dual calculation obtains the Catmull–Clark subdivision, three will have the generalization of biquartic B-spline surfaces, and so on. The difference between these two methods is that the former respectively constructs average operators for odd degree surfaces and even degree surfaces, and the latter executes the average processing by the quadtree data structure.

Ivrissimtzis [35] classifies subdivision schemes as $XY(n, m)$, where X denotes T, Q, or H, Y denotes P, D, or M, T is the triangle subdivision, Q is the quadrilateral subdivision, H is the hexagonal subdivision, P is the primary subdivision, D is the dual subdivision, M is the mixed subdivision of the primary subdivision and the dual subdivision, and (n, m) is a pair of integers. Assume that M^1 is the Lave tiling obtained by subdividing the Lave tiling M^0 once. $\overline{\boldsymbol{O}'\boldsymbol{P}'_1}$ is the generating vector. Let the minimum length of M^1 be 1. M^1 is a submesh of M^0. In the mesh M^0, set $\boldsymbol{O}' = (0, 0)$ if the used subdivision scheme is TP, TM, QP, QM, HP, or HM. If the used subdivision scheme is TD, QD, or HD, $\boldsymbol{O}'$ is respectively set as (1/3, 1/3), (1/2, 1/2), (1/3, 1/3). So, $\boldsymbol{O}'$ is a vertex of M^0. Set $\boldsymbol{P}'_1 = \boldsymbol{O}' + (n, m)$. If $\overline{\boldsymbol{O}'\boldsymbol{P}'_1}$ is a edge of M_0, the subdivision scheme is called as $XY(n, m)$. Typically, Doo–Sabin subdivision is $QD(2, 0)$. Catmull–Clark subdivision is $QP(2, 0)$. Loop subdivision is $TP(2, 0)$. $\sqrt{3}$ subdivision is $TP(1, 1)$. The work of [35] shows that this very low split speed of mesh elements usually results in the expense of symmetry and uniformity of meshes.

Remarks

This chapter introduces several representative subdivision schemes. Though there are also other subdivision schemes, such as the simplest subdivision [24], variational subdivisions [82], composite primal/dual $\sqrt{3}$ Subdivisions [39], and the $\sqrt{2}$ subdivision [46], etc., we can have a comprehensive understand for subdivision schemes by these subdivision schemes discussed in this chapter. A subdivision scheme is composed by two parts: topology rules and geometric rules. Subdivision schemes usually are classified based on these two kinds of rules. They are important ideas to construct subdivision schemes and to classify subdivision schemes based on refinements of

Lave tilings. By the classification of primal/dual subdivisions, the uniform B-spline surfaces of arbitrary degree can be generalized to any two-manifold mesh.

Exercises

1. Write program codes for the Doo–Sabin subdivision scheme.
2. Write program codes for the Catmull–Clark subdivision scheme.
3. Write program codes for the Loop subdivision scheme.
4. Write program codes for the honeycomb subdivision scheme.

Chapter 4
Analysis of Subdivision Surface

A mesh sequence is generated when a given mesh is recursively subdivided. Consequently, two questions are always asked: Is the mesh sequence convergent? How many is the continuity order of the limit surface if the mesh sequence is convergent? The two questions will be answered in this chapter. Subdivision rules can be expressed by subdivision matrix since these rules are linear. The discrete Fourier transform is a good tool to calculate eigenvalues and eigenvectors of subdivision matrix. The eigenvalues and the eigenvectors are necessary to answer the two questions. Using the eigenvalues and eigenvectors, the Stam's analytic evaluation approach is discussed in this chapter. Such approach makes the subdivision surface be evaluated as same as B-spline surfaces. The C-C subdivision is taken as an example for discussion in this chapter.

4.1 Subdivision Matrix

Let M^{k+1} be a subdivision result of M^k, express the vertices of M^k as $\boldsymbol{V}^k$, and then observe the geometric rules of C-C subdivision; since these rules are linear in points, the subdivision operator can be expressed as a matrix form:

$$\overrightarrow{\boldsymbol{V}}_g^{k+1} = \boldsymbol{M}_g \overrightarrow{\boldsymbol{V}}_{\mathrm{g}}^{k}$$

where $\overrightarrow{\boldsymbol{V}}_{\mathrm{g}}^{k} = [\boldsymbol{V}_0^k, \boldsymbol{V}_1^k, \cdots, \boldsymbol{V}_i^k, \cdots, \boldsymbol{V}_m^k]^{\mathrm{T}}$, m is the number of vertices of the mesh M^k. $\boldsymbol{V}_i^k$ is a vertex of the mesh M^k.

$\boldsymbol{M}_g$ is a global matrix in the above matrix form. It is very difficult to analyze the property of C-C subdivision surfaces by the global matrix. By investigating the geometric rules of C-C subdivision, we can see the vertices of 1-neighborhood of a vertex $\boldsymbol{V}$ of M^{k+1} are determined by vertices of 1-neighborhood of $\boldsymbol{V}$ of M^k.

W. Liao et al., *Subdivision Surface Modeling Technology*,
DOI 10.1007/978-981-10-3515-9_4

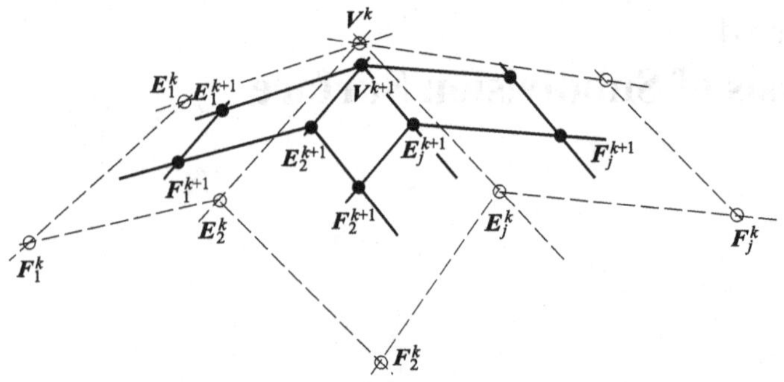

Fig. 4.1 Subdivision of 1-neighborhood of a vertex

Additionally, it can also be found that there are no same vertices on the 1-neighborhood of two different new vertices of M^{k+1}. Consequently, the C-C subdivision can be expressed by a local subdivision matrix.

In order to make the local subdivision matrix as simple as possible and in view of the finding that a random mesh becomes a quadrilateral mesh after one original C-C subdivision step, in this chapter the quadrilateral meshes are only investigated. It should also be noticed that the subdivision mesh M^k will not be added new extra-ordinary point after one initial subdivision step. Consequently, we use $\boldsymbol{E}$ and $\boldsymbol{F}$ to denote the edge point and the face point on the 1-neighborhood of $\boldsymbol{V}$ (see Fig. 4.1). In this book, we also denote the new vertex point, the new edge point, and the new face point as $\boldsymbol{V}, \boldsymbol{E}$, and $\boldsymbol{F}$, respectively.

Now, we research how to define the local subdivision matrix for C-C subdivision. For simplicity, we, respectively, denote the vertices of 1-neighborhood of $\boldsymbol{V}^k$ and $\boldsymbol{V}^{k+1}$ as $\boldsymbol{V}, \boldsymbol{E}, \boldsymbol{F}$ and $\boldsymbol{V}', \boldsymbol{E}', \boldsymbol{F}'$. Based on the subdivision rules given by Catmull and Clark, Ball and Story [15] give the subdivision formula as follows:

$$\boldsymbol{V}' = \alpha_n \boldsymbol{V} + \beta_n \left(\frac{1}{n} \sum_{j=1}^{n} \boldsymbol{E}_j \right) + \gamma_n \left(\frac{1}{n} \sum_{j=1}^{n} \boldsymbol{F}_j \right) \tag{4.1}$$

where $\beta_n = 3/(2n), \gamma_n = 1/(4n), \alpha_n = 1 - \beta_n - \gamma_n$. These weights can be derived by the C-C subdivision rules.
For each $j = 1, 2, \ldots, n$

$$\boldsymbol{E}_j' = \frac{3}{8}(\boldsymbol{V} + \boldsymbol{E}_j) + \frac{1}{16}(\boldsymbol{E}_{j-1} + \boldsymbol{F}_{j-1} + \boldsymbol{F}_j + \boldsymbol{E}_{j+1}) \tag{4.2}$$

$$\boldsymbol{F}_j' = \frac{1}{4}(\boldsymbol{V} + \boldsymbol{E}_j + \boldsymbol{F}_j + \boldsymbol{E}_{j+1}) \tag{4.3}$$

Let $\{\boldsymbol{V}, \overrightarrow{\boldsymbol{E}}, \overrightarrow{\boldsymbol{F}}\} = \{\boldsymbol{V}\} \cup \{\boldsymbol{E}_j | j = 1, 2, \ldots, n\} \cup \{\boldsymbol{F}_j | j = 1, 2, \ldots, n\}$ and

$$\begin{bmatrix} \boldsymbol{V}' \\ \overrightarrow{\boldsymbol{E}'} \\ \overrightarrow{\boldsymbol{F}'} \end{bmatrix} = \boldsymbol{M} \begin{bmatrix} \boldsymbol{V} \\ \overrightarrow{\boldsymbol{E}} \\ \overrightarrow{\boldsymbol{F}} \end{bmatrix} \tag{4.4}$$

Based on the subdivision formulas (4.1)~(4.3), we can define the local subdivision matrix $\boldsymbol{M}$ as follows:

$$\left[\begin{array}{c|ccccc|ccccc} \alpha & \beta/n & \beta/n & \beta/n & \cdots & \beta/n & \gamma/n & \gamma/n & \gamma/n & \cdots & \gamma/n \\ \hline 3/8 & 3/8 & 1/16 & 0 & \cdots & 1/16 & 1/16 & 1/16 & 0 & \cdots & 1/16 \\ 3/8 & 1/16 & 3/8 & 1/16 & \cdots & 0 & 1/16 & 1/16 & 1/16 & \cdots & 0 \\ 3/8 & 0 & 1/16 & 3/8 & \cdots & 0 & 0 & 1/16 & 1/16 & \cdots & 0 \\ \vdots & \vdots & \vdots & \vdots & \vdots & \vdots & \vdots & \vdots & \vdots & \vdots & \vdots \\ 3/8 & 1/16 & 0 & 0 & \cdots & 3/8 & 1/16 & 0 & 0 & \cdots & 1/16 \\ \hline 1/4 & 1/4 & 1/4 & 0 & \cdots & 0 & 1/4 & 0 & 0 & \cdots & 0 \\ 1/4 & 0 & 1/4 & 1/4 & \cdots & 0 & 0 & 1/4 & 0 & \cdots & 0 \\ 1/4 & 0 & 0 & 1/4 & \cdots & 0 & 0 & 0 & 1/4 & \cdots & 1/4 \\ \vdots & \vdots & \vdots & \vdots & \vdots & \vdots & \vdots & \vdots & \vdots & \vdots & \vdots \\ 1/4 & 1/4 & 0 & 0 & \cdots & 1/4 & 0 & 0 & 0 & \cdots & 1/4 \end{array}\right] \tag{4.5}$$

A subdivision scheme should be convergent, i.e., for a mesh sequence $M^k (k = 0, 1, 2, \ldots)$, there is an unique limit surface:

$$\boldsymbol{S} = \lim_{k \to \infty} M^k$$

Equivalently, if we let

$$\begin{bmatrix} \boldsymbol{V}^{k+1} \\ \boldsymbol{E}^{k+1} \\ \boldsymbol{F}^{k+1} \end{bmatrix} = \boldsymbol{M}^k \begin{bmatrix} \boldsymbol{V}^1 \\ \boldsymbol{E}^1 \\ \boldsymbol{F}^1 \end{bmatrix},$$

We should have

$$\boldsymbol{V}^{k+1} \to \boldsymbol{V}^{\infty}, \quad \boldsymbol{E}_j^{k+1} \to \boldsymbol{V}^{\infty}, \quad \boldsymbol{F}_j^{k+1} \to \boldsymbol{V}^{\infty} \tag{4.6}$$

Our current questions are as follows: Is the hypothesis (4.6) correct? How do we obtain the coordinates of $\boldsymbol{V}^{\infty}$ if the subdivision mode is convergent? How much is the continuity order of the limit surface $\boldsymbol{S}$? In order to answer these questions, we have to calculate the eigenvalues of the subdivision matrix $\boldsymbol{M}$. The discrete Fourier transform ("DFT" for short hereinafter) is an important tool to calculate the eigenvalues of $\boldsymbol{M}$.

4.2 Discrete Fourier Transform

The DFT was firstly used by Doo and Sabin [11] to calculate the eigenvalues of the C-C subdivision matrix. Later, Ball and Story [15] developed the transform method. We introduce the method of Ball and Story herein in this section. However, we simplify such method by only investigating the 1-neighborhood of the vertex $\boldsymbol{V}$ other than the 2-neighborhood. On the other hand, we only research how to calculate the eigenvalues and eigenvectors herein; but for Ball and Story, they additionally considered the choice of subdivision weights α_n, β_n, and γ_n.

4.2.1 Fundament of DFT

Let $a = \cos\dfrac{2\pi}{n} + i\sin\dfrac{2\pi}{n}$ and $\overline{a} = \cos\dfrac{2\pi}{n} - i\sin\dfrac{2\pi}{n}$. By some simple deductions, we can have the following conclusions:

$$\begin{aligned} 1 + a^j + (a^j)^2 + \cdots + (a^j)^{n-1} &= 1 + \overline{a}^j + (\overline{a}^j)^2 + \cdots + (\overline{a}^j)^{n-1} \\ &= 0, j = 1, 2, \cdots, n-1 \end{aligned} \tag{4.7}$$

$$a^n = \overline{a}^n = 1 \tag{4.8}$$

$$a\overline{a} = 1 \tag{4.9}$$

Let $\boldsymbol{A} = \begin{bmatrix} 1 & a & \cdots & a^{i-1} & \cdots & a^{n-1} \\ 1 & (a^2)^1 & \cdots & (a^2)^{i-1} & \cdots & (a^2)^{n-1} \\ \vdots & \vdots & \vdots & \vdots & \vdots & \vdots \\ 1 & (a^j)^1 & \cdots & (a^j)^{i-1} & \cdots & (a^j)^{n-1} \\ \vdots & \vdots & \vdots & \vdots & \vdots & \vdots \\ 1 & (a^n)^1 & \cdots & (a^n)^{i-1} & \cdots & (a^n)^{n-1} \end{bmatrix}$. Based on the formula (4.7) and (4.8), we have,

$$\boldsymbol{A}^{-1} = \frac{1}{n}\overline{\boldsymbol{A}}^{\mathrm{T}} \tag{4.10}$$

4.2.2 Discrete Fourier Transform

Using the symbols of the Eq. (4.4) for 1-neighborhood of $\boldsymbol{V}$, we define

$$\boldsymbol{E}_j = \sum_{\omega=0}^{n-1} a_{\omega j} \widetilde{\boldsymbol{E}}_{\omega}, \quad \boldsymbol{F}_j = \sum_{\omega=0}^{n-1} a_{\omega j} \widetilde{\boldsymbol{F}}_{\omega}$$

for $j = 1, 2, \ldots, n$, where

$$a_k = \cos \frac{2k\,\pi}{n} + i \sin \frac{2k\,\pi}{n} = c_k + i s_k$$

Consistently, let $\boldsymbol{V} = \widetilde{\boldsymbol{V}}$ and based on these definitions above, we have

$$\begin{bmatrix} \boldsymbol{V} \\ \vec{\boldsymbol{E}} \\ \vec{\boldsymbol{F}} \end{bmatrix} = \begin{bmatrix} 1 & \boldsymbol{O} & \boldsymbol{O} \\ \boldsymbol{O} & \mathbf{A} & \boldsymbol{O} \\ \boldsymbol{O} & \boldsymbol{O} & \mathbf{A} \end{bmatrix} \begin{bmatrix} \widetilde{\boldsymbol{V}} \\ \vec{\widetilde{\boldsymbol{E}}} \\ \vec{\widetilde{\boldsymbol{F}}} \end{bmatrix} \tag{4.11}$$

where

$$\{\widetilde{\boldsymbol{V}}, \vec{\widetilde{\boldsymbol{E}}}, \vec{\widetilde{\boldsymbol{F}}}\} = \{\widetilde{\boldsymbol{V}}\} \cup \{\widetilde{\boldsymbol{E}}_{\omega} | \omega = 0, 1, \ldots, n-1\} \cup \{\widetilde{\boldsymbol{F}}_{\omega} | \omega = 0, 1, \ldots, n-1\}$$

Replacing $[\boldsymbol{V}, \boldsymbol{E}, \boldsymbol{F}]$ in the Eq. (4.4) by the formula (4.11),

$$\begin{bmatrix} \widetilde{\boldsymbol{V}} \\ \vec{\widetilde{\boldsymbol{E}}}' \\ \vec{\widetilde{\boldsymbol{F}}}' \end{bmatrix} = \begin{bmatrix} 1 & \boldsymbol{O} & \boldsymbol{O} \\ \boldsymbol{O} & \mathbf{A} & \boldsymbol{O} \\ \boldsymbol{O} & \boldsymbol{O} & \mathbf{A} \end{bmatrix}^{-1} \boldsymbol{M} \begin{bmatrix} 1 & \boldsymbol{O} & \boldsymbol{O} \\ \boldsymbol{O} & \mathbf{A} & \boldsymbol{O} \\ \boldsymbol{O} & \boldsymbol{O} & \mathbf{A} \end{bmatrix} \begin{bmatrix} \widetilde{\boldsymbol{V}} \\ \vec{\widetilde{\boldsymbol{E}}} \\ \vec{\widetilde{\boldsymbol{F}}} \end{bmatrix}$$

Rewrite the above equation by using the formula (4.10):

$$\begin{bmatrix} \widetilde{\boldsymbol{V}} \\ \vec{\widetilde{\boldsymbol{E}}}' \\ \vec{\widetilde{\boldsymbol{F}}}' \end{bmatrix} = \frac{1}{n} \begin{bmatrix} n & \boldsymbol{O} & \boldsymbol{O} \\ \boldsymbol{O} & \overline{\mathbf{A}}^{\mathrm{T}} & \boldsymbol{O} \\ \boldsymbol{O} & \boldsymbol{O} & \overline{\mathbf{A}}^{\mathrm{T}} \end{bmatrix} \boldsymbol{M} \begin{bmatrix} 1 & \boldsymbol{O} & \boldsymbol{O} \\ \boldsymbol{O} & \mathbf{A} & \boldsymbol{O} \\ \boldsymbol{O} & \boldsymbol{O} & \mathbf{A} \end{bmatrix} \begin{bmatrix} \widetilde{\boldsymbol{V}} \\ \vec{\widetilde{\boldsymbol{E}}} \\ \vec{\widetilde{\boldsymbol{F}}} \end{bmatrix} \tag{4.12}$$

- Corresponding to the blocks of the matrix (4.5), let

$$\boldsymbol{M} = \begin{bmatrix} \boldsymbol{M}_{0,0} & \boldsymbol{M}_{0,1} & \boldsymbol{M}_{0,2} \\ \boldsymbol{M}_{1,0} & \boldsymbol{M}_{1,1} & \boldsymbol{M}_{1,2} \\ \boldsymbol{M}_{2,0} & \boldsymbol{M}_{2,1} & \boldsymbol{M}_{2,2} \end{bmatrix}.$$

We have,

$$\frac{1}{n} \begin{bmatrix} n & \boldsymbol{O} & \boldsymbol{O} \\ \boldsymbol{O} & \overline{\mathbf{A}}^{\mathrm{T}} & \boldsymbol{O} \\ \boldsymbol{O} & \boldsymbol{O} & \overline{\mathbf{A}}^{\mathrm{T}} \end{bmatrix} \boldsymbol{M} \begin{bmatrix} 1 & \boldsymbol{O} & \boldsymbol{O} \\ \boldsymbol{O} & \mathbf{A} & \boldsymbol{O} \\ \boldsymbol{O} & \boldsymbol{O} & \mathbf{A} \end{bmatrix} = \frac{1}{n} \begin{bmatrix} n\boldsymbol{M}_{0,0} & n\boldsymbol{M}_{0,1}\boldsymbol{A} & n\boldsymbol{M}_{0,2}\boldsymbol{A} \\ \overline{\boldsymbol{A}}^{\mathrm{T}}\boldsymbol{M}_{1,0} & \overline{\boldsymbol{A}}^{\mathrm{T}}\boldsymbol{M}_{1,1}\boldsymbol{A} & \overline{\boldsymbol{A}}^{\mathrm{T}}\boldsymbol{M}_{1,2}\boldsymbol{A} \\ \overline{\boldsymbol{A}}^{\mathrm{T}}\boldsymbol{M}_{2,0} & \overline{\boldsymbol{A}}^{\mathrm{T}}\boldsymbol{M}_{2,1}\boldsymbol{A} & \overline{\boldsymbol{A}}^{\mathrm{T}}\boldsymbol{M}_{2,2}\boldsymbol{A} \end{bmatrix} \tag{4.13}$$

The process from the matrix $\boldsymbol{M}$ to the matrix (4.13) is called the Fourier transform. Based on the definition of $\boldsymbol{A}$ and $\boldsymbol{M}_{1,1}$, we have

$$\frac{1}{n}\overline{\boldsymbol{A}}^{\mathrm{T}}\boldsymbol{M}_{1,1}\boldsymbol{A} = \begin{bmatrix} 1/2 & 0 & \cdots & 0 \\ 0 & \frac{1}{8}(3+c_w) & \cdots & 0 \\ \vdots & \cdots & \ddots & 0 \\ 0 & 0 & 0 & \frac{1}{8}(3+c_w) \end{bmatrix}$$

In order to deduce the above conclusion, we only need the matrix multiplication and make the similar calculation for blocks in the matrix (4.13), we have the following formula; if $\omega = 0$,

$$\begin{bmatrix} \widetilde{\boldsymbol{V}}' \\ \widetilde{\boldsymbol{E}}_0' \\ \widetilde{\boldsymbol{F}}_0' \end{bmatrix} = \begin{bmatrix} \alpha & \beta & \gamma \\ 3/8 & 1/2 & 1/8 \\ 1/4 & 1/2 & 1/4 \end{bmatrix} \begin{bmatrix} \widetilde{\boldsymbol{V}} \\ \widetilde{\boldsymbol{E}}_0 \\ \widetilde{\boldsymbol{F}}_0 \end{bmatrix} \tag{4.14}$$

that is, $[\boldsymbol{R}_0'] = [\boldsymbol{T}_0][\boldsymbol{R}_0]$; and when $\omega = 1, 2, \ldots, n-1$,

$$\begin{bmatrix} \widetilde{\boldsymbol{E}}_\omega' \\ \widetilde{\boldsymbol{F}}_\omega' \end{bmatrix} = \begin{bmatrix} \frac{1}{8}(3+c_\omega) & \frac{1}{16}(1+\bar{a}_\omega) \\ \frac{1}{4}(1+a_\omega) & 1/4 \end{bmatrix} \begin{bmatrix} \widetilde{\boldsymbol{E}}_\omega \\ \widetilde{\boldsymbol{F}}_\omega \end{bmatrix} \tag{4.15}$$

that is, $[\boldsymbol{R}_\omega'] = [\boldsymbol{T}_\omega][\boldsymbol{R}_\omega]$.

4.2.3 Eigenvalues and Eigenvectors from DFT

Let $\boldsymbol{G} = \begin{bmatrix} 1 & \boldsymbol{O} & \boldsymbol{O} \\ \boldsymbol{O} & \mathrm{A} & \boldsymbol{O} \\ \boldsymbol{O} & \boldsymbol{O} & \mathrm{A} \end{bmatrix}$. Since $\boldsymbol{G}^{-1}\boldsymbol{M}\boldsymbol{G}$ and $\boldsymbol{M}$ have the same eigenvalues, the eigenvalues of $\boldsymbol{T}_0$ and $\boldsymbol{T}_\omega(\omega = 1, 2, \cdots, n-1)$ are also the eigenvalues of $\boldsymbol{M}$.

Based on the basic matrix permutation theory, there is a primary matrix $\boldsymbol{P}$, so

$$\boldsymbol{P}^{\mathrm{T}}(\boldsymbol{G}^{-1}\boldsymbol{M}\boldsymbol{G})\boldsymbol{P} = \begin{bmatrix} \boldsymbol{T}_0 & & & \\ & \boldsymbol{T}_1 & & \\ & & \ddots & \\ & & & \boldsymbol{T}_{n-1} \end{bmatrix} = \boldsymbol{T} \tag{4.16}$$

Assume that λ is an eigenvalue and $\boldsymbol{R}$ is a corresponding eigenvector of $\boldsymbol{T}$. From the formula (4.16), we have

$$\boldsymbol{M}(\boldsymbol{GPR}) = \lambda(\boldsymbol{GPR}) \tag{4.17}$$

Consequently, $\boldsymbol{GPR}$ is the eigenvector of $\boldsymbol{M}$. Let $\boldsymbol{R}_i (i = 1, \cdots, n)$ be the eigenvectors of $\boldsymbol{P}^{\mathrm{T}}(\boldsymbol{G}^{-1}\boldsymbol{MG})\boldsymbol{P}$, so the $\boldsymbol{R}_i$ is called the right eigenvector. If $\boldsymbol{M}$ is not defective, the right eigenvectors $\boldsymbol{GPR}_i$ form a base. Consequently, we can define the left eigenvectors as follows:

$$\boldsymbol{L}_k\boldsymbol{R}_j = \delta_{kj} \tag{4.18}$$

In view of the formula (4.16),

$$(\boldsymbol{L}_k\boldsymbol{P}^{\mathrm{T}}\boldsymbol{G}^{-1})(\boldsymbol{GPR}_j) = \delta_{kj}$$

Since $\boldsymbol{GPR}_j$ is a right eigenvector of $\boldsymbol{M}$, $\boldsymbol{L}_k\boldsymbol{P}^{\mathrm{T}}\boldsymbol{G}^{-1}$, and is a left eigenvector of $\boldsymbol{M}$, it should be noticed that $[\boldsymbol{O}^{\mathrm{T}}, (\boldsymbol{r}_\omega)^{\mathrm{T}}, \boldsymbol{O}^{\mathrm{T}}]^{\mathrm{T}}$ is a right eigenvector of $\boldsymbol{T}$; provided that $\boldsymbol{r}_\omega$ and $\boldsymbol{l}_\omega$ are, respectively, a right eigenvector, and a left eigenvector of $\boldsymbol{T}_\omega$, $[\boldsymbol{O}, \boldsymbol{l}_\omega, \boldsymbol{O}]$ is a left eigenvector of $\boldsymbol{T}$.

4.3 Eigenvalues Analysis

4.3.1 Calculation of Eigenvalues and Eigenvectors

For the coefficient matrix $\boldsymbol{T}_0$, its characteristic polynomial is:

$$(1-\lambda)\left[\lambda^2 - \left(\alpha - \frac{1}{4}\right)\lambda + \left(\frac{1}{16} - \frac{1}{8}\beta\right)\right] = 0$$

So, we have the eigenvalues of $\boldsymbol{T}_0$ as follows:

$$\lambda = 1, \frac{\left(\alpha - \frac{1}{4}\right) \pm \sqrt{\left(\alpha - \frac{1}{4}\right)^2 + \left(\frac{1}{4} - \frac{1}{2}\beta\right)}}{2} \tag{4.19}$$

For the coefficient matrix $\boldsymbol{T}_\omega$, its characteristic polynomial is:

$$\lambda^2 - \left(\frac{5}{8} + \frac{1}{8}c_\omega\right)\lambda + \frac{1}{16} = 0$$

So, we have the eigenvalues of $\boldsymbol{T}_\omega$ as follows:

$$\lambda_\omega, \lambda'_\omega = \frac{1}{16}\{(c_\omega + 5) \pm [(c_\omega + 1)(c_\omega + 9)]^{1/2}\} \tag{4.20}$$

Consequently, assume that $\lambda_1 \geqslant |\lambda_2| \geqslant \cdots \geqslant |\lambda_n|$ and in view of $\boldsymbol{T}_{n-\omega} = \overline{\boldsymbol{T}}_\omega$, we have

$$\lambda_1 = 1, \lambda_2 = \lambda_3 = \frac{1}{16}\{(c_\omega + 5) + [(c_\omega + 1)(c_\omega + 9)]^{1/2}\} = \frac{4 + A_n}{16} \tag{4.21}$$

where

$$A_n = 1 + \cos\frac{2\pi}{n} + \cos\frac{\pi}{n}\sqrt{2\left(9 + \cos\frac{2\pi}{n}\right)} \tag{4.22}$$

We now investigate the eigenvectors of $\boldsymbol{M}$; their three right eigenvectors corresponding to the eigenvalues λ_1, λ_2 and λ_3 are:

$$[1, 1, 1]^{\mathrm{T}}, [16(\lambda_a)^2 - 12\lambda_a + 1, 6\lambda_a + 1, 4\lambda_a + 1]^{\mathrm{T}} \text{and}$$
$$[16(\lambda_b)^2 - 12\lambda_b + 1, 6\lambda_b + 1, 4\lambda_b + 1]^{\mathrm{T}}$$

Where 1, λ_a and λ_b are three eigenvalues in the formula (4.19). Based on the definition of the left eigenvalue (see the Formula 4.18), we can construct a linear system as follows:

$$\begin{cases} x + y + z = 1 \\ [16(\lambda_a)^2 - 12\lambda_a + 1]x + (6\lambda_a + 1)y + (4\lambda_a + 1)z = 0 \\ [16(\lambda_b)^2 - 12\lambda_b + 1]x + (6\lambda_b + 1)y + (4\lambda_b + 1)z = 0 \end{cases}$$

i.e.

$$\begin{cases} [16(\lambda_a)^2 - 16\lambda_a]x + (2\lambda_a - 2)y = -4\lambda_a - 1 \\ [16(\lambda_b)^2 - 16\lambda_b]x + (2\lambda_b - 2)y = -4\lambda_b - 1 \end{cases}$$

Solving the linear system, we have

$$x = \frac{n}{n+5}, \quad y = \frac{4}{n+5}, \quad z = \frac{1}{n+5}$$

Consequently, the left eigenvector of $\boldsymbol{T}_0$ is

$$l_1 = \left[\frac{n}{n+5}, \frac{4}{n+5}, \frac{1}{n+5}\right]$$

The two right eigenvectors of $\boldsymbol{T}_\omega$ are,

$$[4\lambda_\omega - 1, a_\omega + 1]^{\mathrm{T}}, [4\lambda'_\omega - 1, a_\omega + 1]^{\mathrm{T}},$$

where λ_ω, λ'_ω are the two eigenvalues defined in the Formula (4.20). Similar to the deduction of $\boldsymbol{I}_1$, we can have the left eigenvector corresponding to λ_ω:

$$\boldsymbol{l}_\omega = [A_n, \bar{a}_{\omega+1}] \tag{4.23}$$

In fact by using the formula (4.22), we have

$$A_n = 4(4\lambda_\omega - 1)$$

Consequently, $A_n(4\lambda'_\omega - 1) + (a_\omega + 1)(\bar{a}_\omega + 1) = 4(4\lambda_\omega - 1)(4\lambda'_\omega - 1) + (2 + c_\omega) = 0$.
From $\boldsymbol{l}_1$ and $\boldsymbol{P}^\mathrm{T}$ in the Formula (4.16),

$$\boldsymbol{L}_1\boldsymbol{P}^\mathrm{T} = \left[\frac{n}{n+5}, \frac{4}{n+5}, 0, \ldots, 0, \frac{1}{n+5}, 0, \ldots, 0\right]$$

$$\text{So,}\quad \boldsymbol{L}_1\boldsymbol{P}^\mathrm{T}\boldsymbol{G}^{-1} = \frac{1}{n+5}[n, 4, 0, \ldots, 0, 1, 0, \ldots, 0]\left(\frac{1}{n}\begin{bmatrix} n & \boldsymbol{O} & \boldsymbol{O} \\ \boldsymbol{O} & \bar{\boldsymbol{A}} & \boldsymbol{O} \\ \boldsymbol{O} & \boldsymbol{O} & \bar{\boldsymbol{A}} \end{bmatrix}\right)$$

$$= \frac{1}{n(n+5)}[n^2, 4, \ldots, 4, 1, \ldots, 1] \tag{4.24}$$

In order to use the denotations in Sect. 4.2.3, based on the Formula (4.23) and let

$$\boldsymbol{L}_2\boldsymbol{P}^\mathrm{T} = [0, \underbrace{0, A_n, 0, \ldots, 0}_{n}, \underbrace{0, \bar{a}_1 + 1, 0, \ldots, 0}_{n}]$$

$$\boldsymbol{L}_3\boldsymbol{P}^\mathrm{T} = [0, \underbrace{0, 0, 0, \ldots, A_n}_{n}, \underbrace{0, 0, 0, \ldots, a_1 + 1}_{n}]$$

$$\begin{aligned} \boldsymbol{L}_2\boldsymbol{P}^\mathrm{T}\boldsymbol{G}^{-1} &= [0, A_n\bar{a}_1, A_n(\bar{a}_1)^2, \ldots, A_n(\bar{a}_1)^n, (\bar{a}_1 + 1)\bar{a}_1, \\ &\quad (\bar{a}_1 + 1)(\bar{a}_1)^2, \ldots, (\bar{a}_1 + 1)(\bar{a}_1)^n] \\ \boldsymbol{L}_3\boldsymbol{P}^\mathrm{T}\boldsymbol{G}^{-1} &= [0, A_na_1, A_n(a_1)^2, \ldots, A_n(a_1)^n, (a_1 + 1)a_1, \\ &\quad (a_1 + 1)(a_1)^2, \ldots, (a_1 + 1)(a_1)^n] \end{aligned}$$

Let $\boldsymbol{L} = \boldsymbol{L}_2\boldsymbol{P}^\mathrm{T}\boldsymbol{G}^{-1}$ and $\overline{\boldsymbol{L}} = \boldsymbol{L}_3\boldsymbol{P}^\mathrm{T}\boldsymbol{G}^{-1}$, and their right eigenvectors are $\boldsymbol{R}$ and $\overline{\boldsymbol{R}}$, thence it is easy to know that $\boldsymbol{L} + \overline{\boldsymbol{L}}$ and $\boldsymbol{L} - \overline{\boldsymbol{L}}$ are the left eigenvectors on $\boldsymbol{R} + \overline{\boldsymbol{R}}$ and $\boldsymbol{R} - \overline{\boldsymbol{R}}$, respectively. Consequently, the left eigenvectors of matrix $\boldsymbol{M}$ corresponding to λ_2 and λ_3 is as follows:

$$\begin{aligned} &\left[0, A_n \cos\frac{2\pi}{n}, \ldots, A_n \cos\frac{2\pi j}{n}, \ldots, A_n \cos\frac{2\pi n}{n}, \cos\frac{2\pi}{n} + \cos\frac{2\pi(1+1)}{n}, \ldots,\right. \\ &\left.\cos\frac{2\pi j}{n} + \cos\frac{2\pi(j+1)}{n}, \ldots, \cos\frac{2\pi n}{n} + \cos\frac{2\pi(n+1)}{n}\right] \end{aligned} \tag{4.25}$$

$$\left[0, A_n \sin\frac{2\pi}{n}, \ldots, A_n \sin\frac{2\pi j}{n}, \ldots, A_n \sin\frac{2\pi n}{n}, \sin\frac{2\pi}{n} + \sin\frac{2\pi(1+1)}{n}, \ldots,\right.$$
$$\left.\sin\frac{2\pi j}{n} + \sin\frac{2\pi(j+1)}{n}, \ldots, \sin\frac{2\pi n}{n} + \sin\frac{2\pi(n+1)}{n}\right] \tag{4.26}$$

In the following sections, some denotations are simplified. Let $\lambda_1 \geqslant \lambda_2 = \lambda_3 \geqslant |\lambda_4| \geqslant \ldots \geqslant |\lambda_i| \geqslant \ldots \geqslant |\lambda_n|$ be the eigenvalues of the local subdivision matrix $\boldsymbol{M}$. Corresponding to λ_i, $\boldsymbol{L}_i$ and $\boldsymbol{R}_i$ are, respectively, the left and right eigenvector of $\boldsymbol{M}$.

4.3.2 Property of the Limit Surface

In view that M^0 is an initial mesh and $M^k (k = 0, 1, 2, \ldots)$ is a mesh sequence generated by Catmull–Clark subdivision. Assume that $\boldsymbol{v}^1$ is a vertex point of M^1 and $\boldsymbol{e}_j^1$ and $\boldsymbol{f}_j^1$ are, respectively, the edge point and face point of 1-neighborhood of $\boldsymbol{v}^1$. Let $\boldsymbol{v}^k$ be the new vertex point corresponding to $\boldsymbol{v}^1$ in the mesh M^k, and $\boldsymbol{e}_j^k$ and $\boldsymbol{f}_j^k$ be the edge point and face point of 1-neighborhood of $\boldsymbol{v}^k$. That is, we convert the capital letters in Sect. 4.1 to the corresponding small letters. Let $\vec{\boldsymbol{V}}^k = [\boldsymbol{v}^k, \boldsymbol{e}_1^k, \ldots, \boldsymbol{e}_n^k, \boldsymbol{f}_1^k, \ldots, \boldsymbol{f}_n^k]^{\mathrm{T}}$.

Based on the choice of subdivision weights α_n, β_n and γ_n (see Formula 4.1), the local subdivision matrix $\boldsymbol{M}$ is not degraded and its right eigenvector $\boldsymbol{R}_i$ forms a base. The 1-neighborhood of $\boldsymbol{v}^1$ can be expanded uniquely as:

$$\vec{\boldsymbol{V}}^1 = \boldsymbol{c}_1\boldsymbol{R}_1 + \cdots + \boldsymbol{c}_m\boldsymbol{R}_m \tag{4.27}$$

where $\boldsymbol{c}_i$ is the geometric position vector, *i.e.*, $\boldsymbol{c}_i = [c_{ix}, c_{iy}, c_{iz}]$. Using the definition of the left eigenvector, we have,

$$\boldsymbol{c}_i = \boldsymbol{L}_i\boldsymbol{V}^1 \tag{4.28}$$

In view of the formula (4.27),

$$\boldsymbol{V}^{k+1} = \boldsymbol{M}^k\boldsymbol{V}^1 = \lambda_1^k\boldsymbol{c}_1\boldsymbol{R}_1 + \cdots + \lambda_m^k\boldsymbol{c}_m\boldsymbol{R}_m \tag{4.29}$$

In view that $\lambda_1 = 1$, $|\lambda_i| < 1$ and $\boldsymbol{R}_1 = [1, \ldots, 1]$, we can see that

$$\boldsymbol{V}^{\infty} = \boldsymbol{c}_1\boldsymbol{R}_1 = [\boldsymbol{c}_1, \ldots, \boldsymbol{c}_1]$$

Consequently, using (4.24) and (4.28),

$$\boldsymbol{v}^{\infty} = \boldsymbol{e}_j^{\infty} = \boldsymbol{f}_j^{\infty} = \frac{n^2\boldsymbol{v}^1 + 4\sum_{j=1}^{n}\boldsymbol{e}_j^1 + \sum_{j=1}^{n}\boldsymbol{f}_j^1}{n(n+5)} \tag{4.30}$$

where $j = 1, \ldots, n$. We now investigate the smoothness of the limit surface G^1. Using $\boldsymbol{p}_s^k$ to denote the $\boldsymbol{e}_j^k$ or $\boldsymbol{f}_j^k$, $s = 2, \ldots, 2n+1$, so,

$$\begin{aligned}\boldsymbol{p}_s^k - \boldsymbol{v}^\infty &= (\lambda_1^k r_{1,s}\boldsymbol{c}_1 + \lambda_2^k r_{2,s}\boldsymbol{c}_2 + \lambda_3^k r_{3,s}\boldsymbol{c}_3 + \cdots + \lambda_m^k r_{m,s}\boldsymbol{c}_m) - \boldsymbol{c}_1 \\ &= \lambda_2^k (r_{2,s}\boldsymbol{c}_2 + r_{3,s}\boldsymbol{c}_3) + \lambda_4^k r_{4,s}\boldsymbol{c}_4 + \cdots \end{aligned} \quad (4.31)$$

In view that $1 > \lambda_2 = \lambda_3 > |\lambda_i| (i > 3)$, we have

$$\begin{aligned}\lim_{n\to\infty} \frac{\boldsymbol{p}_s^k - \boldsymbol{v}^\infty}{\|\boldsymbol{p}_s^k - \boldsymbol{v}^\infty\|} &= \lim_{n\to\infty} \frac{(r_{2,s}\boldsymbol{c}_2 + r_{3,s}\boldsymbol{c}_3) + (\lambda_4/\lambda_2)^k r_{4,s}\boldsymbol{c}_4 + \cdots}{\|(r_{2,s}\boldsymbol{c}_2 + r_{3,s}\boldsymbol{c}_3) + (\lambda_4/\lambda_2)^k r_{4,s}\boldsymbol{c}_4 + \cdots\|} \\ &= \frac{r_{2,s}\boldsymbol{c}_2 + r_{3,s}\boldsymbol{c}_3}{\|r_{2,s}\boldsymbol{c}_2 + r_{3,s}\boldsymbol{c}_3\|} \end{aligned} \quad (4.32)$$

All $\lim_{n\to\infty} \frac{\boldsymbol{e}_j^k - \boldsymbol{v}^\infty}{\|\boldsymbol{e}_j^k - \boldsymbol{v}^\infty\|}$ and $\lim_{n\to\infty} \frac{\boldsymbol{f}_j^k - \boldsymbol{v}^\infty}{\|\boldsymbol{f}_j^k - \boldsymbol{v}^\infty\|} (j = 1, \ldots, n)$ must, therefore, lie in the plane spanned by $\boldsymbol{c}_2$ and $\boldsymbol{c}_3$. Consequently, the normal vector of the tangent plane is $\boldsymbol{c} \times \boldsymbol{c}_3$; if the limit surface $\boldsymbol{v}^\infty$ is G^1 smooth outside, based on (4.25), (4.26), and (4.28), so:

$$\begin{aligned}\boldsymbol{c}_2 &= \sum_{j=1}^{n} A_n \cos\left(\frac{2\pi j}{n}\right)\boldsymbol{e}_j^1 + \left(\cos\left(\frac{2\pi j}{n}\right) + \cos\left(\frac{2\pi(j+1)}{n}\right)\right)\boldsymbol{f}_j^1 \\ \boldsymbol{c}_3 &= \sum_{j=1}^{n} A_n \sin\left(\frac{2\pi j}{n}\right)\boldsymbol{e}_j^1 + \left(\sin\left(\frac{2\pi j}{n}\right) + \sin\left(\frac{2\pi(j+1)}{n}\right)\right)\boldsymbol{f}_j^1 \end{aligned} \quad (4.33)$$

In the reference [44], $\boldsymbol{c}_3$ is given as

$$\sum_{j=1}^{n} A_n \cos\left(\frac{2\pi j}{n}\right)\boldsymbol{e}_{j+1}^1 + \left(\cos\left(\frac{2\pi j}{n}\right) + \cos\left(\frac{2\pi(j+1)}{n}\right)\right)\boldsymbol{f}_{j+1}^1 \quad (4.34)$$

We denote the above expression as $\boldsymbol{c}_3'$ to prove the consistency of the formula (4.33) and (4.34). By replacing $j+1$ by j, we have

$$\boldsymbol{c}_3' = \sum_{j=1}^{n} A_n \cos\frac{2\pi(j-1)}{n}\boldsymbol{e}_j^1 + \left(\cos\frac{2\pi(j-1)}{n} + \cos\frac{2\pi((j+1)-1)}{n}\right)\boldsymbol{f}_j^1$$

In view that $\cos\frac{2\pi(j-1)}{n} = \cos\frac{2\pi j}{n}\cos\frac{2\pi}{n} + \sin\frac{2\pi j}{n}\sin\frac{2\pi}{n}$ and

$$\boldsymbol{c}_3' = \cos\frac{2\pi}{n}\boldsymbol{c}_2 + \sin\frac{2\pi}{n}\boldsymbol{c}_3,$$

so,

$$c_2 \times c_3' = \sin \frac{2\pi}{n} c_2 \times c_3$$

It is only a necessary condition of G^1 continuity of the limit surface that $v^k, e_1^k, \ldots, e_n^k, f_1^k, \ldots, f_n^k$ are convergent to a common plane, which will be explained in the Sect. 4.4.

4.3.3 Another Example for Eigenvalues Analysis

We have given the $\sqrt{3}$ subdivision mode in Sect. 3.4.3. Reader probably puzzles how to get the weight value α_n. In this section, an answer shall be made for the puzzle (the explanation in this subsection comes from Kobbelt [26]). Based on $\sqrt{3}$ subdivision rules, all vertices of the 1- neighborhood of v can be arranged into a vector group $[v, v_0, \cdots, v_{n-1}]$. So the following subdivision matrix can be obtained:

$$M = \frac{1}{3}\begin{bmatrix} u & v & v & v & \cdots & v \\ 1 & 1 & 1 & 0 & \cdots & 0 \\ 1 & 0 & 1 & 1 & \cdots & 0 \\ \vdots & \vdots & \vdots & \vdots & \cdots & \vdots \\ 1 & 0 & 0 & 0 & \cdots & 1 \\ 1 & 1 & 0 & 0 & \cdots & 1 \end{bmatrix}$$

where $u = 3(1 - \alpha_n)$ and $v = 3\alpha_n/n$. However, when analyzing the eigenstructure of this matrix, it is found unsuitable for the construction of a convergent subdivision mode. The reason for this defect is the rotation around v, and the rotation is caused by the application of M; furthermore, such rotation makes all eigenvalues of M become the complex number, see Fig. 4.2 as follows.

From Sect. 3.4.3, we know the twice application on the $\sqrt{3}$ subdivision operator corresponds to a ternary splintering. Therefore, instead of analyzing one single subdivision step, two successive steps can be combined since the 1-neighborhood of v^2 becomes consistent with the original configuration generated by the rotation around v after the M is applied. The backward rotation can be written as a simple permutation matrix:

$$P = \begin{bmatrix} 1 & 0 & \cdots & 0 & 0 \\ 0 & 0 & \cdots & 0 & 1 \\ 0 & 1 & \cdots & 0 & 0 \\ \vdots & \vdots & \cdots & \vdots & \vdots \\ 0 & 0 & \cdots & 1 & 0 \end{bmatrix}$$

Fig. 4.2 Application of the subdivision matrix M causes a rotation around v since the neighboring vertices are replaced by the centers of the adjacent triangles

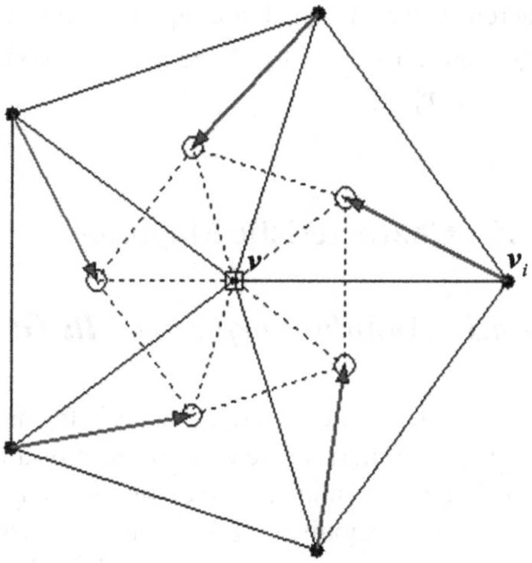

The resultant matrix $\boldsymbol{M}' = \boldsymbol{P}\boldsymbol{M}^2$ now has the correct eigenstructure for the analysis. Its eigenvalues are as follows:

$$\frac{1}{9}\left[9, (2-3\alpha_n)^2, 2+2\cos\left(2\,\pi\,\frac{1}{n}\right), \cdots, 2+2\cos\left(2\,\pi\,\frac{n-1}{n}\right)\right]$$

Sorting these eigenvalues by decreasing modulus, we have

$$\lambda_1 = 1, \lambda_2 = \lambda_3 = \left[2+2\cos\left(2\,\pi\,\frac{1}{n}\right)\right]/9, \lambda_4 = (2-3\alpha_n)^2/9 \tag{4.35}$$

According to (4.31), the necessary condition of G^1 continuity of the limit surface is:

$$\lambda_2 > |\lambda_i| (i \geqslant 4)$$

From (4.35), a natural choice for α_n is as follows:

$$\lambda_4 = (\lambda_2)^2 < \lambda_2$$

so,

$$\left(\frac{2}{3} - \alpha_n\right)^2 = \left(\frac{2+2\cos(2\,\pi\,/n)}{9}\right)^2$$

which leads to

$$\alpha_n = \frac{4 - 2\cos(2\,\pi\,/n)}{9}$$

herein solve the quadratic equation, make $\alpha_n \in [0, 1]$ and $\alpha_n \in [0, 1]$, and make the new vertex point v^{k+1} be a convex combination of the old vertices $\overrightarrow{V}^k = [v^k, v_0^k, \ldots, v_{n-1}^k]$.

4.4 Characteristic Mapping

4.4.1 Annular Surface and Its Gradualness

For simplicity, we investigate a mesh, that is, the 3-neighborhood of an extraordinary vertex. The mesh is the control mesh of a bi-cubic annular surface. We denote the mesh and the annular surface by M^0 and S^0. The extraordinary vertex is denoted as v^0. After subdividing $M^0 k$ times by using the C-C subdivision scheme, the 3-neighborhood of v^k is denoted as M^k and its annular surface as S^k. Figure 4.3 gives explanations for these denotations.

From Fig. 4.3, we can see that a hole is formed by the annular surface S^0. In the process of subdivision, the hole is gradually filled by S^k. When the subdivision process is convergent, there is an unique limit point p, therefore

$$\lim_{k \to \infty} p_k = p \tag{4.36}$$

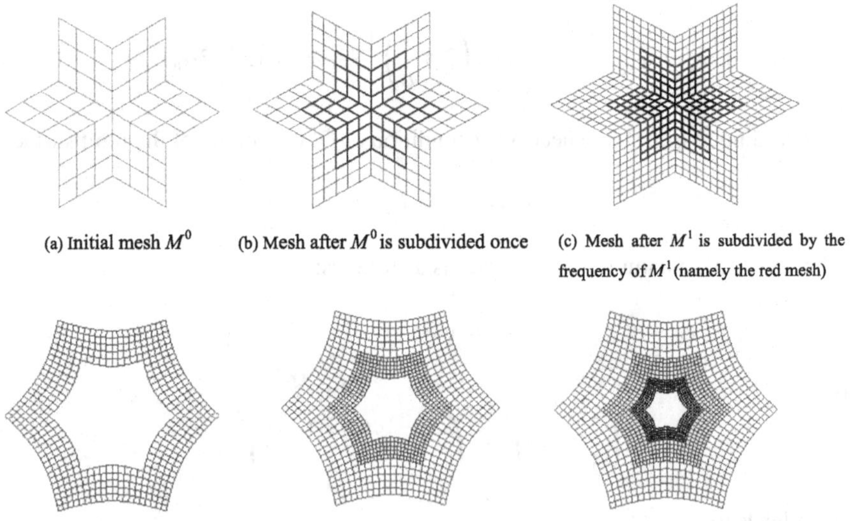

(a) Initial mesh M^0 (b) Mesh after M^0 is subdivided once (c) Mesh after M^1 is subdivided by the frequency of M^1 (namely the red mesh)

(d) Annular surface S^0 (e) Annular surface S^0, S^1 (f) Annular surface S^0, S^1, S^2

Fig. 4.3 S^k is a uniform bi-cubic surface and its control mesh is M^k

for any sequence of points $\{p_k | k = 0, 1, 2, \cdots\} \subset S^k$. The formula (4.36) is a sufficient and necessary condition of the subdivision process convergence. We have discussed the convergence of subdivision schemes in the above section. The description (4.36) on convergence is stricter than the description (4.29) in the above section, since the 1-neighborhood of extraordinary vertices is only investigated there.

It is known that

$$1 = \lambda_1 > \lambda_2 = \lambda_3 > |\lambda_i| (i \geqslant 4)$$

is only a necessary condition of G^1 continuity of the limit surfaces in above section. A sufficient condition of G^1 continuity of limit surfaces is given in this section according to Reif's work [17]. Firstly, we give some basic concepts as follows.

Definition 4.1 The surface $\lim_{n\to\infty} \bigcup_{k=0}^{n} \boldsymbol{S}^k$ is called tangent plane continuous, if the subdivision process converges and if there is an unique limit $\boldsymbol{n}(\boldsymbol{p})$for any normal vector sequence, so:

$$\lim_{k\to\infty} \boldsymbol{n}(\boldsymbol{p}_k) = \boldsymbol{n}(\boldsymbol{p}) \tag{4.37}$$

where $\boldsymbol{p}_k \in S^k$. $\boldsymbol{n}(\boldsymbol{p})$ is just the name of the limit of normal vectors, but not necessarily the normal vector of the $\lim_{n\to\infty} \bigcup_{k=0}^{n} S^k$ at $\boldsymbol{p}$, which might not even exit.

Definition 4.2 The surface $\boldsymbol{S}$ is called regular at $\boldsymbol{P}$, if there is a regular parameterization of $\boldsymbol{S}$ at $\boldsymbol{p}$:

$$\boldsymbol{S} : U \to \boldsymbol{S}(U) \tag{4.38}$$

where U is the neighborhood of (0,0) and $\boldsymbol{S}(0, 0) = \boldsymbol{p}$. A regular parameterization of S is continuous, differentiable, one-to-one, and has a Jacobi matrix of maximum rank. Let $\boldsymbol{S}(U) = (x(u, v), y(u, v), z(u, v))$ and $(u, v) \in U$, the condition is the matrix, i.e.,

$$\begin{bmatrix} \partial x/\partial u & \partial x/\partial v \\ \partial y/\partial u & \partial y/\partial v \\ \partial z/\partial u & \partial z/\partial v \end{bmatrix} \tag{4.39}$$

It has the maximal rank 2.

According to the definition of C^1 continuity in reference [8], if $\boldsymbol{S}$ is regular at $\boldsymbol{p}$, $\boldsymbol{S}$ shall $\boldsymbol{S}$ is C^1 continuous at $\boldsymbol{p}$. Consequently, $\boldsymbol{S}$ is also G^1 continuous at $\boldsymbol{p}$. It is important to see that G^1 continuity is essentially better than the tangent plane continuity. To demonstrate the difference, it is needed to consider the surface $\boldsymbol{S}$ shown in Fig. 4.4. This surface S is the tangent plane and is continuous, where the $\lim_{r\to 0} \boldsymbol{n}(r, t) = [0.0, 1]$, but not regular at the origin since the projection of $\boldsymbol{S}$ on the xy-plane is not injective. In fact, the xy-plane is the tangent plane of $\boldsymbol{S}$. If there is a regular parameterization (4.38) of $\boldsymbol{S}$ at $\boldsymbol{p}$, we can consider that the region is on the tangent plane. Consequently,

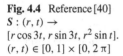

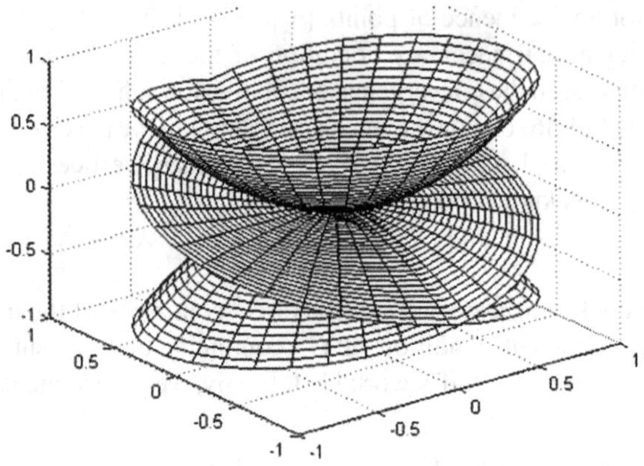

Fig. 4.4 Reference [40] $S : (r, t) \to [r \cos 3t, r \sin 3t, r^2 \sin t]$. $(r, t) \in [0, 1] \times [0, 2\pi]$

there is a one-to-one mapping between $S(U)$ and a region on the tangent plane. Using a parameter transformation, we can make the projection of $S(U)$ on the xy-plane is a one-to-one mapping.

4.4.2 Definition of Characteristic Mapping

Investigate Figs. 4.3a and 4.3d. Assume that the valence of the extraordinary vertex is n. So, there are $3n$ uniform bi-cubic B-spline patches in Fig. 4.3d; they are arranged as S_j^0. Let $\{V_i^0 | i = 0, \ldots, m\}$ be an array of vertices in Fig. 4.3d, the S_j^0 can be expressed in the form as follows:

$$S_j^0 : (u, v) \to S_j^0(u, v) = \sum_{i=0}^{m} b_{i,j}(u, v) V_i^0, \ (u, v) \in [0, 1] \times [0, 1]$$

or in the form of vector as follows:

$$S_j^0(u, v) = \boldsymbol{b}_j(u, v) \vec{V}^0$$

with a row vector $\boldsymbol{b}_j(u, v)$ of a function and a column vector $\vec{V}^0$ consisting of points in R^3. The function $b_{i,j}(u, v)$ satisfies with the following conditions:

$$b_{i,j}(u, v) \geqslant 0$$

$$\sum_{i=0}^{m} b_{i,j}(u, v) = 1$$

For Catmull–Clark subdivision, they are constructed by the uniform bi-cubic B-spline primary functions. Let $\overrightarrow{V}^k$ be the vertex column vector of the mesh M^k and $\boldsymbol{M}$ is the subdivision matrix of $\overrightarrow{V}^k$, *i.e.*,

$$\overrightarrow{V}^k = \boldsymbol{M}^k \overrightarrow{V}^0 \tag{4.40}$$

It should be noticed that the primary function does not depend on k, and

$$\boldsymbol{S}_j^k(u, v) = \boldsymbol{b}_j(u, v)\overrightarrow{V}^k$$

For simplicity, we denote a parameter group of a point of $\boldsymbol{S}_j^k$ as (u, v, j). All parameter groups of $\boldsymbol{S}_j^k(j = 1, \cdots 3n)$ form a set Ω. So, a point of $\boldsymbol{S}^k$ corresponds to an unique parameter group. The function $b_{i,j}(u, v)$ can also be written as $b_i(u, v, j)$. Using these denotations, we can define the characteristic mapping for the subdivision scheme matrix $\boldsymbol{M}$.

Definition 4.3 For a subdivision matrix $\boldsymbol{M}$ with $1 = \lambda_1 > \lambda_2 = \lambda_3 > |\lambda_i|(i \geqslant 4)$, the characteristic mapping $\boldsymbol{\Psi} : \Omega \to \mathrm{R}^2$ is defined by

$$\boldsymbol{\Psi} : (u, v, j) \to \boldsymbol{b}(u, v, j)\boldsymbol{R} = \boldsymbol{b}(u, v, j)[\boldsymbol{r}_2, \boldsymbol{r}_3]$$

where $\boldsymbol{R}$ is a $m\times 2$ matrix with the vectors $\boldsymbol{r}_2$ and $\boldsymbol{r}_3$ as its columns and $\boldsymbol{\Psi} = [\Psi_1, \Psi_2]$. $\boldsymbol{\Psi}$ is called regular, if

$$\Delta(u, v, j) = \begin{vmatrix} \partial\Psi_1(u, v, j)/\partial u & \partial\Psi_1(u, v, j)/\partial v \\ \partial\Psi_2(u, v, j)/\partial u & \partial\Psi_2(u, v, j)/\partial v \end{vmatrix} \neq 0 \tag{4.41}$$

for all $(u, v, j) \in \Omega$.

Note that the characteristic mapping can be viewed as a two-dimensional spline function with two-dimensional control vertices given by rows of the matrix $\boldsymbol{R}$. Of course, the characteristic mapping is subject to the ambiguity in the choice of the vectors $\boldsymbol{r}_2$ and $\boldsymbol{r}_3$. However, all of its crucial properties are well defined. Injectivity and regularity of the characteristic mapping do not depend on the particular choice of the vectors $\boldsymbol{r}_2$ and $\boldsymbol{r}_3$, which will be proved in Theorem 4.1 in the next subsection.

Why do we define the characteristic mapping? To answer the question, the formula (4.31) has to be investigated. Though it only refers to 1-neighborhood of the extraordinary vertex $\boldsymbol{v}$, other than 3-neighborhood, the case is not an obstacle to analyze the meaning of the characteristic mapping; this is just because the 3-neighborhood only leads to a larger subdivision matrix. Let $\boldsymbol{p}^k = \boldsymbol{p}_s^k - \boldsymbol{v}^\infty$ and $\lambda_2 = \lambda_3 = \lambda$, (4.31), so:

$$\frac{\boldsymbol{p}^k}{\lambda^k} = r_{2,s}\boldsymbol{c}_2 + r_{3,s}\boldsymbol{c}_3 + \left(\frac{\lambda_4}{\lambda}\right)^k r_{4,s}\boldsymbol{c}_4 + \cdots \tag{4.41$'$}$$

The formula means that, up to a scaling by λ^k, the 1-neighborhood approaches a fixed configuration. This configuration is determined by $\boldsymbol{r}_2$ and $\boldsymbol{r}_3$, which depends only on the subdivision scheme; as for r_2 and r_3, both of them depend on the initial control mesh.

$\boldsymbol{p}^k$ for sufficiently large k is a linear combination of $\boldsymbol{c}_2$ and $\boldsymbol{c}_3$, up to a vanishing term. This indicates that $\boldsymbol{c}_2$ and $\boldsymbol{c}_3$ span the tangent plane. Also note that if we apply an affine transform, taking $\boldsymbol{c}_2$ and $\boldsymbol{c}_3$ as the coordinate vectors $\boldsymbol{e}_1$ and $\boldsymbol{e}_2$ in the plane, then after it is up to a vanishing term, the scaled configuration will be independent of the initial control mesh. The transformed configuration consists of 2D vertices with coordinates $(r_{2,s}, r_{3,s})$, $s = 1, 2, \ldots$, which depend on the subdivision matrix. Note that $(r_{2,s}, r_{3,s})$, $s = 1, 2, \ldots$ form the matrix $[\boldsymbol{r}_2, \boldsymbol{r}_3]$. The observation can be regarded as a reason that the characteristic mapping is defined.

4.4.3 Characteristic Mapping and Continuity of Subdivision Surfaces

Theorem 4.1 *(1) Injective and regularity of the characteristic mapping do not depend on the particular choice of the vectors $\boldsymbol{r}_2$ and $\boldsymbol{r}_3$. (2) If the characteristic mapping $\boldsymbol{\Psi}$ is regular, then the Δ defined by (4.41) has*

$$\mu(\Delta) = \inf_{\Omega} |\Delta| > 0$$

Proof (1) Let $\widetilde{\boldsymbol{r}}_2$ and $\widetilde{\boldsymbol{r}}_3$ be another two right eigenvectors of the subdivision matrix $\boldsymbol{M}$ in (4.40). They span the same linear space as $\boldsymbol{r}_2$ and $\boldsymbol{r}_3$. Consequently,

$$\widetilde{\boldsymbol{r}}_2 = t_{1,1}\boldsymbol{r}_2 + t_{1,2}\boldsymbol{r}_3$$
$$\widetilde{\boldsymbol{r}}_3 = t_{2,1}\boldsymbol{r}_2 + t_{2,2}\boldsymbol{r}_3$$

Let $\widetilde{\boldsymbol{R}} = [\widetilde{\boldsymbol{r}}_2, \widetilde{\boldsymbol{r}}_3]$ and $\boldsymbol{T} = \begin{bmatrix} t_{1,1} & t_{1,2} \\ t_{2,1} & t_{2,2} \end{bmatrix}$, we have

$$\widetilde{\boldsymbol{R}} = \boldsymbol{R}\boldsymbol{T}$$

This implies

$$\widetilde{\boldsymbol{\Psi}}(u, v, j) = \boldsymbol{b}(u, v, j)\widetilde{\boldsymbol{R}} = \boldsymbol{\Psi}(u, v, j)\boldsymbol{T}$$

So,

$$\widetilde{\Delta}(u, v, j) = \Delta(u, v, j) \det \boldsymbol{T} \tag{4.42}$$

Note that $\det \boldsymbol{T} \neq 0$ and the Formula (4.40) in Definition 4.3, the first statement follows from (4.41).

Proof (2) Assume that $\mu(\Delta) = \inf_{\Omega} |\Delta| = 0$. Since $\boldsymbol{\Psi}(u, v, j)(j = 1, \ldots, 3n)$ forms a C^2 continuity annular surface, we can pick a point series $\{\boldsymbol{\Psi}(u_k, v_k, j)\} \subset \boldsymbol{\Psi}(u, v, j)$ for a fixed j and the series satisfies:

$$\mu(\Delta) = \inf_{\Omega} |\Delta| = 0$$

Note that if $(u_k, v_k) \subset [0, 1] \times [0, 1]$ and $[0, 1] \times [0, 1]$ is a closed set, there should be a parameter group (u, v, j) making

$$\Delta(u, v, j) = 0$$

The above formula is contrary to Formula (4.41). Consequently, the hypothesis is not correct.

We now prove the theorem on convergence of subdivision scheme. Though we have obtained the same conclusion in Sect. 4.3, the following proof process is stricter since it is given according to annular surfaces and the Formula (4.36).

Theorem 4.2 *A subdivision scheme converges, if* $1 = \lambda_1 > |\lambda_i| (i \geqslant 2)$

Proof All rows of the subdivision matrix $\boldsymbol{M}$ sum up to 1. So, $\lambda_1 = 1$ is always an eigenvalue of $\boldsymbol{M}$ and $\boldsymbol{r}_1 = [1, \ldots, 1]^{\mathrm{T}}$ is the corresponding eigenvector. We can write the sequence of vectors of control vertices in the form as follows:

$$\boldsymbol{V}^0 = \boldsymbol{c}_1 \boldsymbol{r}_1 + \boldsymbol{c}_2 \boldsymbol{r}_2 + \cdots + \boldsymbol{c}_m \boldsymbol{r}_m, \quad \boldsymbol{V}^k = \boldsymbol{c}_1 \boldsymbol{r}_1 + \boldsymbol{o}(1)$$

So,

$$\boldsymbol{S}_j^0(u, v) = \sum_{i=1}^{m} \boldsymbol{b}(u, v, j) \boldsymbol{c}_i \boldsymbol{r}_i$$

$$\boldsymbol{S}_j^k(u, v) = \boldsymbol{b}(u, v, j) \boldsymbol{c}_1 \boldsymbol{r}_1 + \boldsymbol{o}(1) = \boldsymbol{c}_1 + \boldsymbol{o}(1)$$

where we have used the property that the function in $\boldsymbol{b}(u, v, j)$ forms a partition of unity. Since the order function $\boldsymbol{o}(1)$ converges uniformly in (u, v), we finally obtain

$$\lim_{k \to \infty} \boldsymbol{S}_j^k(u, v) = \boldsymbol{c}_1$$

and $\boldsymbol{p} = \boldsymbol{c}_1$ is the limit point in the formula (4.36).

The next theorem gives a sufficient condition for the tangent plane continuity of a subdivision scheme. Though the similar conclusions can be applied to other subdivision schemes, we describe the theorem using C-C subdivision; such example can make readers understand better.

Theorem 4.3 *If $\lambda_2 = \lambda_3, 1 > |\lambda_2| > |\lambda_i| (i \geqslant 4)$ are the real eigenvalue with algebraic and geometric multiplicity 2 and its characteristic mapping is regular, so $S = \lim_{k\to\infty} M'^k$ is tangent plane continuous for almost every initial vector $\boldsymbol{V}^0$ of control vertices of the mesh M^0. M^0 is a mesh of k-neighborhood of an extraordinary vertex. M'^k is the mesh that is obtained by subdividing $M^0 k$ times. M^0 is sufficiently larger to be regarded as a control mesh an annular surface.*

Proof Similar to the formula (4.34), we have

$$\boldsymbol{V}^k = \boldsymbol{c}_1\boldsymbol{r}_1 + \lambda_2^k(\boldsymbol{c}_2\boldsymbol{r}_2 + \boldsymbol{c}_3\boldsymbol{r}_3) + \boldsymbol{o}(\lambda_2^k)$$

and

$$\boldsymbol{S}_j^k(u, v) = \boldsymbol{c}_1 + \lambda_2^k\boldsymbol{b}(u, v, j)(\boldsymbol{c}_2\boldsymbol{r}_2 + \boldsymbol{c}_3\boldsymbol{r}_3) + \boldsymbol{o}(\lambda_2^k)$$

So,

$$\boldsymbol{S}_{u,j}^k = \frac{\partial \boldsymbol{S}_j^k(u, v)}{\partial u} = \boldsymbol{c}_1 + \lambda_2^k\boldsymbol{b}_u(u, v, j)(\boldsymbol{c}_2\boldsymbol{r}_2 + \boldsymbol{c}_3\boldsymbol{r}_3) + \boldsymbol{o}(\lambda_2^k)$$
$$\boldsymbol{S}_{v,j}^k = \frac{\partial \boldsymbol{S}_j^k(u, v)}{\partial v} = \boldsymbol{c}_1 + \lambda_2^k\boldsymbol{b}_v(u, v, j)(\boldsymbol{c}_2\boldsymbol{r}_2 + \boldsymbol{c}_3\boldsymbol{r}_3) + \boldsymbol{o}(\lambda_2^k)$$

For the cross product of partial derivatives, we see

$$\frac{\partial \boldsymbol{S}_j^k(u, v)}{\partial u} \times \frac{\partial \boldsymbol{S}_j^k(u, v)}{\partial v} = \lambda_2^{2k}\Delta(u, v, j)\boldsymbol{c}_2 \times \boldsymbol{c}_3 + \boldsymbol{o}(1)$$

$\Delta(u, v, j)$ is nonzero since the characteristic mapping is regular. According to $\lambda_2 = \lambda_3, 1 > |\lambda_2| > |\lambda_i| (\geqslant 4)$, $\boldsymbol{c}_2 \times \boldsymbol{c}_3$ has nonzero norm for almost every choice of control vertices $\boldsymbol{V}^0$. So, the normalized normal vector is given as

$$\boldsymbol{n}^k(u, v, j) = \frac{\boldsymbol{S}_{u,j}^k \times \boldsymbol{S}_{v,j}^k}{\|\boldsymbol{S}_{u,j}^k \times \boldsymbol{S}_{v,j}^k\|} = \frac{\boldsymbol{c}_2 \times \boldsymbol{c}_3}{\|\boldsymbol{c}_2 \times \boldsymbol{c}_3\|} + \frac{\boldsymbol{o}(1)}{|\Delta(u, v, j)|\|\boldsymbol{c}_2 \times \boldsymbol{c}_3\|}$$

Note that

$$\mu(\Delta) > 0, \lim_{k\to\infty} \boldsymbol{n}^k(u, v, j) = \frac{\boldsymbol{c}_2 \times \boldsymbol{c}_3}{\|\boldsymbol{c}_2 \times \boldsymbol{c}_3\|}. \tag{4.43}$$

We have given a similar process to deduce the formula to calculate the normal in Sect. 4.2. However, the above deduction process is stricter mathematically because only 1-neighborhood is considered in Sect. 4.2.

Theorem 4.4 *If $\lambda_2 = \lambda_3, 1 > |\lambda_2| > |\lambda_i| (i \geqslant 4)$ are the real eigenvalue with algebraic and geometric multiplicity 2 and its characteristic mapping is regular and injective, so $S = \lim_{k\to\infty} M'^k$ is regular at $\boldsymbol{v}^\infty$ for almost every initial vector $\vec{\boldsymbol{V}}^0$ of* *control vertices of the mesh* M^0. *Consequently,* $S = \lim_{k\to\infty} M'^k$ *is also* G^1 *continuous. Denotations in the theorem have the meaning as same as those in Theorem 4.2.*

It is not preferred to make overmuch discussion on the theorem proof in this book. Readers who are interested in the proof process can go to the Ref. [17] for details.

From Theorems 4.2 and 4.3, we can decide the continuity of subdivision surfaces by taking advantage of the property of the characteristic mapping. The characteristic mapping is a good tool to analyze the smoothness of subdivision surfaces [8, 102, 103]. References [102, 103] give the condition of G^k continuity of subdivision surfaces according to the characteristic mapping. Using the condition of G^k continuity, references [33] give a G^2 subdivision algorithm by modifying the C-C subdivision scheme. Reif [17] is the first person to find the importance of the characteristic mapping. Our discussion in this section is given according to Refs. [8, 102].

It has to be executed in three steps to analyze the continuity of subdivision surfaces by using the characteristic mapping: (1) Determine the size of the neighborhood of the investigated vertex and construct a subdivision matrix according to the determined neighborhood. The size of the neighborhood is determined by the degree of generalized spline surfaces, and the neighborhood mesh should generate an annular spline surface; (2) calculate eigenvalues and eigenvectors of the subdivision matrix by the DFT and construct the characteristic map; (3) analyze the regularity of the characteristic mapping and determine whether the characteristic mapping is injective. As an example, Prautzsch [102] examines some subdivision schemes that are generalizations of biquadratic B-spline surfaces. The annular spline surface encloses a hexagonal hole. Note that we investigate k-neighborhood of extraordinary vertices for primary subdivision schemes and k-neighborhood of extraordinary faces for dual subdivision schemes. Examples in Prautzsch [102] investigate k-neighborhood of extraordinary faces. Georg [104] gives an entire analysis for the convergence and continuity of loop subdivision scheme by using the characteristic mapping.

It is trivial to calculate eigenvalues and eigenvectors of large subdivision matrix and analyze the properties of the characteristic mapping. This book will only validate the regularity and the injective property of the characteristic mapping of C-C subdivision in numeric.

It is a good idea to give a preparatory estimation for G^1 continuity by 1-neighborhood. It is enough to give an analysis for convergence by 1-neighborhood, since the eigenvalues of the subdivision matrix of 1-neighborhood are also the eigenvalues of the subdivision matrix of k-neighborhood. $\sqrt{3}$ subdivision surfaces are not generalizations of spline surfaces. The exact condition on the eigenvectors and the injecticity of the corresponding characteristic mapping are quite difficult to check strictly. The numerical verification is used by sketching the iso-parameter lines of the characteristic map in Ref. [26].

4.5 Parameter Evaluation

There is a strong belief that subdivision surfaces cannot be evaluated exactly like NURBS surfaces for arbitrary parameter values since subdivision surfaces do not have expressions near the extraordinary vertices. In order to evaluate subdivision surfaces, we have to subdivide control meshes again and again, which is called the iterative evaluation method. Stam [40] breaks this belief and provides a non-iterative technique that efficiently evaluates Catmull–Clark subdivision surfaces and their derivatives up to any order. The time cost of their technique is comparable to the evaluation of a bi-cubic spline surface. Except for C-C subdivision surfaces, Stam [40] still gives a similar parameter evaluation approach for Loop subdivision surfaces. The two approaches are similar. Using Stam's approaches for reference, Wang [42] also gives a parameter evaluation approach for nonuniform subdivision surfaces. The parameter evaluation method discussed in this section is presented by Stam [40].

4.5.1 Notations and Assumpsits

Because the control vertex topology structure near an extraordinary vertex is not a simple rectangular grid, all faces that contain extraordinary vertices cannot be evaluated as uniform B-splines patches. We assume that the initial mesh has been subdivided at least twice, isolating the extraordinary vertices so that each face is a quadrilateral and contains at most one extraordinary vertex, which is shown in Fig. 4.5.

The task of this section is only to demonstrate how to evaluate a patch $\boldsymbol{S}(u, v)$ corresponding to a face with just one extraordinary vertex, such as the region near vertex 1 in Fig. 4.6. Let us denote the valence of that extraordinary vertex by n. Our task is find a expression for $\boldsymbol{S}(u, v)$ defined over the unit square $\Omega = [0, 1] \times [0, 1]$. The patch can be evaluated directly in terms of the $K = 2N + 8$ vertices. In the following discussion, we assume that the surface point corresponding to the

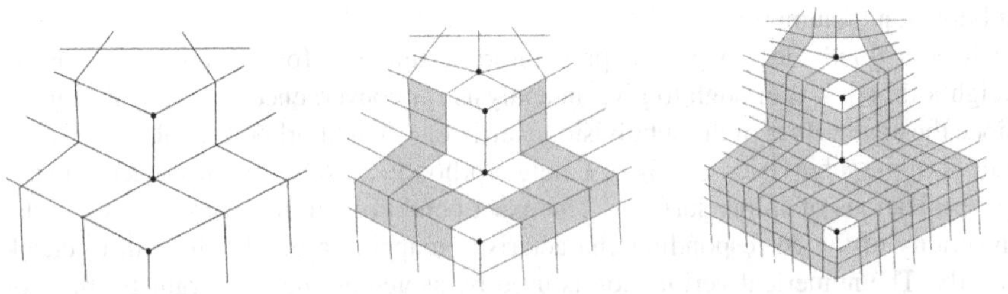

(a) Initial mesh (b) Mesh after one subdivision step (c) Mesh after two subdivision step

Fig. 4.5 Reference [40] Initial mesh and two levels of subdivision. The shaded faces correspond to regular bi-cubic B-spline patches. The dots are extraordinary vertices

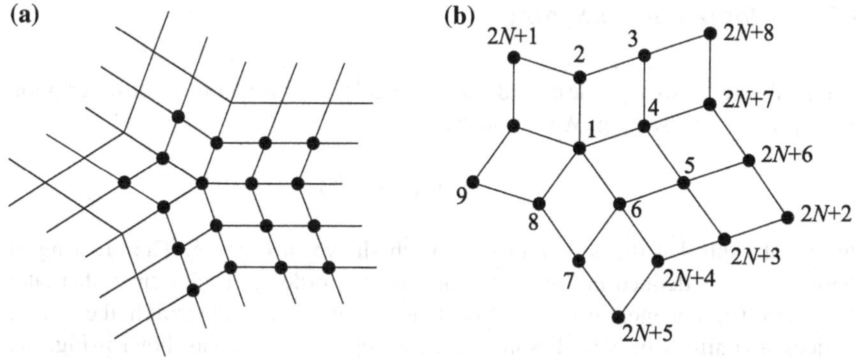

Fig. 4.6 Patch $S(u, v)$ near an extraordinary vertex with its control vertices. Vertex 1 is an extraordinary vertex of valence $N = 5$. **a** submesh with control vertices of a patch $S(u, v)$. The patch can be regard as a correspondence of the shaded face. That is, when the submesh is subdivided, the shaded face generates the patch; **b** indices of control vertices of the patch

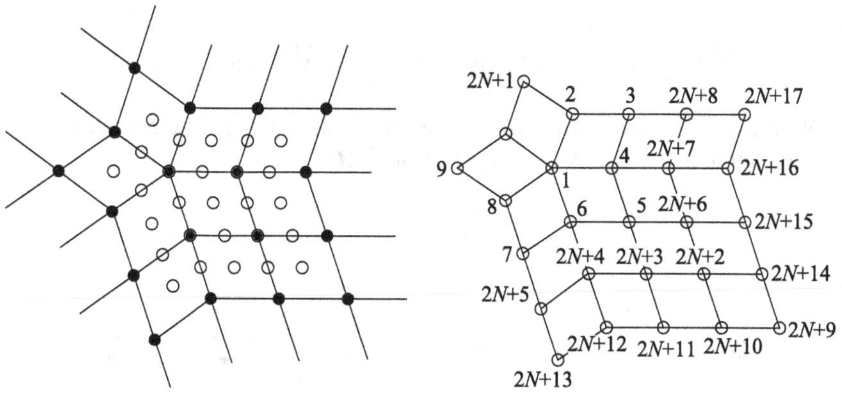

(a) Relations between new vertices and the old mesh (b) Indices of new control vertices

Fig. 4.7 Subdivide a mesh with an extraordinary vertex once

extraordinary vertex is $S(0, 0)$ and the orientation of Ω is chosen such that $s_u \times S_v$ points to outside of the surface.

After subdividing the submesh in Fig. 4.6 b once, a new submesh shown by Fig. 4.7 is then obtained. Note that the boundary vertices and boundary edges do not correspond to new vertices of the submesh in Fig. 4.7 b. From Fig. 4.7 b, it can be found that the old patch is divided into four subpatches, there three of them are regular and the rest one is non-regular. The regular ones are the uniform B-spline patches, and the non-regular one is the patch with an extraordinary point. It is also found that the control meshes of the new non-regular patch and the one of the old non-regular patch have the same topology.

4.5.2 Mathematical Setting

In this subsection, we provide a vivid mathematical setting for the notation description in the previous subsection. We denote by

$$(V^0)^{\mathrm{T}} = (v_1^0, \cdots, v_K^0)$$

the initial control vertices defining the patch shown in Fig. 4.6. The ordering of these vertices is defined in Fig. 4.6b. This peculiar ordering is chosen so that later deductions become more tractable. Note that the vertices do not result in the control vertices of a uniform bi-cubic B-spline patch, except when $N = 4$ as shown in Fig. 4.8.

By subdivision, a new set of $T = K + 9$ vertices is generated, and these vertices are shown as the circles in Fig. 4.7. Subsets of these new vertices are the control vertices of three uniform B-spline patches. Therefore, three-quarter patch is parameterized, and could be evaluated as simple bi-cubic B-spline patches (see Fig. 4.8 and top left of Fig. 4.9). We denote this new set of vertices by

$$(\vec{V}^1)^{\mathrm{T}} = (v_1^1, \ldots, v_K^1) \text{ and } (\vec{\vec{V}}^1)^{\mathrm{T}} = ((\vec{V}^1)^{\mathrm{T}}, v_{K+1}^1, \ldots, v_T^1)$$

With these vectors about vertices, the subdivision step can be represented as a multiplication by an $K \times K$ subdivision matrix $\boldsymbol{M}$:

$$\vec{V}^1 = \boldsymbol{M}\vec{V}^0 \tag{4.44}$$

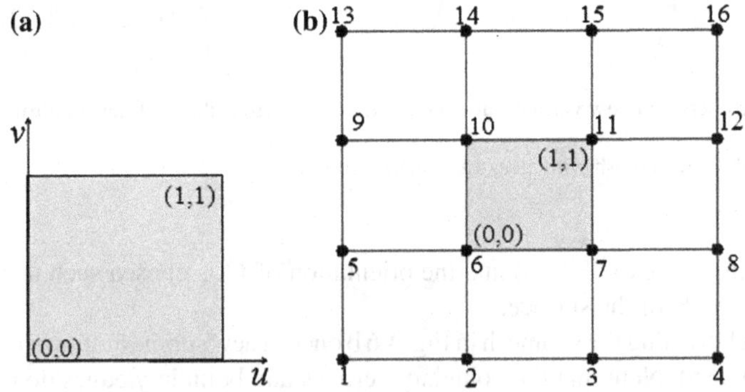

Fig. 4.8 **a** Parameter region of a uniform bi-cubic surface patch; **b** indices of control vertices of the uniform bi-cubic patch. The shaded face can be regarded as a correspondence of the parameter region. It can be also regarded as a correspondence of the patch. That is, when the mesh in Fig. **b** is subdivided, the shaded face generated the patch

Due to the peculiar ordering that we have chosen for the vertices, the extended subdivision matrix has the following block structure:

$$\boldsymbol{M} = \begin{pmatrix} \boldsymbol{M}_{1,1} & \boldsymbol{O} \\ \boldsymbol{M}_{2,1} & \boldsymbol{M}_{2,2} \end{pmatrix} \tag{4.45}$$

where $\boldsymbol{M}_{1,1}$ is the $(2N+1) \times (2N+1)$ subdivision matrix discussed in Sects. 4.2 and 4.3. The remaining two matrices correspond to the midpoint knot insertion rules for B-splines. Their exact definition can be found in Sect. 4.5.6. The additional points used to evaluate the three B-spline patches can be defined using a bigger matrix $\overline{\boldsymbol{M}}$ of the size $T \times K$:

$$\vec{\overline{\boldsymbol{V}}}^{1} = \overline{\boldsymbol{M}} \vec{\boldsymbol{V}}^{0}$$

where

$$\overline{\boldsymbol{M}} = \begin{pmatrix} \boldsymbol{M}_{1,1} & \boldsymbol{O} \\ \boldsymbol{M}_{2,1} & \boldsymbol{M}_{2,2} \\ \boldsymbol{M}_{3,1} & \boldsymbol{M}_{3,2} \end{pmatrix}$$

The matrix $\boldsymbol{M}_{3,1}$ and $\boldsymbol{M}_{3,2}$ are defined in Sect. 4.5.6. The subdivision step defined by Eq. (4.44) can be repeated so as to create an infinite sequence of control vertices:

$$\boldsymbol{V}^{k} = \boldsymbol{M}\boldsymbol{V}^{k-1} = \boldsymbol{M}^{k}\boldsymbol{V}^{0} \quad \text{and} \quad \overline{\boldsymbol{V}}^{k} = \overline{\boldsymbol{M}}\boldsymbol{V}^{k-1} = \overline{\boldsymbol{M}}\boldsymbol{M}^{k-1}\boldsymbol{V}^{0}$$

As shown in Fig. 4.7, on each level $k > 1$, three subsets of the vertices of $\overline{\boldsymbol{V}}^{k}$ are sets of the control vertices of three B-spline patches. Each set can be defined by selecting 16 control vertices from $\overline{\boldsymbol{V}}^{k}$:

$$\boldsymbol{C}_j^k = \boldsymbol{P}_j \overline{\boldsymbol{V}}^{k} (j = 1, 2, 3)$$

where $\boldsymbol{P}_j$ is a $16 \times M$ "picking" matrix. Let $\boldsymbol{b}(u, v)$ be a vector containing a cubic B-spline basis functions (see Sect. 4.5.7). If the control vertices are ordered as shown on the left of Fig. 4.8, then the patch corresponding to each matrix of control vertices is defined as

$$\boldsymbol{s}_j^k(u, v) = (\boldsymbol{C}_j^k)^{\mathrm{T}} \boldsymbol{b}(u, v) = (\overline{\boldsymbol{V}}^{k})^{\mathrm{T}} \boldsymbol{P}_j^{\mathrm{T}} \boldsymbol{b}(u, v) \tag{4.46}$$

where $(u, v) \in \Omega = [0, 1] \times [0, 1], j = 1, 2, 3$.

Using the ordering convention for control vertices in Fig. 4.8, the picking matrix can be defined to obtain vertices shown in Fig. 4.9. Each row of $\boldsymbol{P}_j$ is filled with zeros except for a one in the column corresponding to the index shown in Fig. 4.9. More details will be described in Sect. 4.5.7.

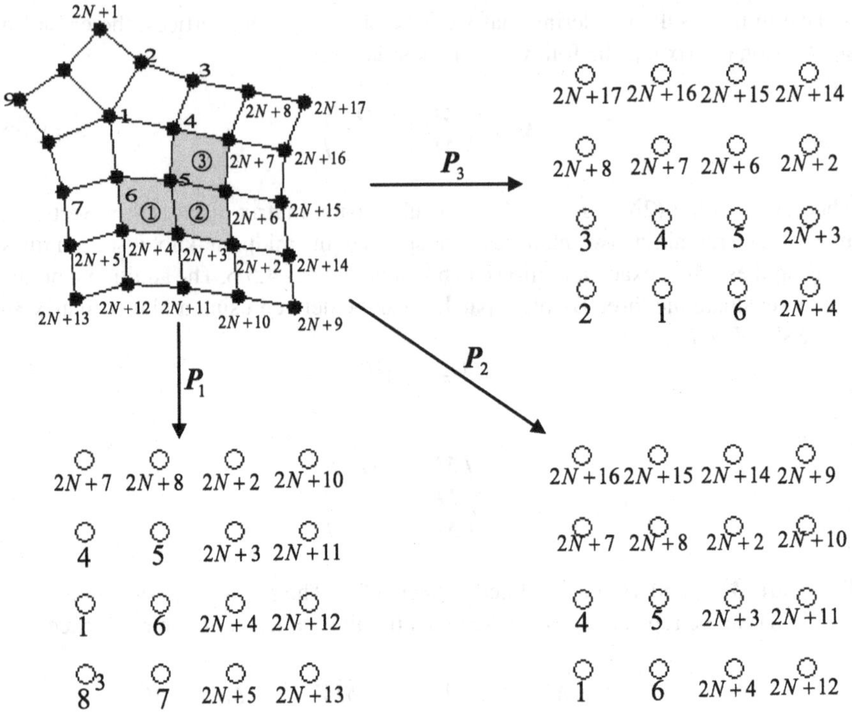

Fig. 4.9 Indices of the control vertices of the three bi-cubic B-spline patches obtained from $\overline{V}^k$

The infinite sequence of uniform B-spline patches defined by Eq. (4.46) forms "patch" $s(u, v)$ that corresponds to the shaded face in Fig. 4.6, when "stitched together." More formally, let us partition the unit square into an infinite set of tiles $\{\Omega_j^k\}, k \geqslant 1, j = 1, 2, 3$, as shown in Fig. 4.10. Each tile with index k is four times smaller than the tiles with index $k - 1$. More precisely:

$$\Omega_1^k = \left[\frac{1}{2^n}, \frac{1}{2^{n-1}}\right] \times \left[0, \frac{1}{2^n}\right]$$

$$\Omega_2^k = \left[\frac{1}{2^n}, \frac{1}{2^{n-1}}\right] \times \left[\frac{1}{2^n}, \frac{1}{2^{n-1}}\right]$$

$$\Omega_3^k = \left[0, \frac{1}{2^n}\right] \times \left[\frac{1}{2^n}, \frac{1}{2^{n-1}}\right]$$

A parametrization of $s(u, v)$ can be constructed by restricting $s(u, v)$ to tile Ω_j^k. In this case, $s(u, v)$ is the B-spline patch defined by the control vertices C_j^k

$$s(u, v)|\Omega_j^k = s_j^k(t_j^k(u, v)) \tag{4.47}$$

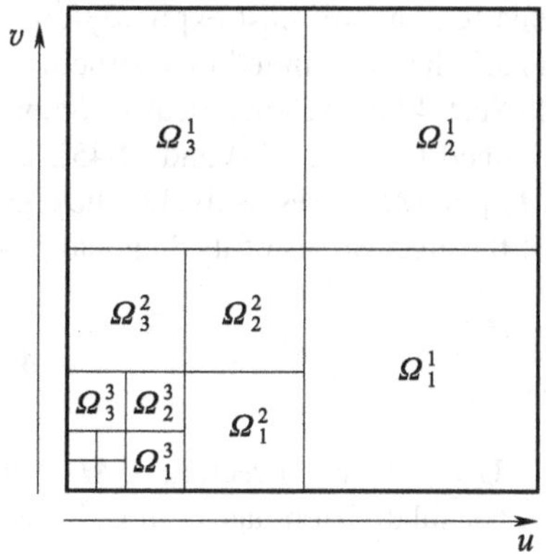

Fig. 4.10 Reference [40] Partition of the unit square into an infinite family of tiles

The transformation t_j^k maps the tile Ω_j^k onto the unit square Ω:

$$t_1^k(u, v) = (2^k u - 1, 2^k v)$$
$$t_2^k(u, v) = (2^k u - 1, 2^k v - 1)$$
$$t_3^k(u, v) = (2^k u, 2^k v - 1)$$

Equation (4.47) gives an actual parameterization for the surface. However, it is very costly to evaluate, since it involves $k - 1$ multiplications of a $K \times K$ matrix $\boldsymbol{M}$. The evaluation can be simplified considerably by computing the eigenstructure of $\boldsymbol{M}$.

4.5.3 Eigenstructure of Subdivision Matrix

The eigenstructure of the subdivision matrix $\boldsymbol{M}$ is defined as the set of its eigenvalues and eigenvectors. In our discussion, assume that the matrix $\boldsymbol{M}$ is non-defective for vertices with any valence. Consequently, there always exists K linearly independent eigenvectors. We denote this eigenstructure by $(\boldsymbol{\Lambda}, \boldsymbol{T})$, where $\boldsymbol{\Lambda}$ is the diagonal matrix containing the eigenvalues of M, and $\boldsymbol{T}$ is an invertible matrix whose columns are the corresponding eigenvectors. The computation of the eigenstructure is then equivalent to the solution of the following matrix equation:

$$\boldsymbol{M}\boldsymbol{T} = \boldsymbol{T}\boldsymbol{\Lambda} \tag{4.48}$$

where the ith diagonal element of $\boldsymbol{\Lambda}$ is an eigenvalue with a corresponding eigenvector equal to the ith column of the matrix $\boldsymbol{T}(i = 1, \ldots, \mathrm{K})$. There are many numerical algorithms which can compute their solutions for such equations. Unfortunately for our purposes, these numerical routines do not always return the correct eigenstructure. For example, in some cases the solver returns complex eigenvalues. For this reason, we must explicitly compute the eigenstructure. Since the subdivision matrix has a definite block structure, our computation can be done in several steps. In Sect. 4.5.6, we analytically compute the eigenstructures of $\boldsymbol{M}_{1,1}$ and $\boldsymbol{M}_{2,2}$ that are defined in Eqs. (4.44) and (4.45). Let $(\boldsymbol{\Sigma}, \boldsymbol{U}_0)$ and $(\boldsymbol{\Delta}, \boldsymbol{W}_1)$ be eigenstructures of $\boldsymbol{M}_{1,1}$ and $\boldsymbol{M}_{2,2}$, respectively. The eigenvalues of the subdivision matrix are the union of the eigenvalues of its diagonal blocks:

$$\boldsymbol{\Lambda} = \begin{pmatrix} \boldsymbol{\Sigma} & \boldsymbol{O} \\ \boldsymbol{O} & \boldsymbol{\Delta} \end{pmatrix}$$

Using the eigenvectors of $\boldsymbol{M}_{1,1}$ and $\boldsymbol{M}_{2,2}$, it can be proven that the eigenvectors of the subdivision matrix must have the following form:

$$\boldsymbol{T} = \begin{pmatrix} \boldsymbol{U}_0 & \boldsymbol{O} \\ \boldsymbol{U}_1 & \boldsymbol{W}_1 \end{pmatrix}$$

The matrix $\boldsymbol{U}_1$ is unknown and is determined from Eq. (4.48). If we replace the matrix $\boldsymbol{\Lambda}, \boldsymbol{T}$ and $\boldsymbol{M}$ by their block representations, we obtain the following matrix equation:

$$\boldsymbol{M}_{2,1}\boldsymbol{U}_0 + \boldsymbol{M}_{2,2}\boldsymbol{U}_1 = \boldsymbol{U}_1\boldsymbol{\Sigma} \tag{4.49}$$

$\boldsymbol{U}_1$ is computed by solving $2N+1$ linear systems of Eq. (4.48). However, we find that the linear system is singular when the valence of the extraordinary vertex is $N = 4n$.In order to solve the linear system, we make a QR decomposition for its coefficient matrix. By the method, $\boldsymbol{U}_0$ can be computed. So, the inverse of our eigenvector matrix is equal to

$$\boldsymbol{T}^{-1} = \begin{pmatrix} \boldsymbol{U}_0^{-1} & \boldsymbol{O} \\ -\boldsymbol{W}_1^{-1}\boldsymbol{U}_1\boldsymbol{U}_0^{-1} & \boldsymbol{W}_1^{-1} \end{pmatrix}$$

where both $\boldsymbol{U}_0$ and $\boldsymbol{U}_1$ can be inverted exactly (see Sect. 4.5.6). This fact allows us to rewrite Eq. (4.48):

$$\boldsymbol{M} = \boldsymbol{T}\boldsymbol{\Lambda}\boldsymbol{T}^{-1} \tag{4.50}$$

4.5.4 Eigenbases and Evaluation for Non-regular Patches

The decomposition formula (4.50) is the crucial result that we use in constructing a fast evaluation scheme of the non-regular patches. Using the formula, the subdivided control vertices at level k are now equal to

$$\overline{\boldsymbol{V}}^k = \overline{\boldsymbol{M}}\boldsymbol{M}^{k-1}\boldsymbol{V}^0 = \overline{\boldsymbol{M}}\boldsymbol{T}\boldsymbol{\Lambda}^{k-1}\boldsymbol{T}^{-1}\boldsymbol{V}^0 = \overline{\boldsymbol{M}}\boldsymbol{T}\boldsymbol{\Lambda}^{k-1}\widehat{\boldsymbol{V}}^0$$

where $\widehat{\boldsymbol{V}}^0 = \boldsymbol{T}^{-1}\boldsymbol{V}^0$. Using this new expression for the control vertices at the kth level of subdivision, Eq. (4.45) can be rewritten in the following form:

$$\boldsymbol{s}_j^k(u, v) = \widehat{\boldsymbol{V}}^0 \boldsymbol{\Lambda}^{k-1} (\boldsymbol{P}_j \overline{\boldsymbol{M}} \boldsymbol{T})^{\mathrm{T}} \boldsymbol{b}(u, v)$$

We observe that the most right terms in the equation are independent of the control vertices and the power k. Therefore, we can precompute this expression and define the following three vectors:

$$\boldsymbol{x}(u, v, j) = (\boldsymbol{P}_j \overline{\boldsymbol{M}} \boldsymbol{T})^{\mathrm{T}} \boldsymbol{b}(u, v) \quad j = 1, 2, 3 \tag{4.51}$$

The components of these three vectors correspond to a set of K bi-cubic splines. $K = 2N + 8$ control vertices defining the patch (See Fig. 4.8). In Sect. 4.5.7, we show how to compute these splines. Note that the splines $x_i(u, v, j)$ depend only on the valence of the extraordinary vertex.

Consequently, we can rewrite the equation for each patch more compactly as:

$$\boldsymbol{s}_j^k(u, v) = \widehat{\boldsymbol{V}}^0 \boldsymbol{\Lambda}^{k-1} \boldsymbol{x}(u, v, j)$$

Let $\widehat{\boldsymbol{v}}_i = (\widehat{x}_i, \widehat{y}_i, \widehat{z}_i)$ denote the rows of $\widehat{\boldsymbol{V}}^0$. According to the Eq. (4.47), the patch can be evaluated as

$$\boldsymbol{s}(u, v)|\Omega_j^k = \sum_{i=1}^{K} (\lambda_i)^{k-1} x_i(\boldsymbol{t}_j^k(u, v), j) \widehat{\boldsymbol{v}}_i \tag{4.52}$$

Therefore, in order to evaluate the patch, we must first compute the new vertices $\widehat{\boldsymbol{v}}_i$ (only once for a given mesh). Next, for each evaluation we determine k and compute λ^{k-1}. Since all but the first of the eigenvalues are smaller than one, contribution of λ decreases as k increases. Thus, for large k, *i.e.*, for surface points near the extraordinary vertex, only a few terms make a significant contribution. In fact for $(u, v) = (0, 0)$, the surface point is $\widehat{\boldsymbol{v}}_i$, which agrees with the definition of a limit point in formula (4.30).

The bi-cubic spline functions $\boldsymbol{x}(u, v, j)$ can be also used to define a set of *eigenbasis functions* about the subdivision. For a given eigenvalue λ_i, we define the function ϕ_i by its restrictions on the domains Ω_j^k as follows:

$$\phi_i(u, v)|\Omega_j^k = (\lambda_i)^{k-1} x_i(\boldsymbol{t}_j^k(u, v), j)$$

where $i = 1, \ldots, K$. By the above definition, these functions satisfy the following scaling relation:

$$\phi_i(u/2, v/2) = \lambda_i \phi_i(u, v) \tag{4.53}$$

In fact, without loss of generality, assume that $(u, v) \in \Omega_1^k$, we have $(u/2, v/2) \in \Omega_1^{k+1}$.

$$\phi_i(u/2, v/2)|\Omega_1^{k+1} = (\lambda_i)^k x_i(t_1^{k+1}(u/2, v/2), 1)$$
$$\because t_1^k(u, v) = (2^k u - 1, 2^k v)$$
$$\therefore t_1^{k+1}(u/2, v/2) = (2^{k+1}u/2 - 1, 2^{k+1}v/2) = (2^k u - 1, 2^k v) = t_1^k(u, v)$$

Consequently,

$$x_i(t_1^{k+1}(u/2, v/2), 1) = x_i(t_1^k(u/2, v/2), 1)$$

So,

$$\phi_i(u/2, v/2)|\Omega_1^{k+1} = \lambda_i \phi_i(u, v)|\Omega_1^{k+1}$$

Since we have defined the eigenbasic function, the evaluation expression of the surface patch given by Eq. (4.52) can be rewritten as:

$$s(u, v) = \sum_{i=1}^{K} \phi_i(u, v)\widehat{v}_i \tag{4.54}$$

This is the key result of Stam's analytic evaluation approach since this equation gives a parametrization for the patch corresponding to any face of the control mesh. Equation (4.22) also allows us to compute derivatives of the patch up to any order. Only the corresponding derivatives of the basis functions appearing in Eq. (4.22) are required. For example, the partial derivative of the ith eigenbasis with respect to u is:

$$\frac{\partial}{\partial u}\phi_i(u, v)|\Omega_j^{k+1} = 2^k(\lambda_i)^{k-1}\frac{\partial}{\partial u}x_i(t_j^k(u, v), j) \tag{4.55}$$

In fact,

$$\begin{aligned}\frac{\partial}{\partial u}\phi_i(u, v)|\Omega_j^{k+1} &= \frac{\partial}{\partial u}(\lambda_i)^{k-1}x_i(t_j^k(u, v), j)\\ &= \frac{\partial}{\partial u}(\lambda_i)^{k-1}x_i(t_j^k(u, v), j)\frac{\partial}{\partial u}t_j^k(u, v)\end{aligned}$$

Consequently, the factor 2^k is equal to the derivative of the affine transformation $t_j^k(u, v)$.

4.5.5 Subdivision Matrix and Their Eigenstructures

We have analyzed the subdivision matrices and their eigenstructures in Sect. 4.3. Those subdivision matrices are constructed for the 1-neighborhood of vertices. However, subdivision matrices in this section are constructed for the 1-neighborhood of a face. According to C-C subdivision rules, the 1-neighborhood forms the control mesh of the patch corresponding to the face. From Figs. 4.6 and 4.7, the 1-neighborhood of a face is also a submesh of the 3-neighborhood of its any vertex. Since we investigate the analytic evaluation method for the patch with an extraordinary vertex, the neighborhood is regarded as a submesh of the 3-neighborhood of the extraordinary vertex of the face. Consequently, these vertices are given indices in Fig. 4.6.

This section analyzes the eigenstructure of the subdivision matrix of 1-neighborhood of faces. The matrix is defined in (4.44) and its block is given in (4.45). According to arrangement of its components, the matrix can be written as:

$$
\boldsymbol{M}_{1,1} = \left(\begin{array}{c|cccccccccc}
a_N & b_N & c_N & b_N & c_N & b_N & \cdots & b_N & c_N & b_N & c_N \\
\hline
d & d & e & e & 0 & 0 & \cdots & 0 & 0 & e & e \\
f & f & f & f & 0 & 0 & \cdots & 0 & 0 & 0 & 0 \\
d & e & e & d & e & e & \cdots & 0 & 0 & 0 & 0 \\
f & 0 & 0 & f & f & f & \cdots & 0 & 0 & 0 & 0 \\
\vdots & & & \vdots & & & \ddots & & \vdots & & \\
d & e & 0 & 0 & 0 & 0 & \cdots & e & e & d & e \\
f & f & 0 & 0 & 0 & 0 & \cdots & 0 & 0 & f & f
\end{array}\right)
$$

where

$$
a_N = 1 - \frac{7}{4N}, b_N = \frac{3}{2N^2}, c_N = \frac{1}{4N^2}, d = 3/8, e = 1/16, f = 1/4
$$

Compared with the matrix defined by (4.4), there only is a difference of arrangement of components because of the difference of arrangement of vertices. Consequently, the two matrices have same eigenvalues and eigenvectors.

According to (4.19) and (4.20), the first block $\boldsymbol{T}_0$ has eigenvalues:

$$
\mu_1 = 1, \mu_2, \mu_3 = \frac{1}{8N}(-7 + 3N \mp \sqrt{49 - 30N + 50N^2}
$$

and eigenvectors

$$
\widehat{\boldsymbol{R}}_0 = \begin{pmatrix} 1 & 16\mu_2^2 - 12\mu_2 + 1 & 16\mu_3^2 - 12\mu_3 + 1 \\ 1 & 6\mu_2 - 1 & 6\mu_3 - 1 \\ 1 & 4\mu_2 + 1 & 4\mu_3 + 1 \end{pmatrix}
$$

The blocks $\boldsymbol{T}_\omega(\omega = 1, 2, \cdots, n-1)$ have eigenvalues:

$$\lambda_\omega^{\mp} = \frac{1}{16}\left(5 + \cos\left(\frac{2\pi\omega}{N}\right) \mp \cos\left(\frac{\pi\omega}{N}\right)\sqrt{18 + \cos\left(\frac{2\pi\omega}{N}\right)}\right)$$

where we have used some trigonometric relations to simplify the resulting expressions. The corresponding eigenvectors of each block are

$$\widehat{\boldsymbol{R}}_\omega = \begin{pmatrix} 4\lambda_\omega^- - 1 & 4\lambda_\omega^+ - 1 \\ 1 + a^\omega & 1 + a^\omega \end{pmatrix}$$

Note that $a = \cos\frac{2\pi}{n} + i\sin\frac{2\pi}{n}$. We have to single out the special case when N is even and $\omega = \mathrm{N}/2$. In this case, the corresponding block is:

$$\widehat{\boldsymbol{R}}_{N/2} = \begin{pmatrix} 1 & 0 \\ 0 & 1 \end{pmatrix}$$

Note that the lower right $2N \times 2N$ block of $\boldsymbol{M}_{1,1}$ has a cyclical structure. In order to make a DFT for the block, we introduce the following $2N \times 2N$ "Fourier matrix":

$$\boldsymbol{F} = \begin{pmatrix} 1 & 0 & 1 & 0 & \cdots & 1 & 0 \\ 0 & 1 & 0 & 1 & \cdots & 0 & 1 \\ 1 & 0 & a^{-1} & 0 & \cdots & a^{-(N-1)} & 0 \\ 0 & 1 & 0 & a^{-1} & & 0 & a^{-(N-1)} \\ & & & & \ddots & \vdots & \\ 1 & 0 & a^{-(N-1)} & 0 & \cdots & a^{-(N-1)^2} & 0 \\ 0 & 1 & 0 & a^{-(N-1)} & \cdots & 0 & a^{-(N-1)^2} \end{pmatrix}$$

Note that the matrix $\boldsymbol{A}$ in subsection 4.2.1. Compared with $\begin{pmatrix} \overline{\boldsymbol{A}}^{\mathrm{T}} & \\ & \overline{\boldsymbol{A}}^{\mathrm{T}} \end{pmatrix}$, $\boldsymbol{F}$ has only a difference on arrangement of components. Let $\boldsymbol{G} = \begin{pmatrix} 1 & \boldsymbol{O} \\ \boldsymbol{O} & \frac{1}{N}\boldsymbol{F} \end{pmatrix}$. We have $\boldsymbol{G}^{-1} = \begin{pmatrix} 1 & \boldsymbol{O} \\ \boldsymbol{O} & \overline{\boldsymbol{F}}^{\mathrm{T}} \end{pmatrix}$. Consequently, $\boldsymbol{G}\boldsymbol{M}_{1,1}\boldsymbol{G}^{-1} = \widehat{\boldsymbol{M}}_{1,1}$ executes a Fourier transform for $\boldsymbol{M}_{1,1}$ and $\widehat{\boldsymbol{M}}_{1,1}$ is a matrix with diagonal blocks:

$$\widehat{\boldsymbol{M}}_{1,1} = \begin{pmatrix} \boldsymbol{T}_0 & & & \\ & \boldsymbol{T}_1 & & \\ & & \ddots & \\ & & & \boldsymbol{T}_{N-1} \end{pmatrix}$$

The eigenvalues of the matrix $\widehat{\boldsymbol{M}}_{1,1}$ are the union of the eigenvalues of its blocks, and its eigenvectors are:

$$\widehat{\boldsymbol{R}} = \begin{pmatrix} \widehat{\boldsymbol{R}}_0 & & & \\ & \widehat{\boldsymbol{R}}_1 & & \\ & & \ddots & \\ & & & \widehat{\boldsymbol{R}}_{N-1} \end{pmatrix}$$

According to (4.17), $\boldsymbol{R} = \boldsymbol{G}^{-1}\widehat{\boldsymbol{R}}$. Consequently, we have computed the eigenvalues and eigenvectors of $\boldsymbol{M}_{1,1}$. However, in this form the eigenvectors are the complex number and most of the eigenvalues are actually of multiplicity 2, since $\lambda_\omega^- = \lambda_{N-\omega}^+$ and $\lambda_\omega^+ = \lambda_{N-\omega}^-$, and we remark these eigenvalues as follows:

$$\mu_4 = \lambda_1^-, \mu_5 = \lambda_1^+, \mu_6 = \lambda_2^-, \mu_7 = \lambda_2^+, \ldots \tag{4.56}$$

Since we have rearranged the eigenvalues, we need to rearrange the eigenvectors. At the same time, we make these eigenvectors real. Let $\boldsymbol{r}_1, \ldots, \boldsymbol{r}_{2N+1}$ be the columns of $\boldsymbol{R}$, then we can construct the columns of a matrix $\boldsymbol{U}_0$ as follows:

$$\begin{aligned} &\boldsymbol{u}_1 = \boldsymbol{r}_1, \boldsymbol{u}_2 = \boldsymbol{r}_2, \boldsymbol{u}_3 = \boldsymbol{r}_3, \\ &\boldsymbol{u}_{2t+2} = \frac{1}{2}(\boldsymbol{r}_{t+3} + \boldsymbol{r}_{2N-t+2}) \text{ and} \\ &\boldsymbol{u}_{2t+3} = \frac{1}{2}(\boldsymbol{r}_{t+3} - \boldsymbol{r}_{2N-t+2}) \end{aligned}$$

More precisely, $\boldsymbol{u}_1, \boldsymbol{u}_2, \boldsymbol{u}_3, \boldsymbol{u}_{2t+2}$, and $\boldsymbol{u}_{2t+3}$ equal to

$$\begin{pmatrix} 1 \\ 1 \\ 1 \\ \vdots \\ 1 \\ 1 \end{pmatrix}, \begin{pmatrix} 16\mu_2^2 - 12\mu_2 + 1 \\ 6\mu_2 - 1 \\ 4\mu_2 + 1 \\ \vdots \\ 6\mu_2 - 1 \\ 4\mu_2 + 1 \end{pmatrix}, \begin{pmatrix} 16\mu_3^2 - 12\mu_3 + 1 \\ 6\mu_3 - 1 \\ 4\mu_3 + 1 \\ \vdots \\ 6\mu_3 - 1 \\ 4\mu_3 + 1 \end{pmatrix}$$

$$\begin{pmatrix} 0 \\ 4\mu_{t+3} - 1 \\ 1 + C_{\gamma(t)} \\ (4\mu_{t+3} - 1)C_{\gamma(t)} \\ C_{\gamma(t)} + C_{2\gamma(t)} \\ \vdots \\ (4\mu_{t+3} - 1)C_{(N-1)\gamma(t)} \\ C_{(N-1)\gamma(t)} + 1 \end{pmatrix}, \begin{pmatrix} 0 \\ 0 \\ S_{\gamma(t)} \\ (4\mu_{t+3} - 1)S_{\gamma(t)} \\ S_{\gamma(t)} + S_{2\gamma(t)} \\ \vdots \\ (4\mu_{t+3} - 1)S_{(N-1)\gamma(t)} \\ S_{(N-1)\gamma(t)} \end{pmatrix}$$

respectively, where $t = 1, \ldots, N_2$, $N_2 = N - 1$ when N is odd and $N_2 = N - 2$ when is even. $\gamma(t) = (t+1)/2$ when t is odd and $\gamma(t) = t/2$ when t is even, and

$$C_k = \cos\left(\frac{2\pi k}{N}\right) \quad \text{and} \quad S_k = \sin\left(\frac{2\pi k}{N}\right)$$

When N is even, the last two eigenvectors are

$$\begin{aligned} \boldsymbol{u}_{2N}^T &= [0, 1, 0, -1, 0, 1, 0, \ldots, -1, 0] \\ \boldsymbol{u}_{2N+1}^T &= [0, 0, 1, 0, -1, 0, 1 \ldots, 0, -1] \end{aligned}$$

Finally, the diagonal matrix of eigenvalues is

$$\boldsymbol{\Sigma} = \mathrm{diag}(1, \mu_2, \mu_3, \mu_4, \mu_4, \ldots, \mu_{N+2}, \mu_{N+2}) \tag{4.57}$$

The remaining blocks of the subdivision matrix $\boldsymbol{M}$ directly follow from the usual B-spline knot insertion rules.

$$\boldsymbol{M}_{2,2} = \begin{pmatrix} c & b & c & 0 & b & c & 0 \\ 0 & e & e & 0 & 0 & 0 & 0 \\ 0 & c & b & c & 0 & 0 & 0 \\ 0 & 0 & e & e & 0 & 0 & 0 \\ 0 & 0 & 0 & 0 & e & e & 0 \\ 0 & 0 & 0 & 0 & c & b & c \\ 0 & 0 & 0 & 0 & 0 & e & e \end{pmatrix}, \quad \boldsymbol{M}_{2,1} = \begin{pmatrix} c & 0 & 0 & b & a & b & 0 & 0 & 0 \\ e & 0 & 0 & e & d & d & 0 & 0 & 0 \\ b & 0 & 0 & c & b & a & b & c & 0 \\ e & 0 & 0 & 0 & 0 & d & d & e & 0 \\ e & 0 & 0 & d & d & e & 0 & 0 & 0 \\ b & c & b & a & b & c & 0 & 0 & 0 \\ e & e & d & d & 0 & 0 & 0 & 0 & 0 \end{pmatrix}$$

where

$$a = 9/16, \quad b = 3/32 \text{ and } c = 1/64$$

For the case $N = 3$, there is no control vertex $\boldsymbol{v}_8 (\boldsymbol{v}_8 = \boldsymbol{v}_2)$, and the second column of the matrix $\boldsymbol{M}_{2,1}$ is equal to $[0, 0, c, e, 0, c, e]^{\mathrm{T}}$, which is shown in Fig. 4.11(a).

The eigenstructure of the matrix $\boldsymbol{M}_{2,2}$ can be computed manually, since this matrix has a simple form. Its eigenvalues are:

$$\Delta = \mathrm{diag}\left(\frac{1}{64}, \frac{1}{8}, \frac{1}{16}, \frac{1}{32}, \frac{1}{8}, \frac{1}{16}, \frac{1}{32}\right)$$

with corresponding eigenvectors:

$$\boldsymbol{W}_1 = \begin{pmatrix} 1 & 1 & 2 & 11 & 1 & 2 & 11 \\ 0 & 1 & 1 & 2 & 0 & 0 & 0 \\ 0 & 1 & 0 & -1 & 0 & 0 & 0 \\ 0 & 1 & -1 & 2 & 0 & 0 & 0 \\ 0 & 0 & 0 & 0 & 1 & 1 & 2 \\ 0 & 0 & 0 & 0 & 1 & 0 & -1 \\ 0 & 0 & 0 & 0 & 1 & -1 & 2 \end{pmatrix}$$

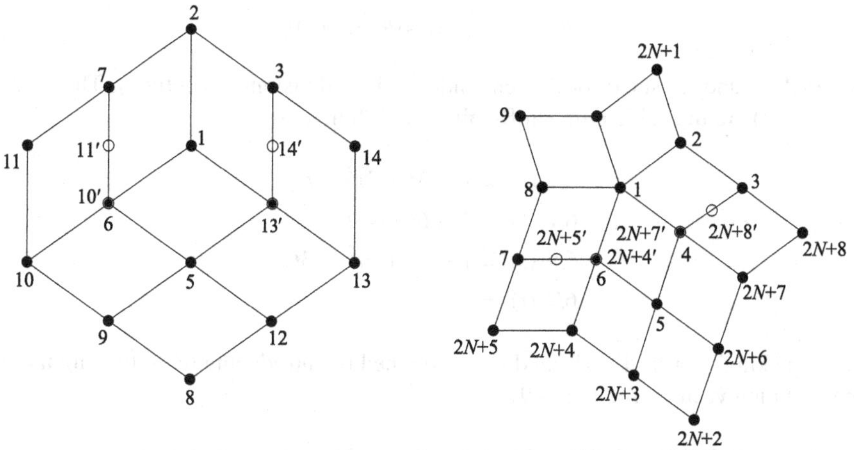

(a) Extraordinary vertex with valence 3 (b) Extraordinay vertex valence 5

Fig. 4.11 Relations between new points and $\boldsymbol{v}_2$, $\boldsymbol{v}_8$ ○ denote new points whose positions we must use $\boldsymbol{v}_2$ or $\boldsymbol{v}_8$ to calculate. Other new points are not shown in this figure

The other two matrices appearing in $\overline{\boldsymbol{M}}$ are:

$$\boldsymbol{M}_{3,1} = \begin{pmatrix} 0&0&0&0&f&0&0&0 \\ 0&0&0&0&d&e&0&0 \\ 0&0&0&0&f&f&0&0 \\ 0&0&0&0&e&d&e&0 \\ 0&0&0&0&0&f&f&0 \\ 0&0&0&e&d&0&0&0 \\ 0&0&0&f&f&0&0&0 \\ 0&0&e&d&e&0&0&0 \\ 0&0&f&f&0&0&0&0 \end{pmatrix}, \quad \boldsymbol{M}_{3,2} = \begin{pmatrix} f&f&0&0&f&0&0 \\ e&d&e&0&e&0&0 \\ 0&f&f&0&0&0&0 \\ 0&e&d&e&0&0&0 \\ 0&0&f&f&0&0&0 \\ e&e&0&0&d&e&0 \\ 0&0&0&0&f&f&0 \\ 0&0&0&0&e&d&e \\ 0&0&0&0&0&f&f \end{pmatrix}$$

It is a trivial and time-consuming task to precompute the eigenstructure of the subdivision matrix $\boldsymbol{M}$. Consequently, it is a good idea to save them in a file to be read in at the start of any application. Stam [40] computed the eigenstructures of the subdivision matrix $\boldsymbol{M}$ up to the valence 50.

4.5.6 Eigenbasis Functions

In this section, we compute the bi-cubic spline pieces $\boldsymbol{x}(u, v, j)$ that are defined in Eq. (4.50). The vector $\boldsymbol{b}(u, v)$ contains the 16 tensor B-spline basis functions, $i = 1, \ldots, 16$, see Fig. 4.8 for indices.

$$b_i(u, v) = N_{(i-1)\%4}(u)N_{(i-1)/4}(v)$$

where "%" and "/" stand for the remainder and the division, respectively. The functions $N_i(t)$ are the cubic uniform B-spline basis functions:

$$\begin{aligned} 6N_0(t) &= 1 - 3t + 3t^2 - t^3 \\ 6N_1(t) &= 4 - 6t^2 + 3t^3 \\ 6N_2(t) &= 1 + 3t + 3t^2 - 3t^3 \\ 6N_3(t) &= t^3 \end{aligned}$$

The "picking" matrix $\boldsymbol{P}_1, \boldsymbol{P}_2$, and $\boldsymbol{P}_3$ are defined by introducing the following three permutation vectors (see Fig. 4.9):

$$\begin{aligned} \boldsymbol{q}^1 &= (8, 7, 2N+5, 2N+13, 1, 6, 2N+4, 2N+12, 4, 5, 2N+3, \\ &\quad 2N+11, 2N+7, 2N+6, 2N+2, 2N+10) \\ \boldsymbol{q}^2 &= (1, 6, 2N+4, 2N+12, 4, 5, 2N+3, 2N+11, 2N+7, 2N+6, \\ &\quad 2N+2, 2N+10, 2N+16, 2N+15, 2N+14, 2N+9) \\ \boldsymbol{q}^3 &= (2, 1, 6, 2N+4, 3, 4, 5, 2N+3, 2N+8, 2N+7, 2N+6, 2N+2, \\ &\quad 2N+17, 2N+16, 2N+15, 2N+14) \end{aligned}$$

Since for the case $N = 3$, the vertices $\boldsymbol{v}_2$ and $\boldsymbol{v}_8$ are the same vertex (see Fig. 4.11). $q_1^1 = 2$ instead of 8 for $N = 3$.

4.5.7 Parameter Evaluation and Characteristic Mapping

Note that the formula (4.49), (4.55) and (4.56). The sixth column and the seventh column of $\overline{\boldsymbol{MT}}$, respectively, correspond to $\boldsymbol{r}_2$ and $\boldsymbol{r}_3$ in Sect. 4.4. In fact, they only contain some components of the two right eigenvectors. The sixth column and the seventh column of $\overline{\boldsymbol{MT}}$, respectively, correspond to the operation: subdivide the mesh formed by $\boldsymbol{t}_6$ and $\boldsymbol{t}_7$ once, where $\boldsymbol{t}_6$ and $\boldsymbol{t}_7$ are the sixth column and the seventh column of $\boldsymbol{T}$.

$\boldsymbol{P}_j(\overline{\boldsymbol{MT}})$ is such an operation, namely to pick the control vertices of a bi-cubic B-spline patch from the subdivided mesh. $(\boldsymbol{P}_j\overline{\boldsymbol{M}}[\boldsymbol{t}_6, \boldsymbol{t}_7])^{\mathrm{T}}\boldsymbol{b}(u, v)$ is a patch of the annular surface determined by the characteristic mapping. Consequently, the eigenbasis functions $\phi_6(u, v)$ and $\phi_7(u, v)$ can define the characteristic mapping:

$$\boldsymbol{\Psi}(u, v) = [x(u, v), y(u, v)] = [\phi_6(u, v), \phi_7(u, v)] \tag{4.58}$$

It should be noticed that the defining region for $\boldsymbol{\Psi}(u, v)$ in formula (4.58) is only a subset of the defining region in Definition 4.3.

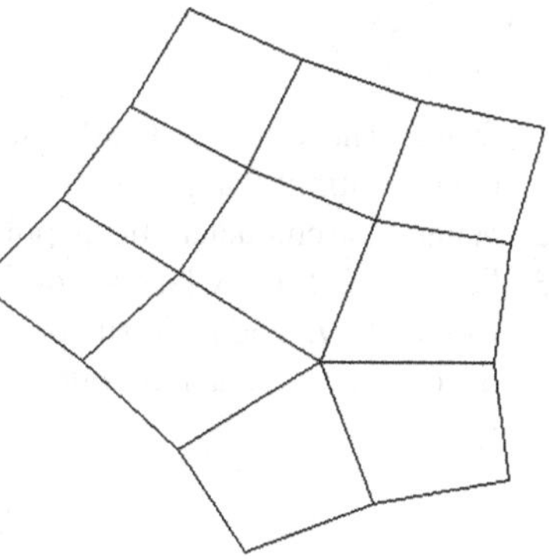

Fig. 4.12 Mesh formed by two eigenvectors corresponding to the subdominating eigenvalue

Figure 4.12 gives the shape of the mesh formed by $\boldsymbol{t}_6$ and $\boldsymbol{t}_7$. The mesh is not self-cross, and sizes of its faces are almost equal. Consequently, the mesh formed by $\overline{\boldsymbol{M}}[\boldsymbol{t}_6, \boldsymbol{t}_7]$ is not self-cross, and sizes of its faces are almost equal, too. According to the symmetry property of the subdivision matrix for k-neighborhood, the regularity and the injective property of the characteristic mapping of C-C subdivision are validated in numeric.

Remarks

Using the C-C subdivision surface as an example, this chapter presents discussions for the DFT in detail. Some different skills are probably used for different subdivision schemes when DFT is executed for subdivision matrices. Compared with the limit normal vector, the characteristic mapping is more precise to distinguish the G^1 continuity of subdivide surfaces. The characteristic mapping is very important to construct rules to distinguish C^k continuity of subdivision surfaces [20, 102, 103]. Based on the characteristic mapping and C^k continuity rules, Refs. [34, 84, 85] give subdivision rules that make subdivision surfaces G^2 continuous at the extraordinary point. Reference [22, 105] also give some methods to adjust subdivision rules around extraordinary vertices. The analytic evaluation approach is discussed in the basis of analysis on the eigenstructures of subdivision matrices. By the analytic evaluation approach, we can also calculate the derivatives at any parameter. Peters et al.[21] give general principles to compute more geometric properties. Bolz et al. [71] sum up various evaluation methods as four types: (I) recursive evaluation; (II) direct evaluation; (III) reduction to the regular setting; and (IV) pretabulated basis function composition. The direct evaluation is the analytic evaluation. The last two types will be discussed in the subsequent chapters.

Exercises

1. Prove: The Catmull–Clark subdivision surface is the surface with the tangent plane continuity at least.
2. What is the characteristic mapping? Why do we define the characteristic mapping?
3. For the Catmull–Clark subdivision surface, Fig. 4.12 is the shape of the mesh formed by t_6 and t_7. Write program codes to compute t_6 and t_7. Draw the figure once more again using your code.

Chapter 5
n-Sided Patches and Subdivision Surfaces

It is the base of surface blending and n-sided holes filling to construct n-sided patches. This chapter discusses three methods to construct n-sided patches. The first one is the manifold method. The control mesh of a manifold patch is constructed on the basis of the 5-ring submesh of an extraordinary vertex of a mesh with C-C subdivision connectivity. The manifold patch constructed in this chapter has uniform cubic B-spline boundaries. It is G^2 continuous everywhere and can G^2 continuously join C-C subdivision surface. The second method and the third method generalize the non-uniform C-C subdivision scheme to open meshes. In these two methods, the key point is how to construct B-spline boundaries of n-sided patches. The second one is called the skirt-removed approach which constructs patches that not interpolate corner vertices of control meshes. The third one is called the corner-vertex-interpolated method which constructs patches that interpolate corner vertices of control meshes. No matter the skirt-removed approach or the corner-vertex-interpolated method, they both construct patches based on subdivision methods.

5.1 Methods for Constructing n-Sided Patches

In the field of GAGD, it has been a difficult problem to construct n-sided patches in order to fill n-sided holes in free-form surfaces or in order to blend several free-form patches. At present, there are many effective methods, and the typical methods can be divided into three kinds: (1) the NURBS method [220]. This method combines several NURBS patches to construct an n-sided patch. Literature [220] uses n bi-septic to construct an n-sided patch which is G^1 continuous almost everywhere. The method is realized through two computation steps, which needs higher computing consumption. In addition, since the purpose of constructing n-sided patches in the method is to blend NURBS patches, these n-sided patches are difficult to integrate with subdivision surfaces whether in construction methods or in surface expression forms. Consequently, we will not discuss the method in this book; (2) the manifold method [221, 222]. The method uses a mesh as the control mesh of n-sided

W. Liao et al., *Subdivision Surface Modeling Technology*,
DOI 10.1007/978-981-10-3515-9_5

patch. Every vertex in the mesh is given a basic function. The linear combination of vertices and basic functions is the expression of the n-sided patches which is going to be constructed. Since the constructed n-sided patch is a 2-dimensional manifold, i.e., the parametric field of every basic function corresponds to a field of a tangent plane space of the patch, the method is called the manifold method in this book. n-sided patches with G^2 continuity can be constructed by using the method; (3) the subdivision method [223–225]. The method firstly constructs a mesh for the patch that is going to be constructed and then obtains the patch by directly subdividing the mesh. Compared with the above two methods, the subdivision method is the simplest one. If we do not allow that the patch containing flat points, the constructed patch is at most G^1 continuous [226], G^1 continuity can meet the requirements of many applications. Furthermore, the subdivision method can greatly enrich the modeling effect due to the appearance of non-uniform subdivision method [228]. Besides the above usual methods, we can also construct n-sided patches by joining Coons patches.

Lastly, we emphasize two concepts: patch and surface. A patch is a finite surface that has deterministic boundaries while a surface may be infinite or closed. In some situations, for example, surface blending and hole filling, we usually call a finite surface as a patch that is used to blend surfaces or fill a hole. If we emphasize that a surface is constituted of several finite surfaces, these finite surfaces are usually called patches. However, in many situations, it is not necessary to distinct a patch from a surface: a patch is also a surface.

5.2 C-C Subdivision Surfaces and G^2 Continuity

It is well known that Catmull–Clark subdivision surface is only G^1 continuous at extraordinary points. Researchers have made a lot of efforts in order to construct G^2 continuous surfaces by using Catmull–Clark subdivision method. Totally there are two types of existing methods: (1) Modify/construct existing subdivision scheme, i.e., by modifying the C-C subdivision method, we make that the limit surface is G^2 continuous everywhere. For example, the literature [226] gives a modified subdivision scheme according to the eigenvalues of the Catmull–Clark matrix. However, although the constructed patch is G^2 continuous everywhere, the extraordinary points of the surface become flat points. Literatures [226, 229] both point out that it is impossible to make the limit surface G^2 continuous and without flat points only by adjusting C-C subdivision scheme. Literature [230] converts the bi-cubic B-spline surface ring which surrounds an n-sided hole into two Bezier surface rings. By the approach, it gets a subdivision method to fill the n-sided hole. By the method, the degree of the patch obtained in the literature is six. Just as the literature [99] has pointed out that there are still flat small pieces in the surface when extraordinary points of the surface are saddle points with higher orders, although the patch is G^2 continuous everywhere. Literature [183] gives subdivision guidelines to construct G^2 continuous patches whose curvature is not zero everywhere. However, we have not found any other method to satisfy the guidelines so far, except the methods in literature

[230] and literature [199]. The method in literature [199] has the same shortcomings as that in literature [99, 230]. (2) Construct n-sided patches. The method is more frequently discussed by researchers. A usual way to construct n-sided patches is to extract the 3-ring around an extraordinary vertex. The 3-ring submesh defines a bi-cubic B-spline surface ring (the surface ring is a part of the limit surface). The surface ring surrounds an n-sided hole, and then, we construct a patch to fill the n-sided hole. Section 5.1 has discussed different kinds of methods to construct n-sided patches. Since it is impossible to make limit surfaces G^2 continuous everywhere and without flat point only by adjusting the C-C subdivision schemes, we attempt to fill these n-sided holes by constructing patches with G^2 continuity. So, we can construct free-form surfaces with arbitrary topology and G^2 continuity.

It is easy to know that we can construct a surface with G^2 continuity based on an initial surface with extraordinary points in the following process: Firstly, we remove a small patch around an extraordinary point and then an n-sided hole appears; Secondly, we fill the n-sided hole by constructing a patch with G^2 continuity. By using the process, Literature [106] modifies the patch around an extraordinary point by the following method:

$$\boldsymbol{q}^*(u, v) = \boldsymbol{q}(u, v)w(u, v) + \boldsymbol{p}(u, v)(1 - w(u, v))$$

where $\boldsymbol{q}(u, v)$ is the part around an extraordinary point after parameterization, $\boldsymbol{p}(u, v)$ is a polynomial patch with low degree, and $w(u, v)$ is a weight function with C^2 continuity. From the boundary of (u, v) to its center, $w(u, v)$ gradually damping from 1 to 0. So, the influence of the extraordinary point can be eliminated by the weight function.

It is easy to know that the manifold method can construct surfaces based on 2-manifold mesh with arbitrary topology. Starting from the property of the manifold method, literature [172] constructs n-sided patches with uniform cubic B-spline boundaries by using control meshes formed by 5-ring submeshes of extraordinary vertices of Catmull–Clark subdivision meshes. After removing the 5-ring submeshes of an extraordinary vertex in a subdivision mesh, an n-sided hole appears on the limit surface. After filling the n-sided hole by using the patch, the whole surface becomes C^2 continuous. In order to reduce computational complexity, literature [172] only extracts the 3-ring submesh of extraordinary vertices and obtains a 5-ring submesh by subdividing the 3-ring submesh. The methods in literatures [106] and [172], respectively, have their advantages and disadvantages and the method of literature [106] makes a surface lost its convex hull property near the extraordinary points, while the method presented by literature [172] abandons the position information of extraordinary vertices of control meshes. This section mainly discusses the method presented by literature [172]. Compared with those subdivision methods to construct patches, the main disadvantage of the manifold method is that it introduces a new approach to construct surfaces near extraordinary points. Consequently, different parts of surfaces probably are constructed by different approaches.

5.3 n-Sided Patches and Catmull–Clark Subdivision

5.3.1 Contribution of Our Method

Our method to construct n-sided patches is presented based on the method in literature [221]. However, if we combine the method of constructing n-sided patches in literature [221] and the C-C subdivision method to construct free-form surfaces with G^2 continuity everywhere, there are at least the following problems to solve or further clarify:

(1) Because of non-uniformity of knot intervals in parameter mesh, although patches constructed by the method can have cubic B-spline boundaries, those boundaries are not uniform cubic B-spline curves. However, regular parts of Catmull–Clark subdivision surfaces are uniform cubic B-spline surfaces. Then, how do we make knot intervals uniform in parameter mesh in order to obtain patches with uniform B-spline boundaries?

(2) When we evaluate points on patches, every coincident vertex should correspond to two charts. According to the mapping rule given in literature [221], there are many coincident vertices on two neighbor sides of control meshes. However, in the process of computation, among those coincident vertices, except 4 vertices should get two charts, the others can only get a chart. So, which chart shall we exactly pick?

(3) In order to obtain patches with B-spline boundaries, we should at least use a 5-ring submesh as a control mesh. Consequently, some initial meshes should be subdivided four times at least, which is a very time-consuming. Furthermore, analysis shows evaluation for patches constructed by the manifold method is also very time-consuming [101]. Then, how do we reduce the number of subdivision times? How do we extract submeshes from subdivision meshes and handle those?

On the basis of solving the above three problems, we use the method to construct n-sided patches in literature [221] to fill the n-sided hole around regular parts of Catmull–Clark subdivision surfaces in order that the resulting surface is G^2 continuous everywhere.

5.3.2 General Steps to Construct n-Sided Patches with Manifold Method

In order to have a comprehensive understanding of the process to construct n-sided patches by using the manifold method, we briefly summarize the construction process in literature [221] as follows:

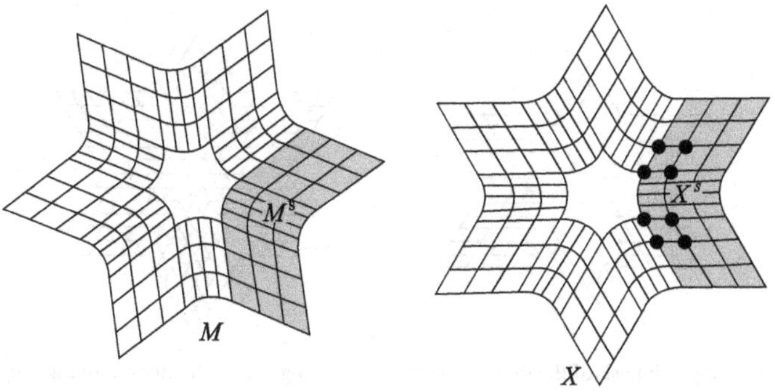

(a) Control mesh (b) Parameter region

Fig. 5.1 Control mesh and parameter region needed to construct n-sided patches

(1) Get the control mesh M, as shown in Fig. 5.1a. The control mesh is constituted of regular layers and an n-sided hole that is surrounded by these regular layers. Let M divided into a series of sides $\overset{s}{M}(s = 0, 1, \ldots, n-1)$;
(2) Define a planar mesh X which has the same topology as that of M, as shown in Fig. 3.1b. We can obtain a new planar mesh by removing the outermost layer faces of X. The new mesh surrounds a region in $\mathbb{R}^2$ space. The region in the parameter region of the patch is going to be constructed. X is also divided into a series of sides $\overset{s}{X}(s = 0, 1, \ldots, n-1)$. Obtain a mesh $\overset{s+}{X}$ by outward extending $\overset{s}{X}$ two face layers and map $\overset{s+}{X}$ into a rectangular mesh $\overset{s+}{\overline{X}}$ in $\mathbb{R}^2$ space;
(3) A chart is defined for every vertex $\boldsymbol{x}_c$ in $\overset{s}{X}$. The chart has 5×5 vertices including *the vertex* $\boldsymbol{x}_c$. We can obtain 5×5 new vertices after mapping the chart into $\overset{s+}{\overline{X}}$. These vertices are the needed knots when we define two cubic B-spline basic functions $N_i(u)$, $N_j(v)$, the multiplication product of these two basic functions and a weight is the base function corresponding to the vertex $\boldsymbol{x}_c$. The basic function is denoted by N_c. In $\overset{s}{X}$, only those vertices marked by black dots correspond to two charts, whose weight is chosen as 1/2, and the other vertices only correspond to a chart whose weight is chosen as 1. So, all the vertices in X (all the vertices in M) at least correspond to a basic function N_c;
(4) Denote the corresponding vertex of $\boldsymbol{x}_c$ in mesh M also as $\boldsymbol{x}_c$, for any a point in the parameters region X, we compute the point on the patch whose control mesh is M as the following:

$$\boldsymbol{p}(\boldsymbol{x}_c) = \sum_{\boldsymbol{x}_c \in X} \boldsymbol{P}_c N_c(\boldsymbol{x}_c)$$

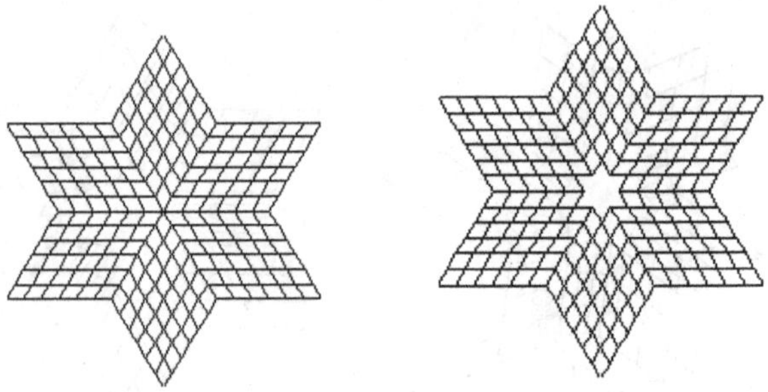

(a) k-ring submesh of the extraordinary vertex (b) k-ring submesh after removing the extraordinary vertex

Fig. 5.2 k-ring submesh of an extraordinary vertex and the control mesh formed by it

5.3.3 Construction of Control Meshes of Manifold Patches

The mesh shown in Fig. 5.2a is the k-ring submesh of an extraordinary vertex (there are no other extraordinary vertices except the middle extraordinary vertex); we can obtain a limit surface if we continue to subdivide the k-ring submesh. The limit surface is a patch denoted as S_1 which is G^1 continuous at the extraordinary points. Obviously, a n-sided hole can be surrounded by a uniform cubic B-spline surfaces ring S_2 whose control mesh is the submesh formed by removing the inner $(k-3)$-ring for the $(k+1)$-ring of an extraordinary vertex. S_1 just fills the n-sided hole and joins S_2 with G^2 continuity. The mesh shown in Fig. 5.2b is the mesh obtained by removing the extraordinary vertex and its neighboring edges, for the k-ring submesh of the extraordinary vertex. In the following discussion, we will construct a surface S with G^2 continuity everywhere, we replace S_1 by S to fill the n-sided hole surrounded by the surface ring S_2, and S joins S_2 with G^2 continuity.

5.3.4 Parameter Region, Normalization Mapping, and Basic Function

The basic methods are used to construct parameter regions, normalization mappings, and basic functions start from literature [221]. There are two motives of still discussing these methods in the section: (1) modify or embody the original methods and (2) give some basic conceptions conveniently for the later discussion. The following discussion is made on the basis of the mesh shown in Fig. 3.2b. We denote the mesh by M.

1. Construction of parameter regions

Let $\alpha = 2\pi/n, x_0 = \cos\alpha + 1$ and $y_0 = \sin\alpha$, we construct a map $\mu(y)$ that satisfies the following conditions:

$$\begin{cases} \mu(y_0) = \mu(-y_0) = x_0 \\ \mu'(y_0) = \mu'(-y_0) = ctg\alpha \\ \mu''(y_0) = \mu''(-y_0) = 0 \end{cases} \tag{5.1}$$

Consequently, we can assume that $\mu(y) = a_0 y^4 + a_1 y^2 + a_2$. By applying the conditions (5.1), there are $a_0 = -ctg\alpha/8y_0^3,\ a_1 = 3ctg\alpha/4y_0,\ a_2 = x_0 - 5\ y_0 ctg\alpha/8$. Therefore, we can define a C^2 continuous function $\overline{\mu}(y)(-ky_0 \leqslant y \leqslant ky_0)$:

$$\overline{\mu}(y) = \begin{cases} (ctg\alpha)y + 1 & y_0 \leqslant y \leqslant ky_0 \\ \mu(y) & -y_0 \leqslant y \leqslant y_0 \\ -(ctg\alpha)y + 1 & -ky_0 \leqslant y \leqslant -y_0 \end{cases} \tag{5.2}$$

So we obtain an initial mesh curve of the mesh X in $\mathbb{R}^2$ space. The mesh X corresponds to the mesh M in $\mathbb{R}^3$ space. Move $\overline{\mu}(y)k - 1$ times in turn toward the positive direction of the coordinated axis. Every time it is moved by one unit length. Together with $\overline{\mu}(y)$, we can obtain *k curves* $\overline{\mu}_j(y)(j = 0, \ldots, k-1)$, by connecting point pairs:

$$(\overline{\mu}_0((k-i)y_0), (k-i)y_0), \quad (\overline{\mu}_{k-1}((k-i)y_0), (k-i)y_0),$$

we obtain line segment $l_i(\boldsymbol{x})(i = 0, \ldots, 2k)$. So, we get the side $\overset{0}{X}$ of the mesh X. Rotate $\overset{0}{X}$ by angle $\alpha s(s = 1, \ldots, n-1)$ in turn and then obtain $\overset{s}{X}$, all $\overset{s}{X}(s = 0, \ldots, n-1)$ from the mesh X. Rotate $\overline{\mu}_{k-2}(y)$ in turn by the angle $\alpha s(s = 1, \ldots, n-1)$ and obtain a curve segment series that surrounds a region. The region is a close set including its boundary curves and it is taken as the parameters region X' of the surface that is going to be constructed, we get $\overset{s+}{X}$ by extending $\overset{s}{X}$ two layers outside. Treat the mesh curve and vertices of $\overset{s+}{X}$ as the mesh curves and vertices of the parameter region X'. For each vertex $\overset{s}{\boldsymbol{x}}_{i,j}$ in $\boldsymbol{X}^s$, we define a chart $\overset{s}{c}_{i,j} \subset \overset{s+}{X}$. There are 5×5 vertices $\overset{s}{\boldsymbol{x}}_{i',j'}(i-2 \leqslant i' \leqslant i+2, j-2 \leqslant j' \leqslant j+2)$ in $\overset{s}{c}_{i,j}$. Assume that the chart is an open set without its boundaries in $\mathbb{R}^2$ space. For mesh X, although $\overset{s}{X}$ and $\overset{s+1}{X}$ have many coincidence vertices, besides vertex $\overset{s}{\boldsymbol{x}}_{k\pm i,j}(i = 1, 2; j = 0, 1; s = 0, \ldots, n-1)$ which corresponds to two charts, the rest vertices, respectively, have a corresponding chart, which is taken as the corresponding chart in $\overset{s}{X}$ or in $\overset{s+1}{X}$, as shown in Fig. 5.3.

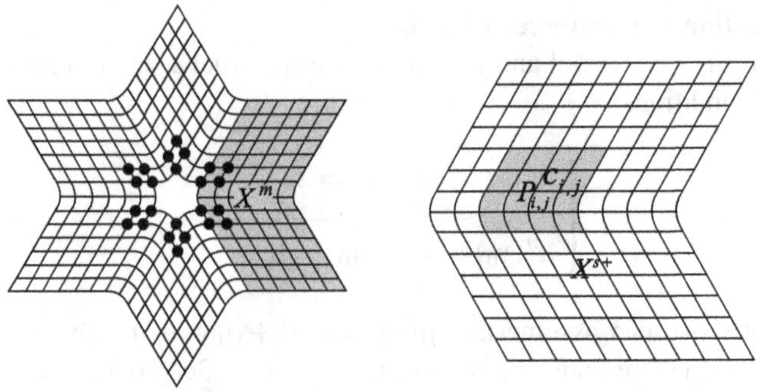

(a) Mesh X and side $\overset{s}{X}$ (b) $\overset{s+}{X}$ and the chart $\overset{s}{c}_{i,j}$ corresponding to $x_{i,j}$

Fig. 5.3 Mesh X, $\overset{s+}{X}$, and $\overset{s}{c}_{i,j}$

2. Normalized mapping and basic function

In order to map $\overset{s+}{X}$ into some rectangular meshes whose edges are, respectively, parallel to x-axis and y-axis, we define a rotate mapping at first:

$$\overset{s}{\varphi} : (x, y) \rightarrow (x \cos(m\alpha) + y \sin(m\alpha), -x \sin(m\alpha) + y \cos(m\alpha))$$

Again define a mapping:

$$\phi : (x, y) \rightarrow (x - \overline{\mu}(y), y/\sin\alpha)$$

Denote $\overset{s}{\phi} = \phi \circ \overset{s}{\varphi}$. Obviously, $\overset{s}{\phi}$ is a C^2 mapping. By using the mapping $\overset{s}{\phi}$, $\overset{s+}{X}$ is mapped into a rectangular mesh $\overset{s+}{\overline{X}}$ in which vertex interval is uniform. The table $\overset{s}{c}_{i,j}$ in $\overset{s+}{X}$ is mapped into the chart $\overset{s}{\overline{c}}_{i,j}$ in $\overset{s+}{\overline{X}}$; in the chart $\overset{s}{c}_{i,j}$, we define the basic function:

$$B_{i,j}(u, v) = N_{i-2,3}(u)N_{j-2,3}(v) \quad (u, v) \in \overset{s}{\overline{c}}_{i,j}$$

In the above formula, $N_{i-2,3}(u)$, $N_{j-2,3}(v)$ are cubic B-spline basic functions, respectively, taking longitudinal vertices and transverse vertices in $\overset{s+}{\overline{X}}$ as their knots vector. For the convenience of discussion, let

$$\overset{s}{B}_{i,j}(\boldsymbol{x}) = B_{i,j} \circ \overset{s}{\phi}(\boldsymbol{x})$$

where $\overset{s}{B}_{i,j}$ is called the basic function corresponding to $\overset{s}{\boldsymbol{x}}_{i,j}$, the function is also C^2 continuous.

Since the vertex intervals of $\overset{s+}{X}$ are uniform, $N_{i-2,3}(u)$, $N_{j-2,3}(v)$ are uniform cubic B-spline basic functions. Consequently, when the mesh M is enough large, we can make the constructed patch S has uniform cubic B-spline boundaries. So far, by constructing the planar mesh X corresponding to the mesh M in Fig. 3.2b, we have solved the first problem in Sect. 5.3.1. Different from the method in literature [221], our method can obtain knots with uniform longitudinal intervals that use y_0 as units in formula (3.2). In addition, the evaluation approach of (x_0, y_0) in this section is equivalent to the approach that the aperture factor γ is evaluated as 2 in literature [221]. The evaluation approach does not conform to the condition satisfied by the aperture factor given in literature [100]:

$$2\gamma \cos\alpha \leqslant [k/2+1] \ (k \text{ is the continuity degree})$$

However, it can be proved that when we take the continuity degree as 2, in order to make any point at least in a chart $\overset{s}{c}_{i,j}$, γ only have to satisfy the condition:

$$\gamma \leqslant 32\cos\alpha/(13\cos^2\alpha + 8\cos\alpha - 5)$$

Unfortunately, here we cannot adjust the size of the aperture factor γ; otherwise, we cannot get patches with uniform B-spline boundaries.

5.3.5 *Related Vertex and Normalized Basic Functions*

Notice that $\overset{s+}{X}$ and $\overset{s+1}{X}$ have many coincident vertices and most vertices of the mesh X can only correspond to one table. Literature [221] presents an approach. The approach firstly abandons tables corresponding to some vertices in X^s and then computes points on patch; here, we do not abandon tables corresponding to any vertices while giving a method that is called as the dynamically selecting vertex method. We first define a concept.

Definition 3.1 Arbitrarily take a point $\boldsymbol{x}$ in the parameter field X'. If there is a vertex $\overset{s}{\boldsymbol{x}}_{i,j}$ in $\overset{s}{X}$ which makes $\boldsymbol{x} \in \overset{s}{c}_{i,j}$, we call that $\overset{s}{\boldsymbol{x}}_{i,j}$ is a related vertex of $\boldsymbol{x}$ in $\overset{s}{X}$.

Assume that all related vertices of $\boldsymbol{x}$ in $\overset{s}{X}$ form a set. Denote the set as $\overset{s}{\Theta}$. Obviously, there are probably coincident vertices in set $\overset{s-1}{\Theta}, \overset{s}{\Theta}, \overset{s+1}{\Theta}$. In order to determine which chart should be taken for these vertices that have only a corresponding table, we give the procedure of dynamically selecting vertex method as follows:

(1) Select a set in $\overset{s}{X} (s = 0, \ldots, n-1)$. Compared with other sets, the set has the most elements. Denote the set as $\overset{s_0}{\Theta}$. If there are several sets that they have the same number of elements, we take the one with the smallest superscript.

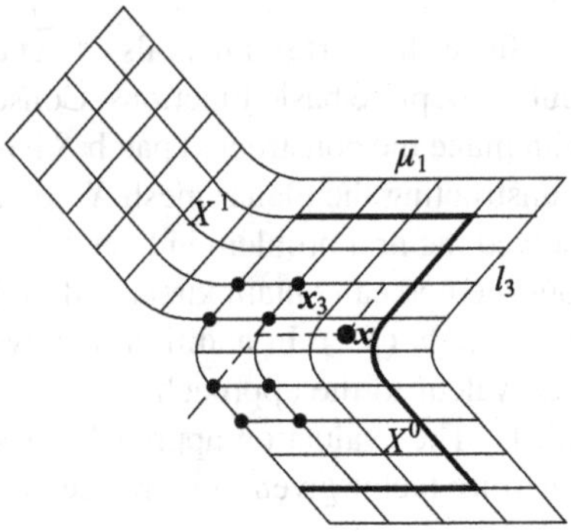

Fig. 5.4 Rationality of dynamically selecting vertex method

(2) Let $s = s_0 + 1$. Starting from $\overset{s}{\Theta}$, we clear those vertices if these vertices have coincident vertices in $\overset{s-1}{\Theta}$. However, if a vertex is a coincident vertex but it has two corresponding tables, we will not delete it. When $s = s_0 - 1$, we have to clear these vertices that have coincident vertices in $\overset{s_0}{\Theta}$.

After selecting vertices by using dynamically selecting vertex method, we obtain sets; these sets are still denote as $\overset{s}{X}(s = 0, \ldots, n-1)$, for $\overset{s}{\Theta}$, if we any pick two sets $\overset{s}{\Theta}$, $\overset{s+1}{\Theta}$, there is no other coincident vertex between the two sets, expect these vertices with two tables. Denote the element in $\overset{s}{\Theta}$ as $\overset{s}{\boldsymbol{x}}_i$: So we can define a normalized basic function:

$$\overset{s}{B}_i(\boldsymbol{x}) = \frac{\overset{s}{w}_i \overset{s}{B}_i(\boldsymbol{x})}{\sum_{s=0}^{n-1} \sum_{i=0}^{k_s} \overset{s}{w}_i \overset{s}{B}_i(\boldsymbol{x})} \quad \text{where } x \in X' \tag{5.3}$$

In the above formula, if $\overset{s}{\boldsymbol{x}}_i$ corresponds to two charts, we take its weight $\overset{s}{w}_i = 1/2$, otherwise take $\overset{s}{w}_i = 1$. Denote the vertex in the mesh M corresponding to $\overset{s}{\boldsymbol{x}}_i$ as $\overset{s}{\boldsymbol{V}}_i$, so the point in the patch S can be computed by the following formula:

$$\boldsymbol{p}(\boldsymbol{x}) = \sum_{s=0}^{n-1} \sum_{i=0}^{k_s} \overset{s}{\boldsymbol{V}}_i \overset{s}{\boldsymbol{B}}_i(\boldsymbol{x}) \quad \text{where } \boldsymbol{x} \in X' \tag{5.4}$$

From formula (5.3), we can find that the resulting surface S has the convexity hull property and the local convexity hull property, that is, S is in the convex hull formed by the whole meshes, and $\boldsymbol{p}(\boldsymbol{U}(\boldsymbol{x}, \varepsilon))$ is in the convex hull formed by vertices $\overset{s}{\boldsymbol{x}}_i(i = 0, \ldots, k_s; s = 0, , \ldots, n-1)$ where $\boldsymbol{U}(\boldsymbol{x}, \varepsilon)$ is a ε neighborhood *of* $\boldsymbol{x}$ in the parameter field X'. So far, we have solved the second problem in Sect. 5.3.1. As for the problem whether there are flat points in the patch S, it is entirely determined by the control mesh M; the conclusion will be taken as Theorem 3.3 of this book.

Again, we analyze the rationality of the dynamically selecting vertex method. Investigate Fig. 5.4, the affine distances of $\boldsymbol{x}$ to two mesh curve $\overline{\mu}_1$ and l_3 are, respectively, (u, v); again, let the affine distances between $\boldsymbol{x}$ and the tangent line of $\overline{\mu}_1$ at $\boldsymbol{x}_3$ be u', obviously, $u \neq u'$. It should be found that (u', v) is the absolute value of coordination of $\boldsymbol{x}$ for $\boldsymbol{x}_3$ when we take the corresponding chart of $\boldsymbol{x}_3$ in $\overset{1}{X}$, consequently, for the parameter value $\boldsymbol{x}$, values of the two normalization basic functions corresponding to $\boldsymbol{x}_3$ are not equal, that is to say, there is a difference when different tables are picked. However, since $\boldsymbol{x}$ is in $\overset{0}{X}$, the dynamically selecting vertex method just satisfies the requirement, which effectively solves the problem that which will we should take for $\boldsymbol{x}_3$. If we further prescribe that $\boldsymbol{x}_3$ should correspond the chart in $\overset{1}{X}$, it is not appropriate the value of the basic function at parameter point $\boldsymbol{x}$. For the property that the resulting surface S has B-spline boundaries by using the dynamically selecting point method, it is the conclusion of the posterior Theorem 3.2.

5.3.6 Properties of the Patch S

The n-sided patches constructed in literature [221] are G^2 continuous and have B-spline boundaries, which have been proved in literature [201]. Now let us prove that the patch S has such properties: (1) G^2 continuity (2) S has B-spline boundaries when we take the k-ring ($k \geqslant 5$) submesh of an extraordinary vertex. Notice that the patch is constructed on the basis of the mesh in Fig. 5.2b and dynamically selecting point method.

Theorem 3.1 *Obtain a mesh M by deleting k-ring submeshes of an extraordinary vertex and edges neighboring to these submeshes for a C-C subdivision mesh, so the resulting surface S constructed by using M as the control mesh and the dynamically selecting vertex method is G^2 continuous everywhere.*

Proof Investigate the continuity of S at the point $\boldsymbol{p}(x)$ when $\boldsymbol{x}$ is in the inner of the region surrounded by $\overset{s}{X}(s = 0, \ldots, n-1)$ (denote the region as X''). When $\boldsymbol{x}$ is in other parts of the parameter region X', the conclusion can be proved by using a similar process. The inner region X'' of the parameter region X' and the related mesh curves in the mesh X are shown in Fig. 5.5.

Firstly, investigate the continuity of the patch S in the quadrilateral region rounded by $\boldsymbol{P}_0\boldsymbol{Q}_1\boldsymbol{Q}_{n-1}\boldsymbol{P}_{n-1}$. The position of a parameter point $\boldsymbol{x}$ in the region can be divided into three cases: ① inside a face surrounded by mesh curve; ② on a mesh curve; and ③ coincident with a mesh vertex.

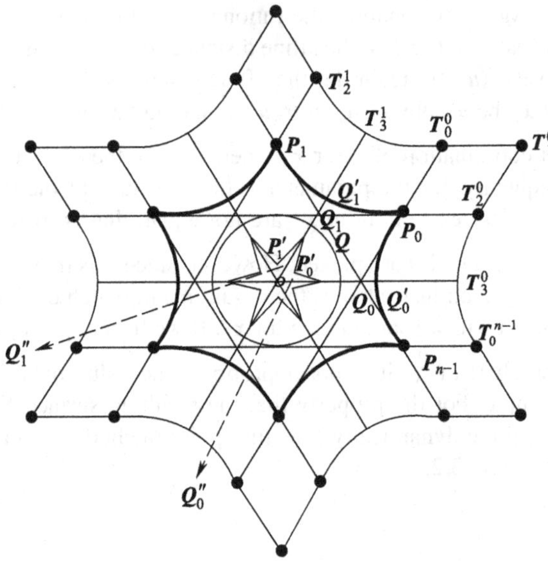

Fig. 5.5 Region X'' and mesh curves related to X'' in X

(1) $\boldsymbol{x}$ inside a face surrounded by mesh curves, assume that vertex sets related to $\boldsymbol{x}$ in $\overset{s}{X}$ be $\overset{s}{\Theta}(s = 0, \ldots, n-1)$, when we select vertices by using dynamically selecting vertex method, all vertices in $\overset{0}{\Theta}$ will be reserved while some vertices in other sets will be selectively deleted. After the operation step, we get some new sets which are still denoted as $\overset{s}{\Theta}(s = 0, \ldots, n-1)$. the vertex number in the set $\overset{s}{\Theta}$ is denoted as k_s, so

$$\boldsymbol{p}(x) = \sum^{n-1} \sum^{k_s} \overset{s}{\boldsymbol{V}}_i \overset{s}{B}_i(\boldsymbol{x})$$

Consider the case that we take parameter value $\boldsymbol{x}'$ in a ε neighborhood $U(\boldsymbol{x}, \varepsilon)$ of $\boldsymbol{x}$. Because related vertices of $\boldsymbol{x}'$ are the same as related vertices of $\boldsymbol{x}$, and because normalized basic function $B_i^m(\boldsymbol{x})$ is C^2 continuous,

$$\lim_{\varepsilon \to 0} \boldsymbol{p}(\boldsymbol{x}') = \boldsymbol{p}(\boldsymbol{x}) \tag{5.5}$$

$$\lim_{\varepsilon \to 0} \left. \frac{d\boldsymbol{p}(\boldsymbol{y})}{d\boldsymbol{y}} \right|_{\boldsymbol{y}=\boldsymbol{x}'} = \left. \frac{d\boldsymbol{p}(\boldsymbol{y})}{d\boldsymbol{y}} \right|_{\boldsymbol{y}=\boldsymbol{x}} \tag{5.6}$$

$$\lim_{\varepsilon \to 0} \left. \frac{d^2\boldsymbol{p}(\boldsymbol{y})}{(d\boldsymbol{y})^2} \right|_{\boldsymbol{y}=\boldsymbol{x}'} = \left. \frac{d^2\boldsymbol{p}(\boldsymbol{y})}{(d\boldsymbol{y})^2} \right|_{\boldsymbol{y}=\boldsymbol{x}} \tag{5.7}$$

(2) $\boldsymbol{x}$ on a mesh curve, without loss of generality, $\boldsymbol{p} \in \boldsymbol{P}_0\boldsymbol{Q}_0$. After selecting vertices by using dynamically selecting vertex method, vertex numbers in sets $\overset{0}{\Theta}$

and $\overset{1}{\Theta}$ are, respectively, 8 and 6 and the other sets are empty. When the parameter $\boldsymbol{x}' \in U(\boldsymbol{x}, \varepsilon) \cap \text{region}\boldsymbol{P}_0 \boldsymbol{Q}_1 \boldsymbol{Q} \boldsymbol{Q}_0 \boldsymbol{Q}_0'$, relating vertices of $\boldsymbol{x}'$ are added vertices $\boldsymbol{P}_1$ and $\boldsymbol{T}_2^1$ in $\overset{1}{X}$, on the basis of element of $\overset{0}{\Theta}$ and $\overset{1}{\Theta}$, vertices $\boldsymbol{P}_1$ and $\boldsymbol{T}_2^1$, respectively, corresponds to a vertex in the mesh M. The corresponding vertices are still denoted as $\boldsymbol{P}_1$, $\boldsymbol{T}_2^1$. Basic functions corresponding to them are written as A_1, A_2^1, so

$$\boldsymbol{p}(\boldsymbol{x}') = \frac{\sum_{s=0}^{n-1}\sum_{i=0}^{k_m} \boldsymbol{V}_i^s \omega_i^s B_i^s(\boldsymbol{x}') + \boldsymbol{P}_1\omega_1 A_1(\boldsymbol{x}') + \boldsymbol{T}_2^1\omega_2^1 A_2^1(\boldsymbol{x}')}{\sum_{s=0}^{n-1}\sum_{i=0}^{k_m} \omega_i^s B_i^s(\boldsymbol{x}') + \omega_1 A_1(\boldsymbol{x}') + \omega_2^1 A_2^1(\boldsymbol{x}')}$$

Since when $\varepsilon \to 0$, $A_1(\boldsymbol{x}')$, $A_2^1(\boldsymbol{x}')$, and their first-order derivatives and second-order derivatives all approach to 0, so the Eqs. (5.5)–(5.7) are correct. By using the same reasoning, we can prove that the Eqs. (5.5)–(5.7) are also correct when $\boldsymbol{x}' \in U(\boldsymbol{x}, \varepsilon)$.

(3) $\boldsymbol{x}$ is coincident with a mesh vertex of X; without loss of generality, assume that it is coincident with $\boldsymbol{Q}_0$. After selecting vertices by using the dynamically selecting vertex method, the vertex numbers in sets P^0 and P^1 are, respectively, 6 and 3 and the other sets are empty. When the parameter $\boldsymbol{x}' \in U(\boldsymbol{x}, \varepsilon) \cap \text{region}\boldsymbol{P}_0 \boldsymbol{Q}_1 \boldsymbol{Q}$ $\boldsymbol{Q}_0 \boldsymbol{Q}_0'$, relating vertices of $\boldsymbol{x}'$ are added vertices $\boldsymbol{P}_1$, $\boldsymbol{T}_2^1$, $\boldsymbol{T}_0^0$, $\boldsymbol{T}_1^0$, $\boldsymbol{P}_1$ in X^0; on the basis of elements of $\overset{0}{\Theta}$ and $\overset{1}{\Theta}$, considering that when $\varepsilon \to 0$, basic functions corresponding to these vertices and their first-order derivatives and second-order derivatives all approach to 0. From the proof way of the case (2), we can know that the Eqs. (5.5)–(5.7) are correct. By using the same reasoning way, we can know that Eqs. (5.5)–(5.7) are still correct when $\boldsymbol{x}' \in U(\boldsymbol{x}, \varepsilon)$.

The above three cases have proven the conclusion that the patch S are C^2 continuous when $\boldsymbol{x}$ is in the quadrilateral region rounded by $\boldsymbol{P}_0 \boldsymbol{Q}_1 \boldsymbol{Q}_{n-1} \boldsymbol{P}_{n-1}$. By using the same reasoning way, it can be known that the patch S is also G^2 continuous when p is in the quadrilateral region rounded by $\boldsymbol{P}_i \boldsymbol{Q}_{i+1} \boldsymbol{Q}_{i-1} \cdot \boldsymbol{P}_{i-1} (i = 1, \ldots, n-1)$. The following process will prove the conclusion that patch S is also C^2 continuous when $\boldsymbol{x} \in \boldsymbol{P}_0' \ldots \boldsymbol{P}_{n-1}'$ rounded region and its boundary curve. In order to prove the conclusion, we only have to prove the conclusion that the patches are C^2 continuous when $\boldsymbol{x}$ is coincident with 0, and when $\boldsymbol{x}$ is at other position, the proved process is same as the above process.

When we take $\boldsymbol{O}$ as the parameter point $\boldsymbol{x}$, the vertex numbers in $\overset{s}{\Theta} (s = 0, \ldots, n-1)$ are all 3. Denote the region rounded by $\boldsymbol{Q}_0'' \boldsymbol{P}_0' \boldsymbol{Q}_1'' \boldsymbol{O}$ as $\boldsymbol{Q}_0'' \boldsymbol{P}_0' \boldsymbol{Q}_1'' \boldsymbol{O}$. When $\boldsymbol{x}' \in U(\boldsymbol{x}, \varepsilon) \cap \boldsymbol{Q}_0'' \boldsymbol{P}_0' \boldsymbol{Q}_1'' \boldsymbol{O}$, the related vertices of $\boldsymbol{x}'$ are added to increased $\boldsymbol{T}_0^0$ in $\overset{0}{X}$ and $\boldsymbol{T}_0^0$ in $\overset{1}{X}$, by using the proof way of the case (2), we can know the Eqs. (5.5)–(5.7) are correct. By using the same reasoning way, we can know that Eqs. (5.5)–(5.7) are still correct when $\boldsymbol{x}' \in U(\boldsymbol{x}, \varepsilon)$.

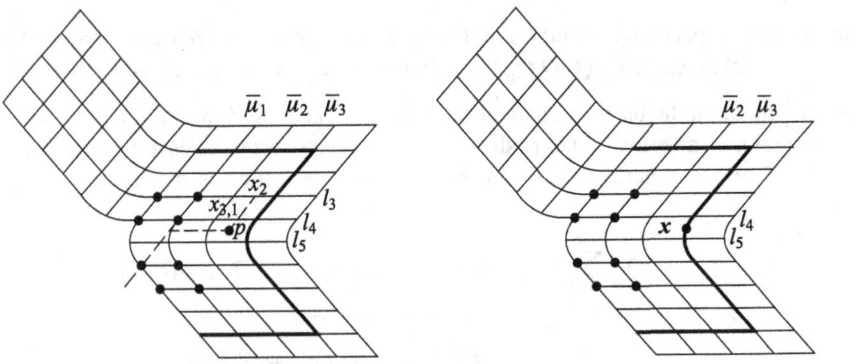

(a) Parameter point x inside the parameter region

(b) Parameter point x on the boundary curve of parameter region

Fig. 5.6 Taking parameter point in different places of parameter region

So, the patch S is C^2 continuous when the parameter point $\boldsymbol{x}$ is inside the region rounded by $\overset{s}{X}(s = 0, \ldots, n-1)$. #

Theorem 3.2 *Taking the k-ring submesh of an extraordinary vertex in a C-C subdivision mesh, obtain a mesh M after deleting the extraordinary vertex and edges neighbor to the vertex, then, if and only if k* $\geqslant$*5, the patch S has uniform cubic B-spline boundaries, and the patch S is constructed by using the method presented in this section and has the control mesh M. These boundary curves are boundary curves of uniform bi-cubic B-spline surface rings whose control meshes are* $\overset{s}{\boldsymbol{V}}_{i,j}(s = 0, \ldots, n-1; i = 0, \ldots, 2k; j = k-4, \ldots, k-1)$.

Proof Firstly, prove that when $k = 5$, inside the patch S, there is no bi-cubic B-spline surface ring whose control vertices are vertices of the mesh M, the boundary of the patch S is constituted of n uniform cubic B-spline curves. Considering that $\overset{s}{X}(s = 1, \ldots, n-1)$ is formed by rotating $\overset{0}{X}$, we only investigate the side $\overset{0}{X}$ of the mesh X. In the mesh $\overset{0}{X}$, only the region surrounded by thick curves is in the parameter region $\boldsymbol{X}'$, as shown in Fig. 5.6a. Pick a point p, in the face surrounded by the mesh curve $\overline{\mu}_2, \overline{\mu}_3, l_4, l_5$. After selecting vertices by using dynamically selecting vertex method, there are 16 elements in the set $\overset{0}{\Theta}$ and 2 elements in the set $\overset{1}{\Theta}$. Every element in the set $\overset{1}{\Theta}$ corresponds to two tables, and the two elements are also elements of $\overset{0}{\Theta}$. For $\boldsymbol{P}_{3,1}$, one of the two coincident elements, when it is regarded as a vertex in $\overset{0}{X}$, let the affine distances, respectively, be (u, v), which are the distances between $\boldsymbol{x}$ and two mesh curves $\overline{\mu}_1$ and l_3, the distances are also the absolute value of coordinates of $\boldsymbol{x}$ to $\boldsymbol{P}_{3,1}$ after applying the mapping $\overset{0}{\phi}$ to $\overset{0}{X}$ and $\boldsymbol{x}$. Again, assume that the absolute value of coordinates of p to $\boldsymbol{P}_{3,1}$ be (u^1, v^1) after mapping $\overset{0}{\phi}$ to $\overset{1}{X}$ and $\boldsymbol{x}$. From Fig. 5.6a, we can know that $v^1 > u$ and $u^1 = v$. Consequently, for the two normalized basic functions corresponding to $\boldsymbol{P}_{2,1}$, there must be: $\overset{0}{B}_{3,1}(\boldsymbol{x}) \neq \overset{1}{B}_{3,1}(\boldsymbol{x})$, so $\boldsymbol{p}(\boldsymbol{x})$ cannot be in the uniform bi-cubic B-spline patch determined by these corresponding vertices of the mesh M. The process has proved the conclusion that when $k = 5$, inside the patch S, there is no uniform bi-cubic B-spline surface ring whose control vertices are vertices of the mesh M. The following process will prove the conclusion that the boundary is constructed of uniform cubic B-spline curves. Take $\boldsymbol{x} \in \overline{\mu}_3 \cap X'$, as shown in Fig. 5.6b. After selecting vertices by using the dynamically selecting point method, there are 9 elements in the set $\overset{0}{\Theta}$ and the other sets are empty. For the 9 elements, each only corresponds to a chart. From the defining way of the normalized basic functions, we can know that $\boldsymbol{p}(\boldsymbol{x})$ must be the uniform bi-cubic B-spline surfaces determined by $\overset{0}{\boldsymbol{V}}_{i,j}(0 \leqslant i \leqslant 10, 1 \leqslant j \leqslant 4)$, so $\boldsymbol{p}(\overline{\mu}_3 \cap X')$ is a uniform cubic B-spline curves.

When $k > 5$, the region denoted by the entire $\overset{0}{X}$ in Fig. 5.6 is in the parameter region. It is easy to know that when take parameter points in parameter region on the right of the mesh curve $\overline{\mu}_3$, these related vertices selected by the dynamically selecting vertex method are all in $\overset{0}{\Theta}$. Furthermore, each of these vertices corresponds to a chart; consequently, the part of the patches corresponds to the parameters must be the bi-cubic B-spline surface whose control mesh is composed of $\underset{i,j}{\overset{0}{V}}(i = 0, \ldots, 2k;\ j = 3, \ldots, k-1)$. So the theorem has been proved. #

Theorem 3.3 *Assume that a patch S is constructed by the method presented by this section. Then, whether the patch S has flat points or is not determined by its control mesh M.*

Proof For formula (5.4), let $\boldsymbol{x} = (x, y)$, arrange these vertices in set $\bigcup_{s=0}^{n-1}\{\overset{s}{\boldsymbol{V}}_i | i = 0, \ldots, k_s\}$ in proper order, if there are identical vertices in two different vertices in the union set. Denote the union set as $\{\boldsymbol{H}_j | j = 0, \ldots, t\}$, B_j is the normalized basic function corresponding to $\boldsymbol{H}_j$. So formula (5.4) becomes:

$$\boldsymbol{p}(x, y) = \sum_{j=0}^{t} \boldsymbol{H}_j B_j(x, y) \tag{5.8}$$

By using the knowledge in the differential geometry, we know that $L_s = \boldsymbol{n} \cdot \dfrac{\partial^2 \boldsymbol{p}}{\partial x^2} / \|\boldsymbol{n}\|$, $M_s = \boldsymbol{n} \cdot \dfrac{\partial^2 \boldsymbol{p}}{\partial x \partial y} / \|\boldsymbol{n}\|$, $N_s = \boldsymbol{n} \cdot \dfrac{\partial^2 \boldsymbol{p}}{\partial y^2} / \|\boldsymbol{n}\|$ are second-class fundamental variables of S, when $L_s = M_s = N_s = 0$, $\boldsymbol{p}(\boldsymbol{x})$ is a flat point, where $\boldsymbol{n} = \dfrac{\partial \boldsymbol{p}}{\partial x} \times \dfrac{\partial \boldsymbol{p}}{\partial y}$ is the normal vector of S at the point $\boldsymbol{p}(\boldsymbol{x})$. Now investigate whether L_s is 0 or not. In fact,

$$\boldsymbol{n} = \frac{\partial \boldsymbol{p}}{\partial x} \times \frac{\partial \boldsymbol{p}}{\partial y} = \left(\sum_{i=0}^{t} H_i \frac{\partial B_i}{\partial x}\right) \times \left(\sum_{j=0}^{t} H_j \frac{\partial B_j}{\partial y}\right)$$
$$= \sum_{i=0}^{t}\sum_{j=0}^{t} (H_i \times H_j) \frac{\partial B_i}{\partial x}\frac{\partial B_j}{\partial y}$$
$$\frac{\partial^2 \boldsymbol{p}}{\partial x^2} = \sum_{k=0}^{t} H_k \frac{\partial^2 B_k}{\partial x^2}$$

Consequently, $\boldsymbol{n} \cdot \frac{\partial^2 \boldsymbol{p}}{\partial x^2} = \sum_{i=0}^{t}\sum_{j=0}^{t}\sum_{k=0}^{t} (\boldsymbol{H}_i, \boldsymbol{H}_j, \boldsymbol{H}_k) \frac{\partial B_i}{\partial x}\frac{\partial B_j}{\partial y}\frac{\partial^2 B_k}{\partial x^2}$.
where $(\cdot,\cdot,\cdot)$ denote the mixed-product operation, when $\boldsymbol{H}_i = \boldsymbol{H}_j$ or $\boldsymbol{H}_j = \boldsymbol{H}_k$ or $\boldsymbol{H}_k = \boldsymbol{H}_i$, there must be $(\boldsymbol{H}_i, \boldsymbol{H}_j, \boldsymbol{H}_k) = 0$. Notice that $\boldsymbol{H}_i \neq \boldsymbol{H}_j \neq \boldsymbol{H}_k \neq \boldsymbol{H}_i$, those $\frac{\partial B_i}{\partial x}\frac{\partial b_j}{\partial y}\frac{\partial^2 B_k}{\partial x^2}$ cannot be all 0, so, whether $\boldsymbol{n} \times \frac{\partial^2 \boldsymbol{p}}{\partial x^2}$ is 0 or not, i.e., whether L_s is 0 or not is decides determined by these mixed products of $(\boldsymbol{H}_i, \boldsymbol{H}_j, \boldsymbol{H}_k)$. By using the same reasoning way, whether the other fundamental variables M_s and N_s are 0 or not are determined by these mixed products of control vertices. The above process proves the conclusion that whether any a point $\boldsymbol{p}(\boldsymbol{x})$ on the patch S is a flat point or not is determined by its control mesh M.

5.3.7 Extracting Submeshes from C-C Subdivision Meshes

In the subsection, we will extract k-ring submeshes of extraordinary vertices of C-C subdivision meshes. After a k-ring submesh is processed, it can be used as the control mesh of patch S discussed in the above subsection. From Theorems 3.1 and 3.3, we can find that when using the control mesh formed by the 5-ring submesh of an extraordinary vertex, the resulting patch S only has uniform cubic B-spline boundaries. However, the patch S can G^2 continuously join with the submesh whose control mesh is the mesh formed by deleting 2-ring submesh and its neighboring edges of the extraordinary vertex for a C-C subdivision mesh on a subdivision level. Notice that the following fact is obvious: Assume that the mesh M is obtained by subdividing M', in the mesh M, the 5-ring submesh of an extraordinary vertex can be obtained by subdividing the 3-ring submesh of the corresponding extraordinary vertex in M'; for the mesh obtained by deleting 2-ring submesh and its neighboring edges in the mesh M, it can be obtained by subdividing a mesh obtained by deleting the corresponding extraordinary vertex and its neighboring edges. Consequently, we can have the following corollary:

Corollary 3.1 *Extracting the 3-ring submesh of an extraordinary vertex from a C-C subdivision mesh M. Obtain the 5-ring submesh of the corresponding extraordinary vertex by subdividing the 3-ring submesh once. Obtain a mesh M by deleting the extraordinary vertex and its neighboring edges in the 5-ring submesh. By using M as the control mesh and using the dynamically selecting vertex method, a patch S can be obtained. The patch $S G^2$ joins with a subdivision surface whose control mesh is the mesh formed by deleting the extraordinary vertex and its neighboring edges in the C-C subdivision mesh.*

So, the three problems in Sect. 5.3.2 have all been solved. For any initial mesh, we can make the case exist by subdividing the initial mesh at most three times: In the 3-ring submesh of an extraordinary vertex of the subdivision mesh (if the 3-ring submesh exists), there is no other extraordinary vertex. Furthermore, if we construct two patches, respectively, based on 3-ring submesh of any two extraordinary vertices, the two patches do not intersect. If they are close to each other, they just G^2 join on their common boundaries.

5.3.8 Examples

In these examples shown in Figs. 5.7 and 5.8, we construct n-sided holes and fill these holes for subdivision meshes that are obtained by subdividing initial meshes three times. For the surface in Fig. 5.7b, it is not entirely surface constructed by using the mesh in Fig. 5.7a as the control mesh. Its outer layer ring is a bi-cubic B-spline surface ring whose control mesh is constructed by deleting the extraordinary vertex and its neighboring edges for the 4-ring submesh of the middle extraordinary vertex. The patch inside the surface ring is the patch S constructed by using the method presented by this section. The most medium part of S corresponds to the parameter

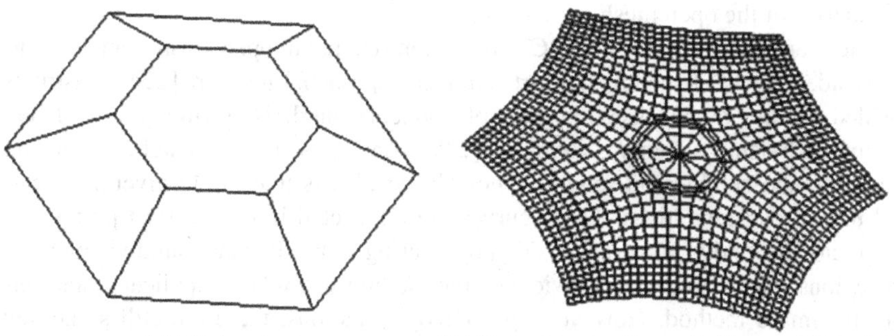

(a) Initial mesh (b) Effects of constructing n-sided hole and fill the hole

Fig. 5.7 n-sided patch and bi-cubic B-spline surface ring G^2 join

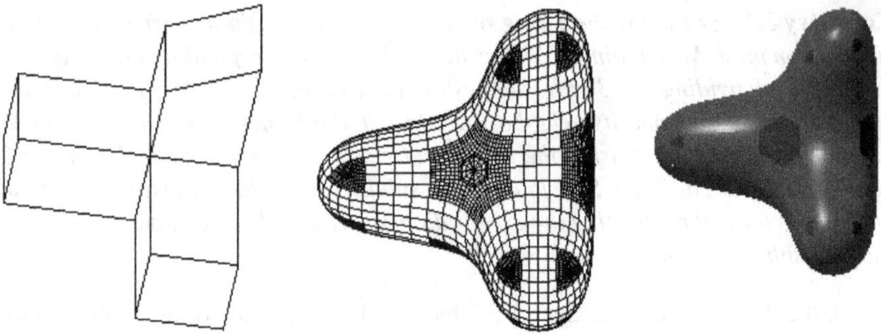

(a) Initial mesh (b) Construct n-sided holes and fill these holes for all extraordinary vertex

Fig. 5.8 Construct surface with entire G^2 continuity

region surrounded by $\overset{s}{X}(s = 1, \ldots, n - 1)$ in the parameter region X'. Figure 5.8b gives the effect by constructing n-sided holes and filling them after subdividing the initial meshes twice.

5.4 Subdivision Modeling, n-Sided Patches and B-Spline Boundaries

So far, among all kinds of modeling methods of n-sided patches, the subdivision method is the simplest one. The earliest subdivision scheme for surface modeling is presented for close meshes or infinite open meshes; in fact, infinite open meshes do not exist. When we apply a subdivision scheme to an open mesh, a patch naturally appears. However, in this case, we have to design new subdivision rules near the boundaries of the open mesh.

Literature [223] applies the C-C subdivision scheme in open meshes and obtaining n-sided patches by using the shirt-removed approach. Literature [224] constructs n-sided patches by constructing a set of boundary subdivision rules for the Doo–Sabin subdivision scheme. Literature [225] constructs n-sided patches by using the combined subdivision scheme whose basic idea is that firstly given n curves and their cross derivatives. These curves are connected in turn at start points and endpoints. Secondly, construct n-sided by altering Catmull–Clark subdivision rules on regions near boundaries. Obviously, the method is a widely applicable and relatively simple method. However, on NURBS situations, there are still some key points that can be improved in these methods. The method presented in literature [223] can only construct patches with uniform cubic B-spline boundaries without non-uniform cubic B-spline boundaries. Literature [224] uses quadratic uniform subdivision, which makes the patches constructed by using its method only have uniform quadratic B-spline boundaries; it is a troublesome task to blend such patches and bi-cubic NURBS patches. The way to process boundaries is similar to the combined subdivision scheme. For literature [225], we can investigate the following contents: at the situations of joining NURBS patches, if these patches do not interpolate corner vertices of their control meshes, two patches can G^2 join when vertex arrays of control meshes have three common rows. If these patches interpolate corner vertices of their control meshes, two patches can G^1 join when vertex arrays of their control meshes have one common row. In the common row and respective second row, the corresponding 3 vertices in a line, The method is simple and obvious. Compared with the joining conditions of NURBS patches, the combined subdivision scheme seems slightly to be trivial. However, if we design such a method: fill n-sided holes or blending NURBS patches by using the n-sided patches presented by literature, the method has such simple and obvious property in the case of joining NURBS patches. Consequently, it is important to research the problem of how to construct n-sided patches by using the non-uniform Catmull–Clark subdivision scheme.

Though the fact is obvious: after extending the initial mesh by a fictitious layer outward, the skirt-removed approach is still fit for the non-uniform C-C subdivision scheme; if we construct n-sided patches by using the subdivision scheme, at least the following problems have to be solved:

(1) Which conditions should the initial mesh satisfy? Since every edge of subdivision meshes have a parameter, which conditions should those parameters satisfy? How should expressions of B-spline boundary be given according to the initial mesh?

(2) For commonly seen NURBS patches, their corner points do interpolate corner vertices of control meshes, their boundary curves are NURBS curves whose control polygons are corresponding boundaries of control meshes, how do we design new subdivision rules on regions near boundaries in order to make n-sided patches have such properties?

This section presents methods to solve the above problems, and these methods are strictly demonstrated, which means that we can construct n-sided patches by using simpler approach and enrich modeling effects at the same time. Though our discussion is made on the basis of non-uniform B-spline surfaces, our method is also fit for NURBS surfaces, because in the rational case we only have to introduce 4-dimensional vertices and project subdivision meshes to the hyperplane $\omega = 1$ after subdividing the meshes in the 4-dimensional space. Thus, it can be seen that the method to construct n-sided patches on an open mesh by using non-uniform C-C subdivision scheme in fact presents an idea for filling n-sided holes in NURBS surfaces or blending several NURBS patches.

5.5 Construct n-Sided Patches by Using Non-uniform C-C Subdivision Scheme and Skirt-Removed Approach

5.5.1 Skirt-Removed Approach

In order to obtain the n-sided patches with uniform B-spline boundaries, literature [212, 223, 227] give the skirt-removed approach under the C-C subdivision scheme. The skirt-removed approach is the key point of the methods to fill holes and blend patches. When we use skirt-removed approach, it is not necessary to discuss the convergence of subdivision meshes and the continuity of the limit surfaces. However, such discussions are necessary when we introduce new boundary subdivision rules. The skirt-removed approach can obtain patches with B-spline boundaries when vertices on several layers near boundaries of control meshes are regular. In this section, we use the method to construct n-sided patches with non-uniform B-spline boundaries under the non-uniform B-spline surfaces.

The idea of skirt-removed approach is very simple. Generally speaking, each time when we subdivide the mesh, the boundary edges do not correspond to new edge points and boundary vertices do not correspond to new vertex points.

Literature [223] calls the C-C subdivision scheme based on the skirt-removed approach as the skirt-removed approach scheme and called the usual C-C subdivision scheme as the standard subdivision scheme. It gives the following properties for the skirt-removed approach:

Property 5.1 The limit surface obtained by using the skirt-removed approach scheme is a part of the limit surface obtained by using the standard subdivision scheme. So the limit surface is C^2 continuous everywhere except that it is C^1 continuous at extraordinary vertices.

Property 5.2 If the initial mesh is regular, the limit surface obtained by using the skirt-removed scheme is bi-cubic B-spline surfaces whose control mesh is the initial mesh.

Figure 5.9 gives examples that construct surfaces by using skirt-removed scheme. Obviously, the boundary curves of the limit surface obtained by using the skirt-removed scheme are not uniform cubic B-spline curve. As for the problem how to construct n-sided patches with non-uniform B-spline boundaries, we will discuss it in the following sections.

5.5.2 Construct n-Sided Patches by Using Non-uniform C-C Subdivision Scheme and Skirt-Removed Approach

From introductions of the C-C subdivision scheme and the skirt-removed approach, it can be known that the skirt-removed approach is also fit for the non-uniform C-C

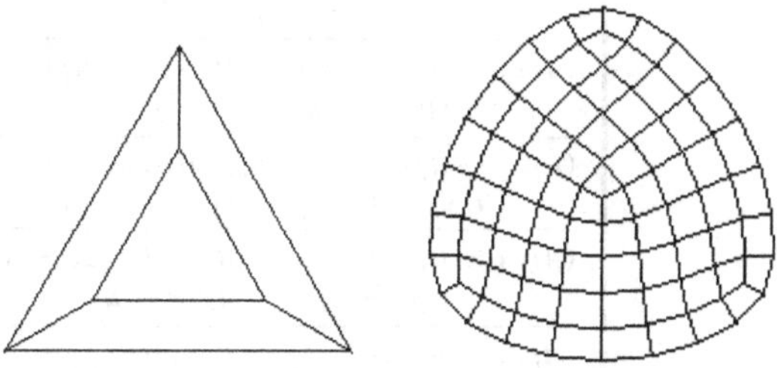

(a) Initial mesh

(b) Subdividing it three times by using the skirt removed scheme

Fig. 5.9 Skirt-removed scheme

subdivision scheme. In order to make the case in the non-uniform C-C subdivision scheme similar to the case in the uniform C-C subdivision scheme, we have to extend the initial mesh M^0 a layer outward and obtain the mesh M^{0+}; such an operation makes the regular vertices in the boundaries become the inner regular vertices. After the operation, we evaluate parameter values for these edges in the extended layer. And these edges neighbor to M^0. After subdivision, we obtain the mesh M^{1+} by deleting the extended layer before subdivision. Again, delete the outmost layer of M^{1+} and the mesh M^1 that we need. Regarding the outermost layers of M^{1+} as the extended layer of M^1, we can subdivide M^1 in the way of subdividing M^0. For convenience, we denote the non-uniform subdivision operator as S_C, denote the skirt-removed operator as R, and denote the combination $R \circ S_C$ of the two operators as S_{RC}. We call the combination operator as the non-uniform skirt-removed scheme.

Note that for B-spline surface without multiple knots, when we insert the middle point of the knot interval for every interval, changes of control mesh are as same as changes of those meshes by using the non-uniform subdivision scheme. Consequently, we give the convention: Except specific explained, for edges in the regular layer near the boundaries and edge neighboring to the regular layer, their parameter values are not 0.

In order to give algorithm to construct patches which do not interpolate mesh corner vertices, we give the following two theorems at first.

Theorem 5.1 *For a cubic B-spline curve* $\boldsymbol{p}(u) = \sum^3 \boldsymbol{P}_i N_i(u)$ *where* $u \in [u_3, u_4]$ *and* $U = [u_0, u_1, \ldots, u_6, u_7]$ *is its knot vector, there exist the following conclusions: if we do not alter the values of* $u_1, \ldots, u_6$*, no matter how we take the values of* u_0 *and* u_7*, the curve will not change; if we give the value of* $u_1, \ldots, u_5$*, and control vertices* $\boldsymbol{P}_0, \boldsymbol{P}_1, \boldsymbol{P}_2$*, no matter how we take the value of* u_0, u_6, u_7 *and* $\boldsymbol{P}_3$*, the endpoint* $\boldsymbol{p}(\boldsymbol{u}_3)$ *of the curve will not change.*

Prove: considering that the segmentation expression of the non-uniform cubic B-spline basic function:

$$N_{i,3}(u) = \begin{cases} \dfrac{(u-u_i)^3}{(u_{i+3}-u_i)(u_{i+2}-u_i)(u_{i+1}-u_i)} & u_i \leqslant u < u_{i+1} \\ \dfrac{(u_{i+2}-u)(u-u_i)^2}{(u_{i+2}-u_i)(u_{i+2}-u_{i+1})(u_{i+3}-u_i)} + \\ \dfrac{(u_{i+3}-u)(u-u_{i+1})(u-u_i)}{(u_{i+3}-u_{i+1})(u_{i+2}-u_{i+1})(u_{i+3}-u_i)} + & u_{i+1} \leqslant u < u_{i+2} \\ \dfrac{(u-u_{i+1})^2(u_{i+4}-u)}{(u_{i+3}-u_{i+1})(u_{i+2}-u_{i+1})(u_{i+4}-u_{i+1})} \\ \dfrac{(u_{i+3}-u)^2(u-u_i)}{(u_{i+3}-u_{i+2})(u_{i+3}-u_{i+1})(u_{i+3}-u_i)} + \\ \dfrac{(u_{i+4}-u)(u_{i+3}-u)(u-u_{i+1})}{(u_{i+4}-u_{i+1})(u_{i+3}-u_{i+1})(u_{i+3}-u_{i+2})} + & u_{i+2} \leqslant u < u_{i+3} \\ \dfrac{(u_{i+4}-u)^2(u-u_{i+2})}{(u_{i+4}-u_{i+1})(u_{i+4}-u_{i+2})(u_{i+3}-u_{i+2})} \\ \dfrac{(u_{i+4}-u)^3}{(u_{i+4}-u_{i+1})(u_{i+4}-u_{i+3})(u_{i+4}-u_{i+2})} & u_{i+2} \leqslant u \leqslant u_{i+3} \\ 0 & u \notin [u_i, u_{i+4}] \end{cases}$$

It is easy to know that the conclusion exists.

Theorem 5.2 *Assume that M be the control mesh of a non-uniform bi-cubic B-spline curve $\mathbf{p}(u, v)$, there are multiple knots at most in the 4 knots, respectively, in the start position and the end position of the two knot vectors of $\mathbf{p}(u, v)$. After evaluating parameter for edges of the control mesh according to knot vectors of $\mathbf{p}(u, v)$, by repeatedly applying the non-uniform C-C subdivision operator and skirt-removed operator to the mesh, the obtained limit surface is $\mathbf{p}(u, v)$, i.e.,*

$$\lim_{j\to\infty} S_{RC}^j(M) = \mathbf{p}(u, v)$$

Proof Apply the knot insertion algorithm [222] to the surface $\mathbf{p}(u, v)$. We insert the middle point for every knot interval. From literatures [40, 222], we can know that the following two facts exist:

(i) After inserting knots, every face, every inner edge, and every inner vertex of M, respectively, corresponds to vertices in the mesh M', these corresponding relating expressions are, respectively, equivalent to formulas (2.1)–(2.3). Let the mesh with matrix form in topology formed by these vertices in M' be M^1;

(ii) According to the relationship between M^1 and the new knot vector after inserting knots, the B-spline surface determined by M^1 and the new knot vector is $\mathbf{p}(u, v)$.

Denote the knot insertion operation as I, let $M = M^0$ and $I(M^0) = M^1$. According to the above second fact, there is $\lim I^j(M^0) = \mathbf{p}(u, v)$;

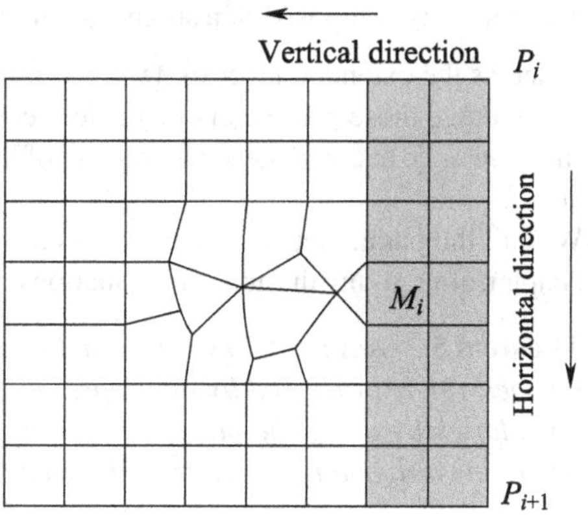

Fig. 5.10 Mesh which the most outside three layers are regular

According to the first fact, there again is $I(M^0) = S_{RC}(M^0)$, so

$$\lim_{j\to\infty} S_{RC}^j(M^0) = \lim_{j\to\infty} I^j(M^0) = \boldsymbol{p}(u, v),$$

i.e.,

$$\lim_{j\to\infty} S_{RC}^j(M) = \boldsymbol{p}(u, v)$$

Now consider the problem how to construct a non-uniform C-C subdivision patch with B-spline boundaries. From properties of B-spline curves, it is well known that if there are no multiple knots in the knot vector, any endpoint of the curve is determined by the first three control vertices of the corresponding end of the control polygon. So, we assume that at least these vertices in the outmost 3 layers of the mesh M should be regular, as shown in Fig. 5.10. According to the assumption, we carry on an average process for parameters corresponding to the edges in the outmost 3 layers of M and edges neighboring to the outmost 3 layers:

(i) Assume that corner vertices of the mesh M be $\boldsymbol{P}_i (i = 0, \ldots, n-1)$ and denote the boundary polygon between $\boldsymbol{P}_i$ and $\boldsymbol{P}_{i+1}$ as L_i, L_i totally has $m (m \geqslant 5)$ vertices. In the mesh M, there is a submesh with matrix form in topology, the submesh is formed by vertices in the outmost layers including L_i. Denote the submesh as $\overset{i}{M}$. Definite two directions in the submesh $\overset{i}{M}$: the row direction (or horizontal direction) in which $\boldsymbol{P}_i$ and $\boldsymbol{P}_{i+1}$ are. The other direction is column direction (or vertical direction). Take $\overset{i}{M}$ and edges neighboring to the submesh from the mesh M. These edges can be classified into two classes: vertical edges (total 3 rows) and horizontal edges (total m-1column, 3 edges in every column). Call $\overset{i}{M}$ as ith side of mesh M. Notice that $\overset{i}{M}$ and faces neighbor to the mesh corresponds to a patch on the limit surface; we call the patch as ith side of the limit surface. In the extended layer of M, there is a

part that M is composed of a submesh with matrix form in topology. The part is also called as the extended layer of $\overset{i}{M}$ in M.
(ii) Average these parameters of vertices edges in the same row in turn;
(iii) If $m > 5$, average these parameters of horizontal edges in jth column ($3 \leqslant j \leqslant m-2$).
We call the parameters average process as the parameters average process I. In this chapter, after giving the above explanations, we can obtain the following conclusions:

Theorem 5.3 *Assume that vertices in the outermost layer of the mesh M are regular, the mesh M is processed by the steps: (1) Average these parameters of edges near boundaries by using the parameters average process I; (2) extend the mesh M a layer outward, and parameters of these edges in the extended layer are equal, these edges neighbor to $\overset{i}{M}$. Under the above conditions, if we apply the non-uniform skirt-removed scheme to the mesh M, the obtained limit surface S has B-spline boundaries.*

Proof Denote the vertex array of $\overset{i}{M}$: as $\{\boldsymbol{P}_{s,t} | s = 0, 1, 2; t = 0, \ldots, m-1\}$, denote parameters of edge $\boldsymbol{P}_{1,t}\boldsymbol{P}_{1,t+1} (t = 0, \ldots, m-2)$ as d_t. Denote the parameters of edges neighboring to $\overset{i}{M}$ in the extended layer of $\overset{i-1}{M}$ as d_0, denote the parameter of edges neighboring to $\overset{i}{M}$ in the extended layer of $\overset{i+1}{M}$ as d_{m-1}. Denote the parameters of edge $\boldsymbol{P}_{s,0}\boldsymbol{P}_{s+1,0} (s = 0, 1)$ as e_s, parameters of edges neighboring to $\overset{i}{M}$ in the extended layer of $\overset{i}{M}$ as e_0, parameters of edges neighboring to $\overset{i}{M}$ in M as e_1. take knot vector $\boldsymbol{U} = [u_0, u_1, \ldots, u_{m+2}, u_{m+3}]$, $\boldsymbol{V} = [v_0, v_1, \ldots, v_5, v_6, v_7]$ where $u_1 = 0$, $u_k = \sum^{k-3} d_j (k = 2, \ldots, m+2)$, u_1, u_{m+3} can be taken as arbitrary values that make $\boldsymbol{U}$ be a non-decrement series. $v_2 = 0$, $v_k = \sum^{k-3} e_j (k = 2, \ldots, 5)$, v_0, v_6, and v_7 are arbitrary real numbers that make $\boldsymbol{V}$ be a non-decrement series. Denote vertices neighboring to $\overset{i}{M}$ in the mesh M as $\boldsymbol{P}_{4,t} (t = 0, \ldots, m-1)$. Consider the mesh formed by $\{\boldsymbol{P}_{s,t} | s = 0, \ldots, 3; t = 0, \ldots, m-1\}$. For horizontal edges in the last row of the mesh formed by $\{\boldsymbol{P}_{s,t} | s = 0, \ldots, 3; t = 0, \ldots, m-1\}$, the parameter of every edge is equal to the parameters of those edges in column in which the edge is. We also denote the mesh formed by $\{\boldsymbol{P}_{s,t} | s = 0, \ldots, 3; t = 0, \ldots, m-1\}$ as $\overset{i}{M}$. Extend $\overset{i}{M}$ a layer downward, the parameter values of the extending edges are v_6–v_5. So, the non-uniform bi-cubic B-spline surface determined by the mesh M, for knot vector $\boldsymbol{U}, \boldsymbol{V}$, is:

$$\boldsymbol{p}(u, v) = \sum^{3} \sum^{m-1} \boldsymbol{P}_{s,t} N_s(v) N_t(u)$$

where $(u, v) \in [u_3, u_m] \times [v_3, v_4]$. Now, we repeatedly apply the operator S_{RC} to the mesh M and $\overset{i}{M}$. Denote the meshes obtained after jointly applying the operator S_{RC}, respectively, as M^j and $\overset{i}{M}{}^j$. Notice that the vertices in the outmost 3 rows of side three rows of M^j are the same as the vertices in the front 3 rows of $\overset{i}{X}{}^j$. According to

Theorem 5.2, when $j \to \infty$, vertices, respectively, in the 3 rows of ith side of M^j for M^j are coincident, in the progress of applying the operator S_{RC}, these vertex rows, respectively, form the boundary curves of ith side of the limit surface of M and a cubic B-spline curve:

$$p(u, v_3) = \sum_{s=0}^{2} \sum_{t=0}^{m-1} P_{s,t} N_s(v_3) N_t(u) \quad u \in [u_3, u_m] \tag{5.9}$$

So the boundary curve of ith side of the limit surface of M is $p(u, v_3)$. Again according to Theorem 3.1, value of u_0, u_{m+3}, v_0, v_6 does not affect the curve $p(u, v_3)$, which is coincident with the uniqueness of limit surface S.

5.6 Non-uniform C-C Subdivision Surface Interpolating Corner Vertices of Control Meshes

The above section illustrates thing that we can obtain limit surfaces with B-spline boundaries by using the non-uniform Catmull–Clark subdivision scheme and the skirt-removed approach. For the initial meshes, the limit surfaces shrink, which is different from the commonly used non-uniform B-spline surfaces— surfaces interpolate corner vertices of control meshes. This section will discuss the problem that how we should design boundaries in order to make a limit surface interpolate corner vertices of its control meshes and its boundaries curve be B-spline curves. When we apply the non-uniform Catmull–Clark subdivision scheme to open mesh, assume that there is a non-uniform cubic B-spline curve:

$$p(u) = \sum_{i=0}^{n-1} P_i N_i(u) \tag{5.10}$$

where $u \in [u_3, u_n]$, $U = [u_0, u_1, \ldots, u_{n+2}, u_{n+3}]$, except the first knot and the last knot with the multiplicity 4, there are no other coincident knots. Obviously, the curve expressed by the expression (5.10) interpolates endpoints of its control polygon. Inspired by the Riesenfeld parameterization methods [21], we give a parameter d_i for every edge $P_i P_{i+1} (i = 0, \ldots, n-2)$ of the control polygon:

$$P_i P_{i+1} \to d_i = u_{i+3} - u_{i+2} \quad i = 0, \ldots, n-2$$

After inserting its middle point for every knot interval by using the knot insertion algorithm, the newly obtained control polygon P^1 has $2n - 3$ vertices, that can be computed by using the following formulas:

$$\begin{cases} \boldsymbol{P}_0^1 = \boldsymbol{P}_0, \boldsymbol{P}_1^1 = (\boldsymbol{P}_0 + \boldsymbol{P}_1)/2, \boldsymbol{P}_{2n-3}^1 = (\boldsymbol{P}_{n-2} + \boldsymbol{P}_{n-1})/2, \boldsymbol{P}_{2n-4}^1 = \boldsymbol{P}_{n-1} \\ \boldsymbol{P}_{2i-1}^1 = \dfrac{(2d_{i+1} + d_i)\boldsymbol{P}_i + (d_i + 2d_{i-1})\boldsymbol{P}_{i+1}}{2(d_{i+1} + d_i + d_{i-1})}, \quad i = 1, \dots, n-3 \\ \boldsymbol{P}_{2i-2}^1 = \dfrac{d_i \boldsymbol{P}_{2i-3}^1 + (d_{i-1} + d_i)\boldsymbol{P}_i + d_{i-1}\boldsymbol{P}_{2i-1}^1}{2(d_i + d_{i-1})} \quad i = 2, \dots, n-3 \end{cases} \tag{5.11}$$

In fact, $\boldsymbol{P}_{2i-1}^1$ is the edge point produced by $\boldsymbol{P}_i\boldsymbol{P}_{i+1}$ and $\boldsymbol{P}_{2i-2}^1$ is the new vertex point produced by $\boldsymbol{P}_i$. Parameters corresponding to every edge of P^1 are given according to the following formulas:

$$\begin{cases} \boldsymbol{P}_{2i}^1 \boldsymbol{P}_{2i+1}^1 = d_{i+1}/2, i = 0, \dots, n-2 \\ \boldsymbol{P}_{2i+1}^1 \boldsymbol{P}_{2i+2}^1 = d_{i+1}/2, i = 0, \dots, n-2 \\ \boldsymbol{P}_1^1 \boldsymbol{P}_2^1 = d_1/2, \boldsymbol{P}_{2n-4}^1 \boldsymbol{P}_{2n-3}^1 = d_{n-1}/2 \end{cases} \tag{5.12}$$

Formulas (5.11) and (5.12) give a subdivision method for the control polygon P. By repeatedly applying the subdivision method to P, the obtained limit curve is the curve $\boldsymbol{p}(u)$ expressed by the expression (5.11).

Again investigate the non-uniform bi-cubic B-spline surface:

$$\boldsymbol{p}(u, v) = \sum_{i=0}^{m-1} \sum_{j=0}^{n-1} \boldsymbol{P}_{i,j} N_i(v) N_j(u) \tag{5.13}$$

where $(u, v) \in [u_3, u_n] \times [v_3, v_m]$, knot vector $\boldsymbol{U} = [u_0, u_2, \dots, u_{n+2}, u_{n+3}]$, $\boldsymbol{V} = [v_0, v_2, \dots, v_{m+2}, v_{m+3}]$. Except the first knot and the last knot with the multiplicity 4, there are no other coincident knots in the knot vectors $\boldsymbol{U}$ and $\boldsymbol{V}$. Let $d_j = u_{j+3} - u_{j+2} (j = 0, \dots, n-2)$, $e_i = v_{i+3} - v_{i+2}$ $(i = 0, \dots, m-2)$. Evaluate parameters for every edge of the control mesh P of the surface expressed by (5.13), assignment according to the following way:

$$\begin{aligned} P_{i,j} P_{i,j+1} &\to d_j, i = 0, \dots, m-1, j = 0, \dots, n-2 \\ P_{i,j} P_{i+1,j} &\to e_i, i = 0, \dots, m-2, j = 0, \dots, n-1 \end{aligned} \tag{5.14}$$

For the control mesh P, we investigate the following subdivision way:

(i) Subdivide every row of P once by the subdivision way defined by Formulas (5.3) and (5.4). Connect those new points from a mesh $P^{1'}$ with the topological matrix form, parameters of vertical edges of $P^{1'}$ in the row direction and equivalent to parameters of vertical edges of P in the corresponding row;

(ii) Subdivide every column of P once by the subdivision way defined by the expressions (5.11) and (5.12), connect all those new points and form a mesh P^1 with the topological matrix form, parameters of horizontal edges of P^1 in the column direction are equivalent to parameters of horizontal edges of $P^{1'}$ in the corresponding column.

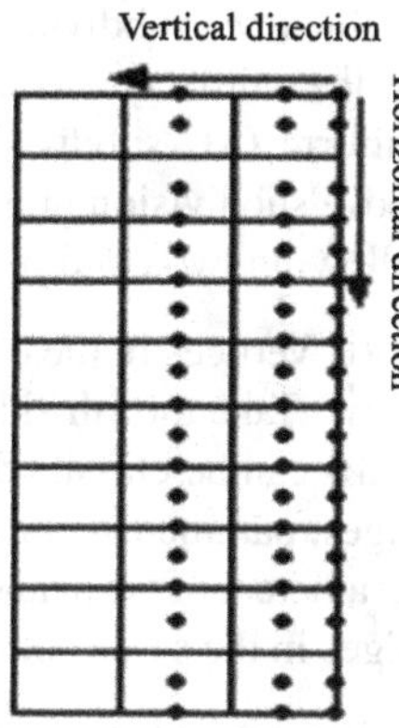

Fig. 5.11 Subdivision operators T on mesh

Denote the subdivision operator as T, i.e., $T(P) = P^1$, so the conclusion is obvious:

$$\lim_{j\to\infty} T^j(P) = \boldsymbol{p}(u, v)$$

From Formula (5.3), it can be known that the conclusion is right: For the submesh obtained by deleting the outmost two layers of the mesh P, its every surface, edge, and vertex has corresponding vertex in the mesh P^1. The corresponding relation formulas are equivalent to Formulas (2.1)–(2.3). Consequently, we can obtain P^1 by subdividing P according to the following ways:

(i) Assume that P has n corner vertices. Take the ith side ($i = 0, \ldots, n-1$) of P, extend P_i a layer downward and obtain a $4 \times k$ mesh with the topology matrix form: still denote the mesh as P_i, parameters of vertical edges in the third row of P_i are the parameters of edges neighboring to P_i in P, parameters of a horizontal edge in the fourth row are equal to the parameters of its preceding 3 edges in the column in which the horizontal edge is;

(ii) Subdivide the $4 \times k$ mesh P_i with the topology matrix form by using the subdivision operator T. However, we only compute the front 3 vertices when we subdivide P_i in column direction, the 3 vertices are, respectively: New point corresponds to the first vertex; new point corresponds to the first and the second edges. After such a subdivision step, we obtain a $3 \times (2k-3)$ mesh P_i^1 with topological matrix form, as shown in Fig. 5.11;

(iii) For vertices in the third row of P_i^1, if we delete the first two vertices and the last two vertices, every vertex successively corresponds to a face or vertical edge in the second row of P_i. Record these corresponding relations;

(iv) For those remained vertices, edge, and faces after deleting vertices in the outermost two layers and their neighboring edges of P compute their corresponding new points according to Formulas (2.3), (2.2), and (2.1), connect these new points and the point recorded in the previous step according to the subdivision topological rules, so the mesh P^1 is obtained.

Since we subdivide the vertices, edges, and faces in the mesh region surrounded by the third layer (including the third layer) of the mesh according to the non-uniform C-C subdivision scheme, when we subdivide the mesh M by using the above subdivision progress, consequently, for an ordinary mesh M, if it satisfies the following two conditions:

(i) Vertices in the outmost three layers are regular;

(ii) Take the ith side $M_i$of M and edges neighbor to M_i. Edges in M_i neighbor to M_i can be classified into two types, 3 rows horizontal edges and 3 rows vertical edges; parameters corresponding to them satisfy the relation Formula (5.14), that is, parameters of vertical edges in the same row are equal and parameters of horizontal edges in the same column are equal.

We can subdivide the mesh M by using the subdivision process. We call the subdivision process as the subdivision operator S_{CT}. According to the above two conditions, for the mesh M whose vertices in the outmost 3 layers are regular and whose every edge has parameters, in order to apply the subdivision operator S_{CT} to the mesh, we have to process these parameters corresponding to edges of the mesh according to the following ways:

(i) Average those parameters corresponding to edges in the outmost 3 regular layers and edges neighboring to these regular layers once by using the parameter average process I;

(ii) For the ith side $\overset{i}{M}(i = 0, \ldots, n-1)$ of M, we add the parameters of the vertical edges in the first row to the parameters of the vertical edges in the second row. At the same time, the parameters of the vertical edges in the first row are set as 0.

Call the above parameters handling process as the parameter average process II. By using these explanations, we can have:

Theorem 5.4 *Assume that vertices in the outmost layers of the mesh M are regular, parameters of edges of the mesh M are handed by using the parameters average process II, when we repeatedly apply the subdivision operator to the mesh M, the limit surface S has the following properties*

(1) Interpolate corner vertices of the mesh M, the boundary curve of the ith side of the limit surface is a uniform B-spline curve.

(2) Assume that we obtain the mesh M' by deleting the outmost layer of the mesh M. Regarding the outmost layer of M as an extension of M', if we obtain a limit surface S' by repeatedly applying S_{RC} operator to M', there is $C(S) = C(S')$ where $C(\cdot)$ denotes the operator taking the continuity degree.

Proof Firstly, prove the first part of the theorem. Denote the vertex array of the mesh $\overset{i}{M}$ as $\{\boldsymbol{P}_{s,t} | s = 0, 1, 2; t = 0, \ldots, m-1\}$. Denote parameters corresponding to edges $\boldsymbol{P}_{0,t}\boldsymbol{P}_{0,t+1}(t = 0, \ldots, m-2)$ as d_t, parameters corresponding to edges $\boldsymbol{P}_{s,1}\boldsymbol{P}_{s+1,1}(s = 0, 1)$ as e_s, parameters corresponding to edges neighboring to $\overset{i}{M}$

in M as e_2. Take knot vector $\boldsymbol{U} = [u_0, u_1, \ldots, u_{m+2}, u_{m+3}]$, $\boldsymbol{V} = [v_0, v_1, \ldots, v_5, v_6]$ where $u_0 = u_1 = u_2 = 0$, $u_k = \sum^{k-3} d_j / \sum^{m-1} d_j (k = 3, \ldots, m+1)$, $u_{m+2} = u_{m+3} = 1$, $v_0 = v_1 = v_2 = 0$, $v_k = \sum^{k-4} e_j$, $(k = 3, \ldots, 5)$, v_6 is taken as an arbitrary real number that makes $\boldsymbol{V}$ be a non-deceasing sequence. Extend M_i two layers downward and obtain a $5 \times m$ mesh, which is still denoted as $\overset{i}{M}$, whose vertex array is denoted as $\{\boldsymbol{P}_{s,t} | s = 0, \ldots, 4; t = 0, \ldots, m-1\}$. Add two vertices v_7, v_8 to $\boldsymbol{V}$, take $v_8 = v_7 = v_6 = v_5$. Obviously, for the surface,

$$\boldsymbol{p}(u, v) = \sum_{s=0}^{4} \sum_{t=0}^{m-1} \boldsymbol{P}_{s,t} N_s(v) N_t(u), \ (u, v) \in [u_3, u_n] \times [v_3, v_5]$$

the boundary curve on it is:

$$\boldsymbol{p}(u, v_3) = \sum_{s=0}^{2} \sum_{t=0}^{m-1} \boldsymbol{P}_{s,t} N_s(v_2) N_t(u) \quad u \in [u_3, u_m] \tag{5.15}$$

When we apply the operator S_{CT} on M, the operator T is also applied to the $5 \times m$ mesh $\overset{i}{M}$. Notice that vertices in the first three rows of the ith side of $S_{CT}^j(M)$ are as same as vertices in the first three rows of $T^j(\overset{i}{M})$. When $j \to \infty$, $T^j(\overset{i}{M})$ forms the surface $\boldsymbol{p}(u, v)$, the boundary curve of the ith side of the limit surface is $\boldsymbol{p}(u, v_3)$ and two deep color regions denote $\overset{i}{M}{}^2$ the submesh of deep color area demined a spline surface.

According to Theorem 3.1, how we pick the value for u_0, u_{m+3}, v_0 and v_6. The curve $\boldsymbol{p}(u, v_3)$ will not be affected, which is coincident with the uniqueness of limit surface S. Now prove the second part of the theorem.

Firstly, we subdivide the mesh M twice by using the operator S_{CT} and we obtain the mesh M^2, obviously, vertices in the outmost layers of the mesh M^2 are regular, as shown in Fig. 5.12.

Similar to the definition of the side of the mesh M. Notice that $\overset{i}{M}$ and its neighboring edges form a $7 \times k$ mesh with topological matrix mesh. Denote the mesh as $\overset{i}{m}{}^2$; from subdivision rules, it is known that parameters of vertical edges in the same row of $\overset{i}{M}{}^2$ are equal and parameters of horizontal edge in the same column of $\overset{i}{M}{}^2$ are equal. Parameters of edges neighboring to $\overset{i}{M}{}^2$ are also equal. Take knot vector $\boldsymbol{U}^2$ and $\boldsymbol{V}^2$ by the method similar to that in the proof of the first part of the theorem. For the convenience of discussion, though $\boldsymbol{V}^2$ only needs 11 knots when we construct a B-spline surface by using $\overset{i}{M}{}^2$ as a control mesh, we take 12 knots here. So, the mesh $\overset{i}{M}{}^2$ and knot vector $\boldsymbol{U}^2$, $\boldsymbol{V}^2$ determine a bi-cubic B-spline surfaces denoted as S_i. Again, take the mesh in the deep color region in Fig. 5.12. According to the

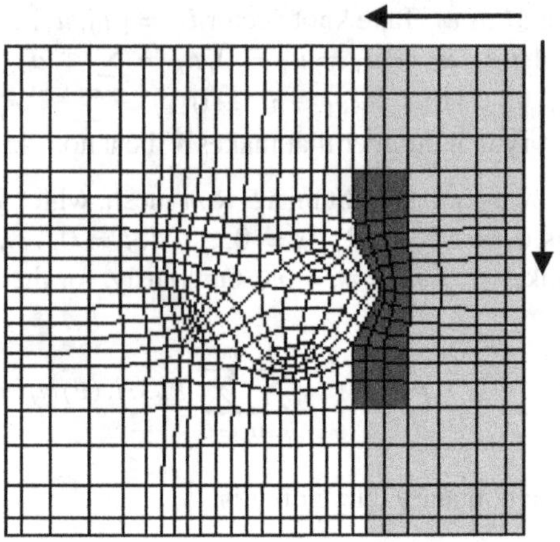

Fig. 5.12 Vertices in the outmost seven layers of M^2 are regular light color layers of region

corresponding relation of vertices in M^2 for B-spline functions, the little mesh also determines a spline surface denoted as S_i', which joins S_i with G^2 continuity. From the used subdivision method, it is known that the boundary on which S_i' and S_i join is the boundary of the ith side of the surface S_i' (though the whole S_i' is not a part of S). Considering that $C(S') \leqslant G^2$ and $C(S_i) = G^2$ according to the appointment in this chapter, there is $C(S) = C(S')$. #

The above theorem not only explains boundary properties of limit surfaces. The limit surfaces can be divided into two parts: B-spline surface ring near the boundaries and the surfaces constructed by using the skirt-removed approach, which in fact converts the continuity analysis problem of the surface S' into continuity analysis problem of the non-uniform Catmull–Clark surface on a closed mesh [63]. At last, we especially point out 2 points:

(1) If we only have to demonstrate the fact that the boundary of ith side of $S_{CT}^j(M)$ is a B-spline curve which interpolates corner vertices of the mesh M, it is not necessary to use the expressions (5.15). We write the expressions (5.15) in order to demonstrate the fact that the cross derivatives on boundary curves of limit surfaces also have properties of B-spline surfaces.

(2) This section requires meshes to satisfy stricter conditions: not only vertices in the outermost are regular, but also there are at least 5 vertices on the boundaries of every side. For a mesh with only two outermost regular layers or sides whose boundaries have a vertex number less than 5, if we do not require the limit surface that interpolate in corner vertices of its control mesh, we may firstly subdivide it once or twice by using the S_{RC} operator and make a parameter average process. If we want the limit to interpolate in corner vertices, we can only use the parameters average process II and S_{CT} operator after subdividing the mesh by introducing other subdivision rules near boundary.

A simplest method is as follows: when we subdivide the mesh, the new vertex points of boundary vertices are themselves, the new edge points of boundary edges and edges with a boundary vertex as endpoints are the average values of two endpoints of every edge, the new face points of every face are averages of all vertices of every face, and other new points are computed according to subdivision formulas of the used subdivision scheme. If we use the method to subdivide a mesh, we only subdivide the mesh twice and the subdivided meshes can satisfy requirements.

5.7 Some Examples to Construct Patches by Using the Non-uniform C-C Subdivision Scheme

Here, we design two examples—quadrilateral mesh and pentagon mesh. In order to illustrate the fact that the patches constructed by using the methods in this chapter have B-spline boundaries, we construct B-spline patches which join the two limit patches by subdividing quadrilateral meshes. The methods by which we construct control meshes of spline surfaces are to extend the ith side $\overset{i}{M}$ of the mesh M outward. For the limit patch obtained by using the S_{RC} operators, extend $\overset{i}{M}$ a layer outward and take the whole side as the control mesh; for the limit patch by using the S_{CT} operators, extend $\overset{i}{M}$ three layers outward and take the mesh formed by the first layer of $\overset{i}{M}$ and the three extended layers as the control mesh. Computing results are shown in Figs. 5.13, 5.14 and 5.15. In these example figures, light color meshes represent initial meshes or B-spline surfaces, and deep color mesh represents subdivision surfaces.

Remarks

Construction of n-sided patches is the fundamental of surface blending. It has important applications in CAD field and is a classical research content in CAGD field. Since the subdivision surface modeling technique has prevailed, the construction of n-sided patches is no more a difficult problem. Around the problem to construct

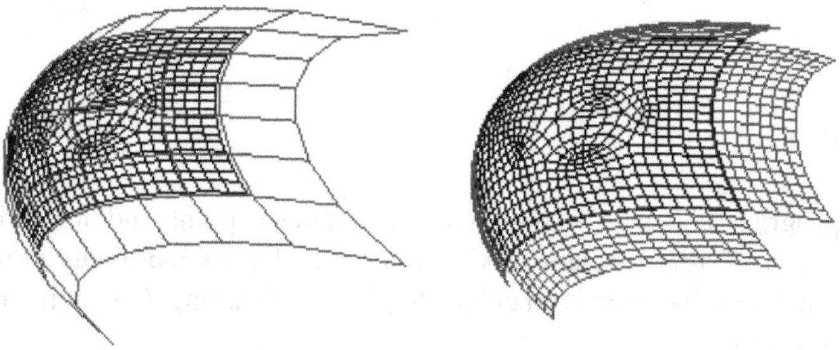

(a) Surface shrinks to its control mesh (b) Surface with B-spline boundaries

Fig. 5.13 Surface by using the non-uniform C-C subdivision scheme and the skirt-removed approach

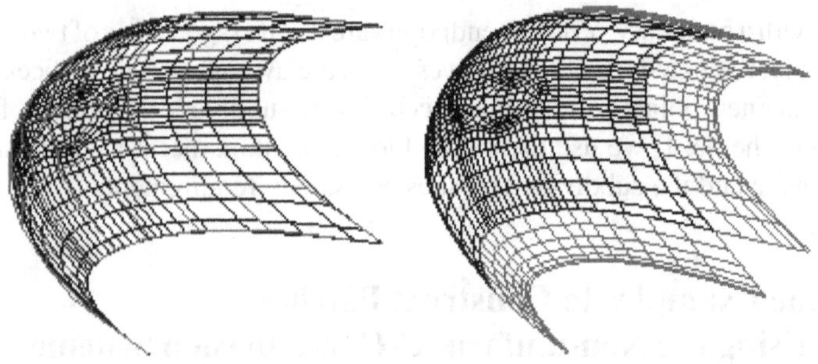

(a) Surface interpolate corner vertices of meshes (b) Surface with B-spline boundary

Fig. 5.14 Non-uniform C-C subdivision surface interpolates corner vertices of its control meshes

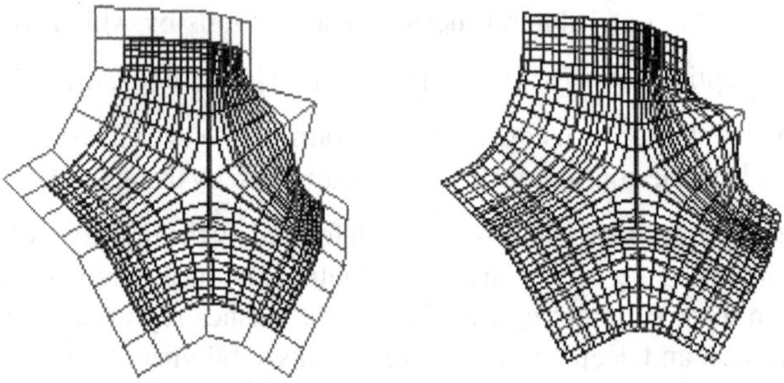

(a) Patches constructed by using the skirt removed approach (b) Surface constructed by introducing boundary subdivision rules

Fig. 5.15 Non-uniform C-C subdivision patches on open mesh

n-sided patches by using the subdivision methods, researchers' efforts focused on two aspects: one is how to construct patches with G^2 continuity and the other is how to blending other surfaces by using subdivision patches, from which many research works and literatures appear. This chapter classifies those methods in related subsections.

Exercises

1. Write program codes to extract an extraordinary point and its k-ring ($k = 1, 2, 3$) from a Catmull–Clark subdivision mesh. The extraordinary point and its 3-ring can determine a uniform cubic B-spline surface ring. Compute the surface ring and render it.
2. Write program codes to subdivide an open mesh using the skirt-removed approach under the Catmull–Clark subdivision scheme.

Chapter 6
Energy Optimization Method and Subdivision Surfaces

An important research content of surface modeling is to construct fairing curves and surfaces, and the optimization modeling is an important method to construct fairing curves and surfaces, so which has been always being focused by many researchers. Starting from the model of Canadian scholar Terzopoulos, namely the optimization model based on physical energy in later 1980s [171, 172], many optimization models based on the geometric sense have bloomed. In some literatures, the two optimization modeling methods are classified into two categories [173]: physical PDE (namely the optimization model based on physical energy) and geometric PDE [174, 175]. Generally, the optimization model based on the geometric sense is more complicated than the physical PDE, and it is usually solved by adopting the variation principle. Since physical PDE is easy to solve and can be used for constructing fairing curve, such method has been always being focused by surface modeling researchers. Early in the middle of 1990s, Halstead et al. [176] had introduced the physical PDE into subdivision surface modeling. In this chapter, the position of control vertices of subdivision surfaces is computed by using the physical PDE in order to improve the automation of surface modeling and enhance the fairing of surfaces, and we discuss the purpose of the physical PDE in subdivision surface modeling from two aspects: surface blending and surface interpolation.

6.1 Blending Uniform Bi-Cubic B-Spline Surfaces

6.1.1 Subdivision Surface Blending of B-Spline Patches

Blending of multiple parameter patches is a problem frequently discussed by researchers in the field of CAD/CG. Referring to the survey [177] of Vida et al. the literature [178] discussed a series of existing methods to construct blending

W. Liao et al., *Subdivision Surface Modeling Technology*,
DOI 10.1007/978-981-10-3515-9_6

patches, namely the rolling ball method, the ridge curve-based method, the cut-based method, the partial differential method, the wire-frame method, and the implicit surface method, and additionally pointed out the difficulties when blending parameter surfaces by using these methods. The literature [179] gives a simpler method to blend the quadric parameter surfaces, where its basic idea is to construct the blending surface by a series of curves produced by the base curves. As the subdivision modeling techniques are being increasingly paid attention to by researchers, the subdivision modeling method gradually becomes an effective method to blend parameter surfaces. The literature [180] gives a method to fill n-sided holes by using the combined subdivision scheme. In some occasions of the surface blending (for instance the vertices of different base surfaces are not coincided), the issue of surface blending can be converted into the problem of filling n-sided holes; therefore, the method of the literature can be used for blending parameter surface patches. When constructing the blending patches by using the combined subdivision scheme, we join the blending surface and the base surface G^1 continuity. The literature [177] constructed the blending patches by using the Doo–Sabin subdivision scheme. Such method is similar to the combined subdivision scheme and the blending surfaces join with the base patches at most with G^1 continuity. The literature [178] blends uniform bi-cubic B-spline surfaces by using the C-C subdivision surface. For this method, the three rows of the control vertices of every base patch were used as some control vertices of the initial control mesh of their blending patch, the balance vertices were identified by adopting other ways to obtain the initial mesh, and finally, the blending surface was obtained by using the skirt-removed approach during subdivision. At this time, the blending surface joins with those surfaces with G^2 continuity. Obviously, if only we determine vertices of initial meshes, the subdivision method is easier to construct blending surfaces than those conventional methods. Consequently, it is a very important problem how to determine an initial subdivision mesh when we construct a blending patch by using a subdivision method. Though many literatures give the topology structure of subdivision mesh at present, the other two methods only can determine the control vertices under user's participation except that literature [181] make vertices of the initial mesh entirely determined in the basis of the joining continuity.

In order to solve the problem above and in view of the discussion on blending the uniform cubic B-spline surfaces and the principle of energy optimization in the literature [178], the chapter gives a method to compute the mesh vertices, and thence, the automatic selection of mesh vertices without man-made intervention is achieved. Moreover due to the energy optimization approach solves based on the energy minimum, the blending surface obtained in such condition when the base surface is given randomly, and the blending surface from which has rather good fairness.

Though the method in this section is designed aiming at blending uniform cubic B-spline surfaces, the principle to design the method is also fit for fairing patches constructed in literature [180, 181] and blending non-uniform B-spline surfaces by using the non-uniform subdivision scheme. In the case of surface fairing, it is only needed to subdivide the initial mesh twice at most, keep the vertices relevant to the boundary unchanged, and then adjust the other vertices. In the case of the surface blending, it is needed to firstly determine the topology position of the control vertices, average the corresponding parameters of mesh, and then establish the geometrical position for the vertices determined by the optimization model. As for the method to average the corresponding parameters of some edges, it is given in the later chapter.

6.1.2 Optimization Model Based on Physical Energy

Terzopoulos' energy model is the optimization model based on the physical energy, and it is used widely at present. The model draws the thin elastic deformation equation in elastic mechanics as reference, its general form is

$$E_{curve} = \int (\alpha \boldsymbol{p}_u^2(u) + \beta \boldsymbol{p}_{uu}^2(u) - 2\zeta(u)\boldsymbol{p}(u))du \tag{6.1}$$

$$\begin{aligned} E_{surface} = \iint & [\alpha_{11}\boldsymbol{p}_u^2(u,v) + 2\alpha_{12}\boldsymbol{p}_u(u,v)\boldsymbol{p}_v(u,v) + \alpha_{22}\boldsymbol{p}_v^2(u,v) \\ & + \beta_{11}\boldsymbol{p}_{uu}^2(u,v) + 2\beta_{12}\boldsymbol{p}_{uv}^2(u,v) + \beta_{22}\boldsymbol{p}_{vv}^2(u,v) \\ & - 2\zeta(u,v)\boldsymbol{p}(u,v)]dudv \end{aligned} \tag{6.2}$$

In above equations, $\boldsymbol{p}$ is the surface(or curve) of the parameter u, v to be solved: $\boldsymbol{p}_u$, $\boldsymbol{p}_v$, $\boldsymbol{p}_{uu}$, and $\boldsymbol{p}_{vv}$ are, respectively, one-order and two-order partial derivatives of curve along the direction u and v direction(or u direction), $\boldsymbol{p}_{uv}$ is a mixed partial derivative, α and β are given parameters, and $\boldsymbol{\zeta}$ is a given vector function that represents the imposed loads. In order to construct the B-spline surfaces(curves), the literature [172] discussed in detail the computation of optimization models and effects of every parameter. The computation approach to solve the control vertices of the subdivision surface in this section takes the method given in literature [178] as the base. As for the bi-cubic B-spline surfaces (or cubic B-spline curves), we can assume their expressions, respectively, as follows:

$$\boldsymbol{p}(u) = \sum_{i=0}^{n-1} \boldsymbol{P}_i N_i(u) \tag{6.3}$$

$$\boldsymbol{p}(u,v) = \sum_{i=0}^{n-1}\sum_{j=0}^{m-1} \boldsymbol{P}_{i,j} N_i(u) N_j(v) \tag{6.4}$$

In above expressions, $\boldsymbol{P}_i$, $\boldsymbol{P}_{i,j}$ are, respectively, control vertices of the corresponding surfaces (curves), and $N_i(u)$ and $N_j(v)$ are, respectively, the cubic B-spline basic functions determined by the knot vectors $[u_0, \ldots, u_{n-1}, \ldots, u_{n+3}]$ and $[v_0, \ldots, v_{m-1}, \ldots, v_{m+3}]$. When the imposed load is 0, the energy of the curve expressed by the equation is:

$$E_{curve} = \sum_{i=0}^{n-1} \boldsymbol{P}_i \sum_{j=0}^{n-1} \boldsymbol{P}_j T_{i,j} \tag{6.5}$$

where

$$T_{i,j} = \int_0^1 (\alpha N_i'(u) N_j'(v) + \beta N_i''(u) N_j''(v)) du \tag{6.6}$$

The energy of the surface expressed by the equation is

$$E_{surface} = \sum_{i=0}^{n-1} \sum_{j=0}^{m-1} \boldsymbol{P}_{i,j} \sum_{k=0}^{n-1} \sum_{l=0}^{m-1} \boldsymbol{P}_{k,l} W_{i,j,k,l} \tag{6.7}$$

where

$$W_{i,j,k,l} = \int_0^1 \int_0^1 \begin{bmatrix} \alpha_{11} N_i'(u) N_j(v) N_k'(u) N_l(v) + 2\alpha_{12} N_i'(u) N_j(v) N_k(u) N_l'(v) + \\ \alpha_{22} N_i(u) N_j'(v) N_k(u) N_l'(v) + \beta_{11} N_i''(u) N_j(v) N_k''(u) N_l(v) + \\ 2\beta_{12} N_i'(u) N_j'(v) N_k'(u) N_l'(v) + \beta_{22} N_i(u) N_j''(v) N_k(u) N_l''(v) \end{bmatrix} dudv \tag{6.8}$$

After imposing some constraints on surfaces (or curves) and constructing the optimization model by taking the Formula (6.5) and (6.7) as the target functions, we can then obtain a needed surface(or curve).

6.1.3 Compute Control Vertices of Subdivision Patches

Hereinafter, we introduce the method to blend parameter surfaces presented by literature [178], as shown in Fig. 6.1. Assume that base surfaces are N uniform bi-cubic B-spline surfaces, control vertices of the B-spline surface are denoted as $\{\overset{k}{\boldsymbol{V}}_{i,j} | i = 0, \ldots, n-1;\ j = 0, \ldots, 2m\}$, and directions of arrows are the directions on which the i and j increases; the vertices marked by black dots are undetermined vertices of mesh (for convenience, we call them as new vertices in latter discussion), the center vertex is denoted as $\boldsymbol{V}_0$, and the new vertices between patches $\overset{s}{S}_p$ and $\overset{s+1}{S}_P$ are denoted as $\{\overset{s}{\boldsymbol{V}}_j | j = 0, \ldots, m-1\}$. The mesh in the dotted-line box, namely the one that is composed of $\{\overset{s}{\boldsymbol{V}}_{i,j} | i = 0, \ldots, n-1;\ j = 0, \ldots, 2m\} (k = 0, \ldots, n-1)$ and $\boldsymbol{V}_0$, $\{\overset{s}{\boldsymbol{V}}_j | j = 0, \ldots, m-1\} (s = 0, \ldots, n-1)$, is thence subdivided by adopting the

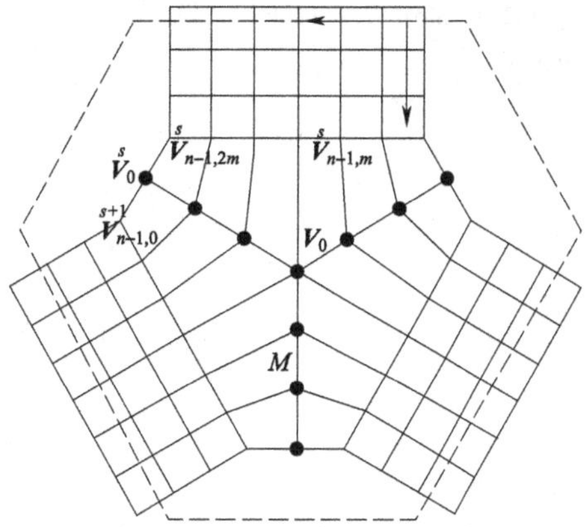

Fig. 6.1 Initial mesh and new vertices

C-C subdivision scheme, and we obtain a limit surface, namely the blending patch that joins with the base surfaces, where the formula to compute new vertices is as follows:

$$
\begin{aligned}
\boldsymbol{V}_0 &= \frac{1}{N}\left(\sum_{s=0}^{N-1}[\overset{s}{\boldsymbol{V}}_{n-1,m} + \alpha(\overset{s}{\boldsymbol{V}}_{n-1,m} - \overset{s}{\boldsymbol{V}}_{n-2,m})]\right) \\
\overset{s}{\boldsymbol{V}}_j &= \frac{1}{2}(\overset{s}{\boldsymbol{V}}_{n-1,2m-j} + \overset{s}{\boldsymbol{V}}_{n-1,j}) + \\
&\quad \overset{s}{\alpha}_j(\overset{s}{\boldsymbol{V}}_{n-1,2m-j} - \overset{s}{\boldsymbol{V}}_{n-2,2m-j} + \overset{s+1}{\boldsymbol{V}}_{n-1,j} - \overset{s+1}{\boldsymbol{V}}_{n-2,j})
\end{aligned}
\tag{6.9}
$$

where $j = 0, \ldots, m-1, s = 0, \ldots, N-1, \overset{s}{\alpha}_j$ is a shape control parameter. In order to be convenient for later discussions, we make the following conventions: Denote $\{\overset{s}{\boldsymbol{V}}_{i,j}|i = n-3, \ldots, n-1;\ j = 0, \ldots, 2m;\ k = 0, \ldots, N-1\}$ as $\{\overset{s}{\boldsymbol{P}}_{i,j}|i = 0, \ldots, 2;\ j = 0 \ldots, 2m;\ k = 0, \ldots, N-1\}$; Denote $\{\overset{s-1}{\boldsymbol{V}}_0, \ldots, \overset{s-1}{\boldsymbol{V}}_{m-1}, \overset{s1}{\boldsymbol{V}}_0, \overset{s}{\boldsymbol{V}}_{m-1}, \ldots, \overset{s}{\boldsymbol{V}}_0\}(k = 0, \ldots, N-1)$ as $\{\overset{s}{\boldsymbol{P}}_{3,j}|j = 0, \ldots, 2m\}(k = 0, \ldots, N-1)$ in turn; unite the vertices array $\{\overset{s}{\boldsymbol{P}}_{i,j}|i = 0, \ldots, 3;\ j = m, \ldots, 2m\}$ and vertex array $\{\overset{s+1}{\boldsymbol{P}}_{i,j}|i = 2, \ldots, 0;\ j = m, \ldots, 0\}$ as $\{\overset{s}{\boldsymbol{Q}}_{i,j}|i = 0, \ldots, 6;\ j = 0, \ldots, m\}$. Denote the middle 5 rows of the vertex array as $\{\overset{s}{\boldsymbol{Q}}{}'_{i,j}|i = 0, \ldots, 4;\ j = 0, \ldots, m\}$, as shown in Fig. 6.2. The vertices array $\{\overset{s}{\boldsymbol{P}}_{i,j}|i = 0, \ldots, 3;\ j = 0, \ldots, 2m\}(k = 0, \ldots, N-1)$ form a mesh that is denoted as the mesh M, that is shown as the mesh in the dot box in Fig. 4.1. Respectively, denote these uniform bi-cubic B-spline surfaces determined by $\{\overset{s}{\boldsymbol{P}}_{i,j}|i = 0, \ldots, 3;\ j = 0, \ldots, 2m\}$, $\{\overset{s}{\boldsymbol{Q}}_{i,j}|i = 0, \ldots, 6;\ j = 0, \ldots, m\}$ and $\{\overset{s}{\boldsymbol{Q}'}_{i,j}|i = 0, \ldots, 4;\ j = 0, \ldots, m\}$ as $\overset{s}{S}_P$, $\overset{k}{S}_Q$ and $\overset{k}{S}_{Q'}$

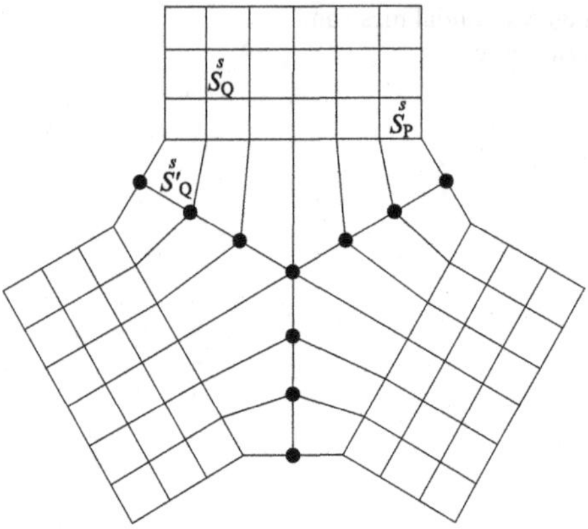

Fig. 6.2 Control vertices of patch $\overset{s}{S}_Q$

6.1.4 *Preliminary Discussion on Selecting New Vertices by Energy Optimization Method*

It is easy to know that $\overset{s}{S}_P(s = 0, \ldots, N-1)$ and $\overset{s}{S}_{Q'}(s = 0, \ldots, N-1)$ form a surface ring (denote it as S). According to C-C subdivision rules, we can know that the surface ring is a part of the limit surface obtained by subdividing the mesh M. Consequently, we can attempt to make the energy of the surface ring S minimum in order to determine the new vertices of mesh M. Denote the energy of the patch $\overset{s}{S}_P$ as $\overset{s}{E}_P$, the energy of the patch $\overset{s}{S}_{Q'}$ as $\overset{s}{E}_{Q'}$, so we can construct the following optimization models:

$$
\begin{aligned}
&\min \sum_{s=0}^{N-1} (\overset{s}{E}_P + \overset{s}{E}_{Q'}) \\
&s.t.\ \overset{s}{\boldsymbol{P}}_{i,j} = \overset{s}{\boldsymbol{V}}_{i+n-3,j} \quad (i = 0, 1, 2;\ j = 0, \ldots, 2m;\ s = 0, \ldots, N-1) \\
&\quad\ \ \overset{s}{\boldsymbol{Q}'}_{i,j} = \overset{s}{\boldsymbol{V}}_{i+n-2,m+j} \quad (i = 0, 1;\ j = 0, \ldots, m;\ s = 0, \ldots, N-1) \\
&\quad\ \ \overset{s}{\boldsymbol{Q}'}_{i,j} = \overset{s}{\boldsymbol{V}}_{i+n-2,m-j} \quad (i = 3, 4;\ j = 0, \ldots, m;\ s = 0, \ldots, N-1) \\
&\quad\ \ \overset{0}{\boldsymbol{P}}_{3,m} = \ldots = \overset{N-1}{\boldsymbol{P}}_{3,m} = \overset{0}{\boldsymbol{Q}'}_{2,0} = \ldots = \overset{N-1}{\boldsymbol{Q}'}_{2,0} \\
&\quad\ \ \overset{s}{\boldsymbol{P}}_{4,m+j} = \overset{s+1}{\boldsymbol{P}}_{4,m-j} = \overset{s}{\boldsymbol{Q}'}_{2,j} (j = 1, \ldots, m;\ k = 0, \ldots, N-1)
\end{aligned}
\tag{6.10}
$$

Obviously, no matter by which way we compute new vertices, edges of the determined mesh M should not cross each other and the obtained surface should satisfy with the fairing criteria. If there is a new vertex set that satisfies the above conditions, we call the new vertex set as an appropriate new vertex set, or as a inappropriate new vertex set otherwise. Unfortunately, the new vertex set computed by using the

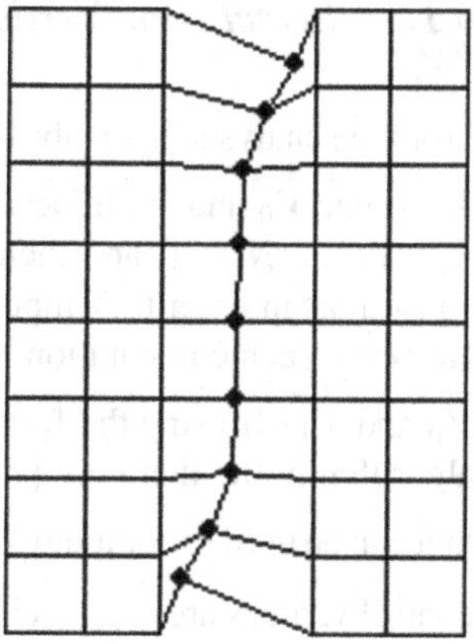

Fig. 6.3 Computed control vertices without constraints

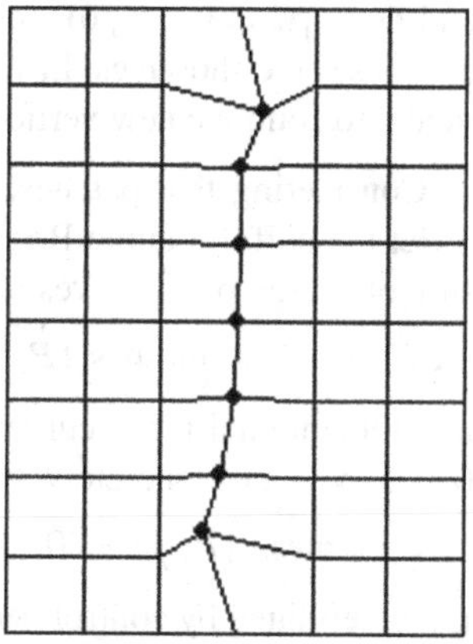

Fig. 6.4 Computed control vertices with constraints

optimization model (6.10) is a inappropriate new vertex set. At least such a new vertex set will not make the constructed surface satisfy with the fairing criteria. The reason of which is that there are no enough constraints. In fact, whichever it is the patch $\overset{s}{S}_P$ or for the patch $\overset{k}{S}_{Q'}$, there is no constraint in column directions (i.e., j directions). In order to explain the problem, we take vertices of the control mesh of a uniform bi-cubic B-spline surface in a plane, as shown in Fig. 6.3, where the vertices in the middle column are undetermined. Vertices marked by black dots are vertices when the formula (6.7) is minimum. When the vertices in the first row and the last row are given, the computed vertices are shown in Fig. 6.4.

6.1.5 Select New Vertices by Energy Optimization Method

From the analysis in the above subsection, it can be seen that we should beforehand determine $\overset{s}{\boldsymbol{V}}_0$ and $\overset{s}{\boldsymbol{V}}_1$ in new vertices at least (or corresponding conditions), where $s = 0, \dots, N-1$, and then add them into the optimization model (6.10) as the constraints in order to compute new vertices by using the optimization model (4.8). However, such computation will bring the following problem: How do we compute $\overset{s}{\boldsymbol{V}}_0$ and $\overset{s}{\boldsymbol{V}}_1$? If using the formula (6.9), our algorithm will have the shortage of the algorithm in the literature [73]. If using the curve-based energy method, the curve that can be used to compute $\overset{s}{\boldsymbol{V}}_j (j = 0, 1)$ is the uniform cubic B-spline curve whose control vertices are $\overset{s}{\boldsymbol{Q}}_{i,m-1} (i = 0, \dots, 6)$, at this time, $\overset{s}{\boldsymbol{V}}_j$ is a linear combination of $\overset{s}{\boldsymbol{Q}}_{i,m-j} (i = 0, \dots, 6, \text{ and } i \neq 3)$. Because the relations between $\overset{s}{\boldsymbol{Q}}_{i,m} (i = 0, \dots, 6)$ and $\overset{s}{\boldsymbol{Q}}_{i,m-1} (i = 0, \dots, 6)$ are not yet considered, the method to determine $\overset{s}{\boldsymbol{V}}_0$ and $\overset{s}{\boldsymbol{V}}_1$ also has some shortages. In view of the reason, this section presents an optimization model to compute new vertices on the basis of the curve-based energy method.

Considering that patches, $\overset{s}{S}_P$ and $\overset{k}{S}_Q$ are constituent parts of the surface ring S, let the uniform cubic B-spline curves in them as possible as fairing, i.e., let the sum of energy of all curves minimum. These curves have two directions: transverse mesh curves of meshes $\{\overset{s}{\boldsymbol{P}}_{i,j} | i = 0, \dots, 3;\ j = 0, \dots, 2m\} (s = 0, \dots, N-1)$ and longitudinal mesh curves of meshes $\{\overset{s}{\boldsymbol{Q}}_{i,j} | i = 0, \dots, 6;\ j = 0, \dots, m\} (s = 0, \dots, N-1)$. Concretely, we classify them as follows:

- for mesh $\{\overset{s}{\boldsymbol{P}}_{i,j} | i = 0, \dots, 3;\ j = 0, \dots, 2m\}$, uniform cubic B-spline curves $\overset{s}{C}_P$ determined by control vertices $\{\overset{s}{\boldsymbol{P}}_{3,j} | j = 0, \dots, 2m\}$.
- for mesh $\{\overset{s}{\boldsymbol{Q}}_{i,j} | i = 0, \dots, 6;\ j = 0, \dots, m\}$, uniform cubic B-spline curves $\overset{s}{C}_{Qj}$ determined by control vertices $\{\overset{s}{\boldsymbol{Q}}_{i,j} | i = 0, \dots, 4\} (j = 0, \dots, m)$.

Might as well, we assume that the energy of $\overset{s}{C}_P$ is $\overset{s}{E}_P$, and the energy of $\overset{s}{C}_{Qj}$ is $\overset{s}{E}_{Qj} (j = 0, \dots, m)$, so we can construct the following optimization model:

$$
\begin{aligned}
\min\ & \sum_{s=0}^{N-1} \left(\overset{s}{E}_P + \sum_{j=0}^{m} \overset{s}{E}_{Qj} \right) \\
s.t.\ & \overset{s}{\boldsymbol{Q}}_{i,j} = \overset{s}{\boldsymbol{V}}_{i+n-3,m+j} (i = 0, 1, 2;\ j = 0, \dots, m;\ k = 0, \dots, N-1) \\
& \overset{s}{\boldsymbol{Q}}_{i,j} = \overset{s}{\boldsymbol{V}}_{-i+n+3,m-j} (i = 4, 5, 6;\ j = 0, \dots, m;\ k = 0, \dots, N-1) \\
& \overset{0}{\boldsymbol{P}}_{3,m} = \dots = \overset{N-1}{\boldsymbol{P}}_{3,m} = \overset{0}{\boldsymbol{Q}}_{3,0} = \dots = \overset{N-1}{\boldsymbol{Q}}_{3,0}
\end{aligned}
$$

$$\overset{s}{\boldsymbol{P}}_{4,m+j} = \overset{s+1}{\boldsymbol{P}}_{4,m-j} = \overset{s}{\boldsymbol{Q}}_{3,j}(j = 1, \ldots, m; s = 0, \ldots, N-1) \tag{6.11}$$

In the above optimization model, there is no any constraint in these column directions on which $\{\overset{s}{\boldsymbol{V}}_j | j = 0, \ldots, m-1\}(s = 0, \ldots, N-1)$ exists. Consequently, here we beforehand determine $\overset{s}{\boldsymbol{V}}_0$ and $\boldsymbol{V}_0(s = 0, \ldots, N-1)$ in order to add constraints to the optimizations model (6.11)

6.1.6 Determine Some New Vertices

For $\overset{s}{\boldsymbol{V}}_0$ on boundaries of the mesh M, it should make the energy of boundary curve $\overset{s}{C}_{Qm}$ minimum. So, we can compute $\overset{s}{\boldsymbol{V}}_0$ by making the value of the following formula minimum:

$$E_{curve} = \sum_{i=0}^{6} \overset{s}{\boldsymbol{Q}}_{i,m} \sum_{j=0}^{6} \overset{s}{\boldsymbol{Q}}_{j,m} T_{i,j} \tag{6.12}$$

where $\overset{s}{\boldsymbol{Q}}_{3,m} = \overset{s}{\boldsymbol{V}}_0$ is unknown variable, Other $\overset{s}{\boldsymbol{Q}}_{i,m}(i = 0, 1, 2, 4, 5, 6)$ are known variables. $T_{i,j}$ is determined by the Formula (6.6). By substituting $\overset{s}{\boldsymbol{Q}}_{3,m}$ by $\overset{s}{\boldsymbol{V}}_0$ in the right part of the Eq. (6.12), so:

$$E_{curve} = \sum_{\substack{i=0\\ i\neq 3}}^{6} \overset{s}{\boldsymbol{Q}}_{i,m} \sum_{\substack{j=0\\ j\neq 3}}^{6} \overset{s}{\boldsymbol{Q}}_{j,m} T_{i,j} + \overset{s}{\boldsymbol{V}}_0 \sum_{\substack{i=0\\ i\neq 3}}^{6} \overset{s}{\boldsymbol{Q}}_{i,m}(T_{i,3} + T_{3,i}) + (\overset{s}{\boldsymbol{V}}_0)^2 T_{3,3}$$

Considering that $\overset{s}{\boldsymbol{V}}_0$ should make E_{curve} minimum, so

$$\frac{dE_{curve}}{d\overset{s}{\boldsymbol{V}}_0} = 2\boldsymbol{V}_0^k S_{3,3} + \sum_{\substack{i=0\\ i\neq 3}}^{6} \overset{s}{\boldsymbol{Q}}_{i,m}(T_{3,i} + T_{i,3}) = 0$$

Consequently, $\overset{s}{\boldsymbol{V}}_0 = -\sum_{\substack{i=0\\ i\neq 3}}^{6} \overset{s}{\boldsymbol{Q}}_{i,m}(T_{3,i} + T_{i,3})/2T_{3,3}$.

For $\boldsymbol{V}_0$, it should make $\sum_{k=0}^{N-1} \overset{s}{E}_{Q0}$ minimum; consequently, by using the way similar to the way to compute $\overset{s}{\boldsymbol{V}}_0$, we have:

$$V_0 = -\sum_{k=0}^{N-1}\sum_{\substack{i=0\\ i\neq 3}}^{6} \overset{s}{Q}_{i,0}(T_{3,i}+T_{i,3})/2NT_{3,3}$$

So, we can compute key vertices and can add constraints to the optimization model (6.11).

6.1.7 Simplify Optimization Model

In order to simplify the optimization model (6.11), we firstly investigate $\overset{s}{E}_P$. Denote $\overset{s}{P}_{3,i}$ as $\overset{s}{P}_i$, we have

$$\overset{s}{E}_P = \sum_{i=0}^{2m}\overset{s}{P}_i\sum_{j=0}^{2m}\overset{s}{P}_i R_{i,j}$$

where $R_{i,j}$ is determined by formula (6.6). substitute $\overset{s}{P}_i = \overset{s-1}{P}_i (i=0,\ldots,m-1)$, $\overset{s}{P}_m = V_0$, and $\overset{s}{P}_i = \overset{s}{V}_{2m-i}(i=m+1,\ldots,2m)$, into the above formula. Consider that V_0 and $\overset{s}{V}_0$ are known variables. By simplifying the expression, we can have:

$$\begin{aligned}
\overset{s}{E}_P = {} & \sum_{i=1}^{m-1}\overset{s-1}{V}_i\sum_{j=1}^{m-1}\overset{s-1}{V}_j R_{i,j} + \sum_{i=1}^{m-1}\overset{s-1}{V}_i\sum_{j=1}^{m-1}\overset{s-1}{V}_j(R_{2m-i,j}+R_{j,2m-i}) + \\
& \sum_{i=1}^{m-1}\overset{s}{V}_i\sum_{j=1}^{m-1}\overset{s}{V}_j R_{2m-i,2m-j} + \sum_{i=1}^{m-1}\overset{s-1}{V}_i[\overset{s-1}{V}_0(R_{0,i}+R_{i,0}) \\
& + V_0(R_{m,i}+R_{i,m}) + \overset{s}{V}_0(R_{2m,i}+R_{i,2m})] + \\
& \sum_{i=1}^{m-1}\overset{s}{V}_i[\overset{s-1}{V}_0(R_{0,2m-i}+R_{2m-i,0}) + \\
& V_0(R_{m,2m-i}+R_{2m-i,m}) + \overset{s}{V}_0(R_{2m,2m-i}+R_{2m-i,2m})] + c
\end{aligned}$$

where c is a constant; for $\overset{s}{E}_{Qj}$, considering that $\overset{s}{Q}_{3,j} = \overset{s}{V}_{m-j}$ is an unknown variable, again,

$$\sum_{i=0}^{m}\overset{s}{E}_{Qi} = \sum_{i=1}^{m-1}\overset{s}{V}_i\sum_{\substack{j=0\\ j\neq 3}}^{6}\overset{s}{Q}_{j,i}(T_{3,j}+T_{j,3}) + \sum_{i=1}^{m-1}(\overset{s}{V}_i)^2 T_{3,3} + c$$

So, if let $\sum_{k=0}^{N-1}\left(\overset{s}{E}_P+\sum_{j=0}^{m}\overset{s}{E}_{Qi}\right)=\sum_{k=0}^{N-1}\overset{s}{\theta}+c$, there is

$$\overset{s}{\theta}=\sum_{i=1}^{m-1}\overset{s}{V}_i\sum_{j=1}^{m-1}\overset{s}{V}_j(R_{j,2m-i}+R_{2m-i,j})+\sum_{i=1}^{m-1}\overset{s}{V}_i\sum_{j=1}^{m-1}$$
$$\overset{s}{V}_j(R_{2m-i,2m-j}+R_{i,j})+\sum_{i=1}^{m-1}(\overset{s}{V}_i)^2T(3,3)+$$
$$\sum_{i=1}^{m-1}\overset{s}{V}_i[V_0(R_{m,2m-i}+R_{2m-i,m}+R_{m,i}+R_{i,m})$$
$$+\overset{s}{V}_0(R_{2m,2m-i}+R_{2m-i,2m}+R_{0,i}+R_{i,0})+$$
$$\overset{s-1}{V}_0(R_{0,2m-i}+R_{2m-i,0})]+\overset{s}{V}_0\sum_{i=1}^{m-1}\overset{s-1}{V}_i(R_{2m+1,i}+R_{i,2m+1})$$
$$+\sum_{i=1}^{m-1}\overset{s}{V}_i\sum_{\substack{j=0\\j\neq 3}}^{6}\overset{s}{Q}_{j,i}(R_{3,j}+R_{j,3})$$

since $\overset{s}{V}_i(i=1,\ldots,m-1;s=0,\ldots,N-1)$, let the value of $\sum_{s=0}^{N-1}\left(\overset{s}{E}_P+\sum_{j=0}^{m}\overset{s}{E}_{Qj}\right)$ minimum, calculate the partial derivate of $\sum_{s=0}^{N-1}\overset{s}{\theta}$ about $\overset{s}{V}_i$, we have:

$$\partial\sum_{k=0}^{N-1}\overset{s}{\theta}/\partial\overset{s}{V}_i=\sum_{j=1}^{m-1}\overset{s-1}{V}_{j-1}(R_{j,2m-i}+R_{2m-i,j})+\sum_{j=1}^{m-1}\overset{s}{V}_j(R_{2m-i,2m-j}$$
$$+R_{i,j}+R_{j,i}+R_{2m-j,2m+i})+2T(3,3)\overset{s}{V}_i$$
$$+\sum_{j=1}^{m-1}\overset{s+1}{V}_j(R_{i,2m-j}+R_{2m-j,i})+\left[\overset{s-1}{V}_0(R_{0,2m-i}+R_{2m-i,0})+\right.$$
$$\overset{s}{V}_0(R_{2m-i,m}+R_{m,2m-i}+R_{0,i}+R_{i,0})+\overset{s+1}{V}_0(R_{2m,i}+R_{i,2m})$$
$$\left.+V_0(R_{m,2m-i}+R_{2m-i,m}+R_{m,i}+R_{i,m})+\sum_{\substack{j=0\\j\neq 3}}^{6}\overset{s}{Q}_{j,i}(T_{3,j}+T_{j,3})\right]=0$$

consequently, $\overset{s}{V}_i(i=1,\ldots,m-1;s=0,\ldots,N-1)$ can be computed by solving a linear system. From the formula (6.6), we can know that the coefficient matrix of the linear system is a symmetrical matrix; therefore, the decomposition method can be used for computing the equation system.

6.1.8 Examples

Compute some examples by using the method described by this section, and then, the results shown in Figs. 6.5, 6.6, 6.7 and 6.8 are obtained. In the examples shown in Figs. 4.6 and 6.5, the base surfaces are symmetrically given. In order to compare effects, respectively, computed by using the method presented in literature [178] and the method presented in this section, the new vertices in Fig. 4.8 are computed by using the formula (6.9), and the base surfaces is as same as the one in Fig. 6.7. It is needed to note the shape-controlled parameter in the example shown in Fig. 6.8 is 0; this is just because it is rather troublesome to adjust these shape-controlled

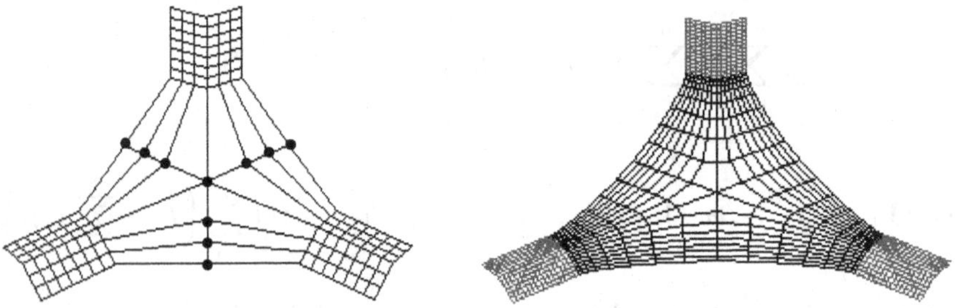

(a) New vertices and initial mesh (b) constructed blending patch

Fig. 6.5 3 base surfaces are given symmetrically

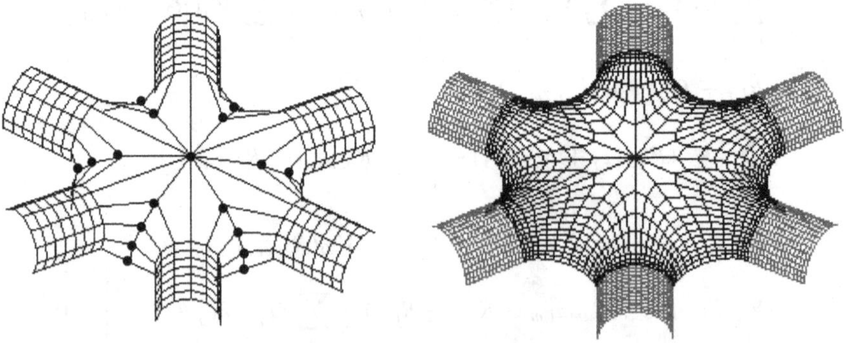

(a) New vertices and initial mesh (b) constructed blending patch

Fig. 6.6 6 base surfaces with the semi-circle cross section are given symmetrically

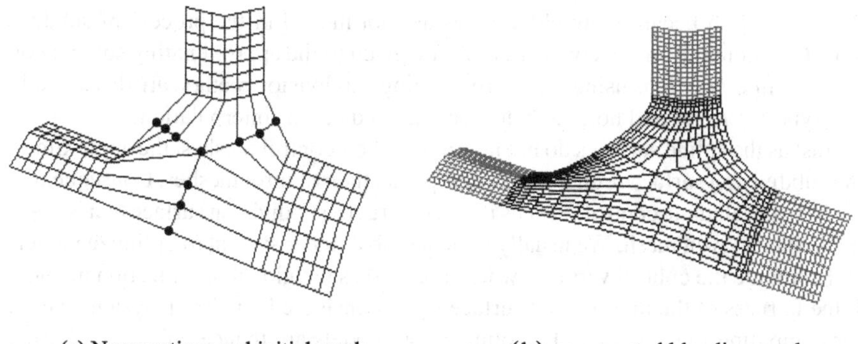

(a) New vertices and initial mesh (b) constructed blending patch

Fig. 6.7 3 base surfaces are given asymmetrically

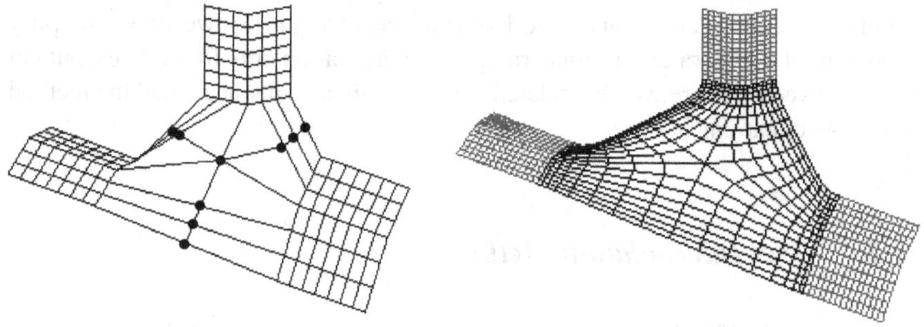

(a) New vertices and initial mesh (b) constructed blending patch

Fig. 6.8 New vertices are computed by using formula (6.11)

parameters and select an appropriate value; moreover, there is no criterion to measure how a shape-controlled parameter is adjusted appropriately.

6.2 Interpolations Using Subdivision Surfaces

6.2.1 Subdivision Surface and Interpolation

Interpolation is a typical problem in CAGD. Interpolation of any topological mesh is a difficult, which is focused by many researchers. The subdivision surface is an efficient tool to solve the problem. There are two major ways to interpolate a given mesh with a subdivision surface: interpolating subdivision [23, 45–47] and approximating subdivision. Comparing with the approximating subdivision method, the interpolating subdivision method is simple since only subdivision rules are recursively executed on a given mesh. However, it is very sensitive to the distortion of the given mesh because any old vertices are not moved in the process of subdivision. Consequently, more concentrations are given to the approximating subdivision method. Interpolations using the approximating subdivision method are discussed in two types: vertices and normals interpolation, and curves interpolation.

Just as the spline surfaces do not interpolate their control meshes, the approximating subdivision surfaces either do not interpolate their control meshes. Consequently, we have to calculate control vertices if we construct a subdivision surface that is interpolated in a given mesh. We usually build a global linear system to optimize models and compute the control vertices, which is called as the global optimization method. If the fairness of the interpolated surface is not considered, the linear system is used for computing the vertices. The optimization models are thence built when fairing interpolatory surfaces are needed. The former is called as the direct interpolation method, and the latter is called as the interpolation method with fairness in this chapter. Those optimization models minimize the physics-based energy. The discussion in this section is based on the work of Halstead et al. [44] However, we simplify the content of the literature by concerning the interpolation only for vertices, but not for normal vector. Moreover, the related equations are also deduced, and the method for solving equations is given.

6.2.2 *Direct Interpolation Method*

According to (4.30), the limit positions of vertices when a quadrilateral mesh is subdivided by the C-C scheme can be calculated as follows:

$$\boldsymbol{v}^{\infty} = \frac{n^2\boldsymbol{v} + 4\sum_{j=0}^{n-1}\boldsymbol{e}_j + \sum_{j=0}^{n-1}\boldsymbol{f}_j}{n(n+5)} \tag{6.13}$$

where $\boldsymbol{e}_j, \boldsymbol{f}_j (j = 0, \ldots, n-1)$ are, respectively, the edge points and face points on the 1-ring of $\boldsymbol{v}$ and n is the valence of $\boldsymbol{v}$, as shown in Fig. 6.9.

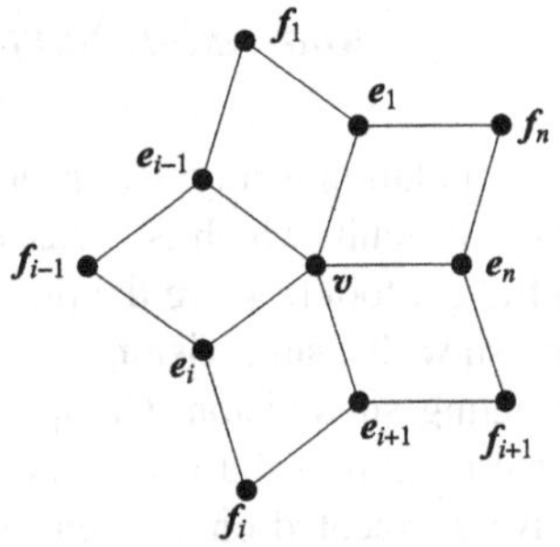

Fig. 6.9 $\boldsymbol{e}_j, \boldsymbol{f}_j (j = 1, \ldots, n)$ are edge points and face points on the 1-ring of $\boldsymbol{v}$ and n is the valence of $\boldsymbol{v}$

Since any two-manifold mesh becomes a quadrilateral mesh after C-C subdivision, we investigate the mesh M^1 so as to build a linear system calculating M^0 by using the formula (6.13). Assume that $\{\boldsymbol{V}_0^1, \ldots, \boldsymbol{V}_{m-1}^1\}$ is the new vertex point set of M^1 and $\{\boldsymbol{q}_0, \ldots, \boldsymbol{q}_{m-1}\}$ is the point set; where I^0 and m is, respectively, the number of the given mesh and the number of vertices of I^0. According to the calculating rules of new points of the C-C scheme, there is a correspondence between $\boldsymbol{V}_i^1$ and $\boldsymbol{q}_i (i = 0, \ldots, m-1)$. Similarly, $\boldsymbol{V}_i^k (i = 0, \ldots, m-1)$ are vertices of $M^k (k \geqslant 0)$, and there is a correspondence between $\boldsymbol{V}_i^k$ and $\boldsymbol{q}_i (i = 0, \ldots, m-1)$. We can build equations by the formula (6.13) if M^∞ is interpolated into I^0:

$$\frac{n_i \boldsymbol{V}_i^1 + 4\sum_j \boldsymbol{E}_{i,j}^1 + \sum_j \boldsymbol{F}_{i,j}^1}{n_i(n_i+5)} = \boldsymbol{q}_i \tag{6.14}$$

$\boldsymbol{V}_i^1$, $\boldsymbol{E}_{i,j}^1$, and $\boldsymbol{F}_{i,j}^1$ can be linearly expressed by $\boldsymbol{V}_0^0, \ldots, \boldsymbol{V}_{m-1}^0$:

$$\begin{aligned} \boldsymbol{v}_i^1 &= \alpha_{i,0}\boldsymbol{V}_0^0 + \ldots + \alpha_{i,m-1}\boldsymbol{v}_{m-1}^0 \\ \boldsymbol{e}_{i,j}^1 &= \beta_{i,j,0}\boldsymbol{V}_0^0 + \ldots + \beta_{i,j,m-1}\boldsymbol{V}_{m-1}^0 \\ \boldsymbol{f}_{i,j}^1 &= \gamma_{i,j,0}\boldsymbol{V}_0^0 + \ldots + \gamma_{i,j,m-1}\boldsymbol{V}_{m-1}^0 \end{aligned} \tag{6.15}$$

Substituting (6.15) for (6.14), so:

$$c_{i,1}\boldsymbol{V}_0^0 + \ldots + c_{i,m-1}\boldsymbol{V}_{m-1}^0 = \boldsymbol{q}_i \tag{6.16}$$

where $i = 1, \ldots, m$. The linear system (4) can be written as

$$\boldsymbol{C}\vec{\boldsymbol{V}} = \vec{\boldsymbol{Q}} \tag{6.17}$$

where $\boldsymbol{C} = \{c_{i,j}\}(i, j{=}1, \ldots, m)$, $\vec{\boldsymbol{V}} = (\boldsymbol{V}_0^0, \ldots, \boldsymbol{V}_{m-1}^0)^{\mathrm{T}}$, and $\vec{\boldsymbol{Q}} = (\boldsymbol{q}_0, \ldots, \boldsymbol{q}_{m-1})^{\mathrm{T}}$. The linear system is possibly a singular or ill-conditioned [44] mesh. The ill-conditioned linear systems are very troublesome because they make us obtain bad control meshes. For example, we obtain the control mesh M^0 shown in Fig. 6.10b via the linear system (6.17); if I^0 is a cubic mesh shown in Fig. 6.10a, M^0 should have the same shape as that of I^0 in this case.

There are probably iterative methods that can obtain "good" results. For example, the method in [61]. In order to overcome the obstacles of singular or ill-conditioned linear systems, we build the optimization model (6.18) instead of using the least-square solution to solve the linear system (6.17). It is unfit to use the least-square solution if the linear system is ill-conditioned, because an ill-conditioned linear system has only an unique solution.

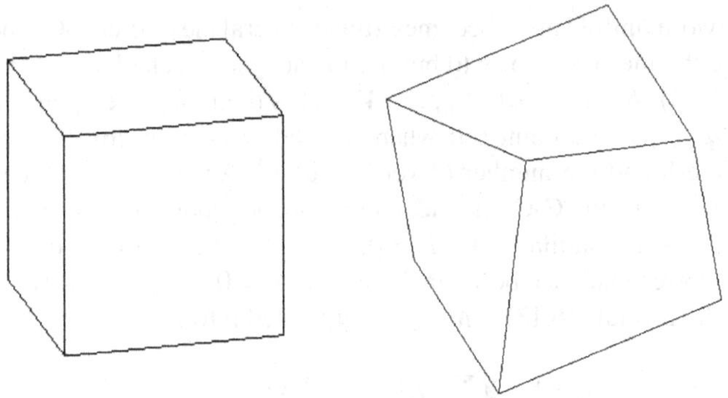
(a) the given mesh I^0
(b) the mesh M^0 computed by the linear system (6.17)
We obtain the result (b) by the Gauss method

Fig. 6.10 Computing effect of an ill-conditioned linear system

$$\begin{aligned} \min \ & (\vec{V} - \vec{Q})^{\mathrm{T}}(\vec{V} - \vec{Q}) \\ \text{s.t.} \ & C\vec{V} = \vec{Q} \end{aligned} \tag{6.18}$$

The optimization model (6.18) is built for computing the desirous shape by investigating the shape of I^0, and we should make the shape of M^0 approach to the shape of I^0 as much as possible. Note that we only want to find an appropriate solution for the linear system (6.17). Consequently, we transform the optimization model (6.18) into the unconstrained optimization model (6.19):

$$\min \ (\vec{V} - \vec{Q})^{\mathrm{T}}(\vec{V} - \vec{Q}) + \mu(C\vec{V} - \vec{Q})^{\mathrm{T}}(C\vec{V} - \vec{Q}) \tag{6.19}$$

The unconstrained optimization model (6.19) can be transformed into the linear system (6.20) by setting its gradient to zero:

$$(\mu C^{\mathrm{T}}C + I)\vec{V} = (\mu C^{\mathrm{T}} + I)\vec{Q} \tag{6.20}$$

where I is a unit matrix. It is easy to know that the solution of Eq. (6.20) can be regarded as the solution of Eq. (6.17) when μ is enough large. The coefficient matrix $\mu C^{\mathrm{T}}C + I$ is symmetric and positive definite. We may get the result as shown in Table 6.1 by calculating M^0 via the linear system 6.20.

In Table 6.1, $(\pm v, \pm v, \pm v)$ are eight vertices of M^0.

Table 6.1 Value of μ and solutions of Eq. (6.16)

μ	10^3	10^4	10^5	10^6	10^7
v	0.998008	0.9998	0.99998	0.999998	1

6.2.3 Interpolation Method with Fairness

Comparing with M^0, $M^k(k \geqslant 1)$ provides extra freedom degrees for the fairness of the interpolatory surface, which enable us to calculate their control vertices by the physics-based energy model (6.21):

$$\begin{aligned} \min \quad & (\vec{V}^k)^{\mathrm{T}} \boldsymbol{E} \vec{V}^k \\ \text{s.t.} \quad & \boldsymbol{C}^k \vec{V}^k = \vec{Q} \end{aligned} \tag{6.21}$$

where $\vec{V}^k = (V_0^k, \ldots, V_{m-1}^k, \ldots, V_{m+n-1}^k)^{\mathrm{T}}$ is the vertex vector of M^k, $m+n$ is the number of vertices of M^k, $\boldsymbol{E} = \{e_{i,j}\}(i = 1, \ldots, m+n;\ j = 1, \ldots, m+n)$ is the energy matrix, and $\vec{Q} = (q_1, \ldots, q_m)^T$ is the vertex vector of I^0. $\boldsymbol{C}^k = \{c_{i,j}^k\}(i = 0, \ldots, m-1;\ j = 1, \ldots, m+n-1)$.

We only can evaluate energy norms for quadrilateral meshes, where any quadrilateral has at most an extraordinary vertex; M^1 does probably not satisfy the condition while any $M^k(k \geqslant 2)$ satisfies the condition, as shown in Fig. 6.11. We can overcome the obstacle by regarding the energy of $M^{k+r}(r \geqslant 0)$ as M^k. In fact,

$$\vec{V}^{k+r} = A^{k,k+r} \vec{V}^k \tag{6.22}$$

where $\boldsymbol{A}^{k,k+r}$ is a coefficient matrix, which is normally expressed by formula 6.22; thence, we have:

$$\begin{aligned} & (\vec{V}^{k+r})^{\mathrm{T}} \boldsymbol{E}^{k+r} \vec{V}^{k+r} \\ = & (\boldsymbol{A}^{k,k+r} \vec{V}^k)^{\mathrm{T}} \boldsymbol{E}^{k+r} (\boldsymbol{A}^{k,k+r} \vec{V}^k)^k \\ = & (\vec{V}^k)^{\mathrm{T}} ((\boldsymbol{A}^{k,k+r})^{\mathrm{T}} \boldsymbol{E}^{k+r} \boldsymbol{A}^{k,k+r}) \vec{V}^k \\ = & (\vec{V}^k)^{\mathrm{T}} \boldsymbol{E}^k \vec{V}^k \end{aligned}$$

Consequently, we can define:

$$\boldsymbol{E}^k = (\boldsymbol{A}^{k,k+r})^{\mathrm{T}} \boldsymbol{E}^{k+r} \boldsymbol{A}^{k,k+r} \tag{6.23}$$

According to Catmull–Clark subdivision rules, it is not a difficult task to find $\boldsymbol{A}^{k,k+r}$ since V_i^{k+1} can be linearly expressed by $\vec{V}^k$. Using the expression (6.23), we can evaluate energy norms for a mesh M by subdividing it several times. The energy norm is a property of meshes in this paper. Our experiences show that it seems to have no effect on resulting surfaces to abandon energies of faces with an extraordinary vertex in the mesh M^{k+r} if there is not other extraordinary vertices on 3-neighborhood of any an extraordinary vertex on in the mesh M^{k+r}. In Fig. 6.11c, the energies of extraordinary faces can be abandoned. So, according to the formula (6.23), we may not consider the evaluation of faces with an extraordinary vertex in the M^k by adding the subdivision time number r. However, the number of faces

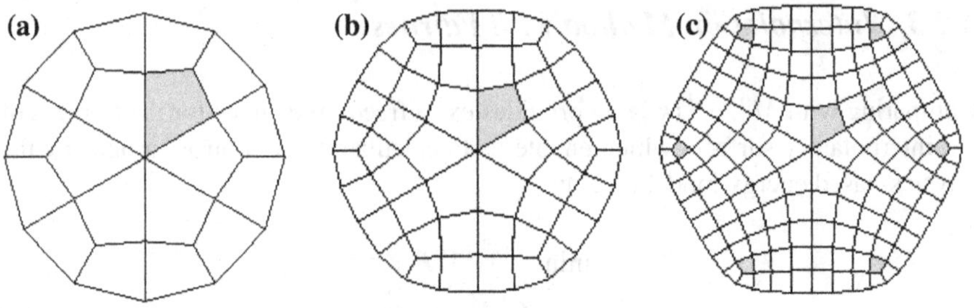

Fig. 6.11 Energy evaluation and subdivision times. The shaded faces are examples of faces with extraordinary vertices. **a** In M^1, internal faces cannot be evaluated energy norms since every of them two extraordinary vertices. **b** In M^2, internal faces with extraordinary vertices can be evaluated energy norms. **c** In M^3, energy of faces with extraordinary vertices can be abandoned when the energy sum of all regular faces can be regarded as the energy of the mesh

increases by 4^r when M^k is subdivided. Consequently, it is necessary to evaluate the energies for faces with an extraordinary vertex in view of time-consuming. How to evaluate the energies for faces with an extraordinary vertex will be discussed in Sect. 5.2.5.

The optimization model can be converted into:

$$\min \quad (\vec{\boldsymbol{V}}_s)^{\mathrm{T}} \boldsymbol{E} \vec{\boldsymbol{V}}_s + \mu(\boldsymbol{C}_s \vec{\boldsymbol{V}}_s - \vec{\boldsymbol{Q}})^{\mathrm{T}} (\boldsymbol{C}_s \vec{\boldsymbol{V}}_s - \vec{\boldsymbol{Q}}) \tag{6.24}$$

where μ is an enough large positive number. So, a linear system can be obtained:

$$(\boldsymbol{E} + \mu \boldsymbol{C}^{\mathrm{T}}_{sSs C_s}) \overleftarrow{\boldsymbol{V}}_s = \mu \boldsymbol{C}^{\mathrm{T}}_{Ss} \vec{\boldsymbol{Q}} \tag{6.25}$$

According to the definition of energy, $\boldsymbol{E}$ is a nonnegative symmetrical matrix, which will be found in Sect. 5.2.4. Note that $\boldsymbol{C}^{\mathrm{T}}_{sSs} \boldsymbol{C}_s$ is a nonnegative and symmetrical matrix. $\boldsymbol{E} + \mu \boldsymbol{C}^{\mathrm{T}}_{sSs} \boldsymbol{C}$ is a nonnegative symmetrical matrix. Generally, $\boldsymbol{E} + \mu \boldsymbol{C}^{\mathrm{T}}_{sSs} \boldsymbol{C}$ is a non-singular matrix. In this case, $\boldsymbol{E} + \mu \boldsymbol{C}^{\mathrm{T}}_{sSs} \boldsymbol{C}_s$ is a positive and symmetrical matrix. So, the Eq. (6.25) can be solved by the relaxed iteration method. The method that solves the optimization model (6.21) by the model (6.24) is called the punishing method. In order to solve the optimization model (6.21), other methods [182] probably can be also used. However, the punishing method is simple and the model (6.21) can be converted to a linear system with a positive and symmetrical coefficient matrix by the punishing method.

6.2.4 SOR Iteration Method to Solve Linear Systems

Assume there is a linear system

$$\boldsymbol{G}\boldsymbol{V} = \boldsymbol{F} \tag{6.26}$$

where $\boldsymbol{G} = [g_{i,j}]$, $\overrightarrow{\boldsymbol{V}} = [\boldsymbol{V}_i]$, $\overrightarrow{\boldsymbol{F}} = [\boldsymbol{F}_i]$, and $\boldsymbol{V}_i$ and $\boldsymbol{F}_i$ are the coordinate vectors of vertices. So, the SOR iteration can be written as:

$$
\begin{aligned}
\boldsymbol{V}_0^{k+1} &= (1-\omega)\boldsymbol{V}_0^k + \frac{\omega}{g_{0,0}}(\boldsymbol{F}_0 - g_{0,1}\boldsymbol{V}_1^k - g_{0,2}\boldsymbol{V}_2^k - \ldots - g_{0,m-1}\boldsymbol{V}_{m-1}^k)\\
\boldsymbol{V}_0^{k+1} &= (1-\omega)\boldsymbol{V}_0^k + \frac{\omega}{g_{1,1}}(\boldsymbol{F}_1 - g_{1,0}\boldsymbol{V}_0^{k+1} - g_{1,2}\boldsymbol{V}_2^k - \ldots - g_{1,m-1}\boldsymbol{V}_{m-1}^k)\\
&\ldots\ldots\\
\boldsymbol{V}_{m-1}^{k+1} &= (1-\omega)\boldsymbol{V}_{m-1}^k + \frac{\omega}{g_{m-1,m-1}}(\boldsymbol{F}_{m-1} - g_{m-1,0}\boldsymbol{V}_0^{k+1} - g_{m,2}\boldsymbol{V}_1^{k+1}\\
&\quad - \ldots - g_{m,m-1}\boldsymbol{V}_{m-1}^{k+1})
\end{aligned}
$$

where $k = 0, 1, \ldots$. ω is called the relaxed factor. If $\boldsymbol{G}$ is a positive symmetrical matrix and when $0 < \omega < 2$, the SOR iteration is convergent. Experientially, we use the $1 \leqslant \omega \leqslant 1.4$ when we solve the Eqs. (6.20) and (6.25). If $\| \overrightarrow{\boldsymbol{V}}^{k+1} - \overrightarrow{\boldsymbol{V}}^{k} \| < \varepsilon$, the iterative process will stop.

It is convenient to handle open meshes by the SOR iteration. We can firstly compute boundary control vertices. In the iterative process, these boundary vertices are fixed. The following next subsection will discussion how to compute these boundary control vertices.

6.2.5 Compute Boundary Vertices

The boundary curves are the quasi-uniform cubic B-spline curves [3]. The boundary curve of an interpolatory surface is probably composed of several quasi-uniform cubic B-spline curve sections. We divide the boundary polygon curves by the corner vertices. The corner vertices are defined in the mesh M^0; the $M^k(k \geqslant 1)$ inherits corner vertices from M^{k-1}. We define the corner vertices as follows:

Definition 5.1 Let $\boldsymbol{V}_i$ be a boundary vertex with valence 2. $\boldsymbol{V}_i$ is called as a corner; if $\angle \boldsymbol{V}_{i-1}\boldsymbol{V}_i\boldsymbol{V}_{i+1} < \delta_c$ (δ_c is a given threshold value). Where $\boldsymbol{V}_{i-1}$ and $\boldsymbol{V}_{i+1}$ are the vertices neighboring to $\boldsymbol{V}_i$ on the boundary polygon.

Experientially, we adopt $\delta_c = 2\pi/3$. Figure 6.12 gives the corner vertices and the quasi-uniform cubic B-spline curves, where the latter form the boundary of the interpolatory surface. We firstly extract all boundaries $L_i (i = 1, \ldots, n)$ of M^0. Every boundary L_i is a closed polygon. In Fig. 6.12, M^0 has only a boundary L_1. Assume that L_i is divided into a polygon series $L_{i,j} (j = 1, \ldots, m)$. Let $\{\boldsymbol{q}_0, \ldots, \boldsymbol{q}_{s-1}\}$ is the points of $L_{i,j}$. we can calculate new boundary $L_{i,j}^M$ via the Eq. (6.27):

$$
\begin{aligned}
&\boldsymbol{V}_0 = \boldsymbol{q}_0\\
&\frac{\boldsymbol{V}_{i-1} + 4\boldsymbol{V}_i + \boldsymbol{V}_{i+1}}{6} = q_i\\
&\boldsymbol{V}_s = \boldsymbol{q}_s
\end{aligned}
\tag{6.27}
$$

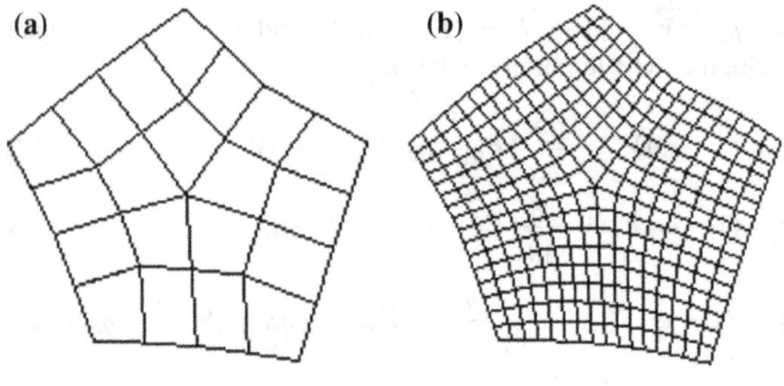

Fig. 6.12 Subdivision of open mesh. a Initial mesh. These 2 valence vertices are all corner vertices; b Interpolatory surface. Its boundary is composed of 5 quasi-uniform cubic B-spline curve sections; the corner vertices are the endpoints of the B-spline curves

According to (2.19), the boundary polygons are subdivided by the following rules:

$$\boldsymbol{V}_{2i}^{k} = \frac{\boldsymbol{V}_{i-1}^{k-1} + 6\boldsymbol{V}_{i}^{k-1} + \boldsymbol{V}_{i+1}^{k-1}}{8}$$
$$\boldsymbol{V}_{2i+1}^{k} = \frac{\boldsymbol{V}_{i}^{k-1} + \boldsymbol{V}_{i+1}^{k-1}}{2} \tag{6.28}$$

6.2.6 Evaluating Energy Norms for Meshes

The energy of a surface contains the energy of a membrane and the energy of a thin plate [44, 107, 108]:

$$E(S) = E_m(S) + E_p(S) \tag{6.29}$$

where $E_m(S) = \iint (\alpha_{11}\boldsymbol{p}_u^2 + 2\alpha_{12}\boldsymbol{p}_u\boldsymbol{p}_v + \alpha_{22}S_v^2)dudv,$

$$E_p(S) = \iint (\beta_{11}\boldsymbol{p}_{uu}^2 + 2\beta_{12}\boldsymbol{p}_{uv}^2 + \beta_{22}\boldsymbol{p}_{vv}^2)dudv,$$

$\boldsymbol{p} = \boldsymbol{p}(u, v)$ is the parameter patch S.

Now, we investigate the uniform bi-cubic B-spline patch:

$$\boldsymbol{p}(u, v) = \sum_{i=0}^{3}\sum_{j=0}^{3} \boldsymbol{P}_{i,j}N_i(u)N_j(v) \tag{6.30}$$

where $(u, v) \in [0, 1] \times [0, 1]$ and $N_i(u)$, $N_j(v)$ $(i, j = 0, 1, 2, 3)$ are the uniform cubic B-spline basic functions.

According to the formula (6.29), the energy of the uniform bi-cubic B-spline patch can be expressed as:

$$E(S) = \sum_{i=0}^{3}\sum_{j=0}^{3} \boldsymbol{P}_{i,j} \sum_{k=0}^{3}\sum_{l=0}^{3} \boldsymbol{P}_{k,l} W_{i,j,k,l}$$

where

$$W_{i,j,k,l} = \int_0^1 \int_0^1 \begin{bmatrix} \alpha_{11} N_i'(u) N_j(v) N_k'(u) N_l(v) + 2\alpha_{12} N_i'(u) N_j(v) N_k(u) N_l'(v) + \\ \alpha_{22} N_i(u) N_j'(v) N_k(u) N_l'(v) + \beta_{11} N_i''(u) N_j(v) N_k''(u) N_l(v) + \\ 2\beta_{12} N_i'(u) N_j'(v) N_k'(u) N_l'(v) + \beta_{22} N_i(u) N_j''(v) N_k(u) N_l''(v) \end{bmatrix} dudv \quad (6.31)$$

The above energy formula can be written as:

$$E(S) = \sum_{i=0}^{15}\sum_{j=0}^{15} \boldsymbol{P}_i \boldsymbol{P}_j W_{i,j} = \vec{\boldsymbol{V}}^{\mathrm{T}} \boldsymbol{E} \vec{\boldsymbol{V}} \quad (6.32)$$

Assume that M is a quadrilateral mesh and any quadrilateral has at most an extraordinary vertex. A quadrilateral is called as the regular face if it has no extraordinary vertices. The non-regular face is called as the extraordinary face. Figure 6.13a gives a regular face and its 1-ring. Using the formula (6.30), a regular face can produce a uniform bi-cubic B-spline patch. In Fig. 5.13a, those vertices are given serial numbers so as to evaluate the energy norm for the uniform bi-cubic B-spline patch by using the formula (6.32). Figure 5.13b gives an extraordinary face and its 1-ring. An extraordinary face can produce a limit patch. However, the limit patch is not a uniform bi-cubic B-spline patch. We will investigate how to evaluate the energy norm for the limit patch. After the 1-ring mesh is subdivided once, there will be three regular faces shown in Fig. 6.13.

Denote the limit patch corresponding to the extraordinary face in Fig. 6.13b as S. According to Fig. 6.13, there is a new extraordinary face after an old extraordinary face is subdivided. We can obtain three uniform bi-cubic B-spline patches: $S_i^k (i = 0, 1, 2)$ after subdividing extraordinary face by the number of k, as shown in Fig. 6.14.

From Fig. 6.15, we have:

$$S = \sum_{k=1}^{\infty} (S_0^k + S_1^k + S_2^k) \quad (6.33)$$

In the expression (6.33), every S_i^k can be evaluated an energy norm by the expression (6.30). Denote the mesh in Fig. 6.13b by M_S and the regular meshes in Fig. 6.15 are, respectively, expressed as $M_{SS,R1}^1$, $M_{SS,R2}^1$, $M_{SS,R3}^1$. The 1-ring mesh of the extraordinary face of M_{SS}^1 has the same topology as M_S, which is denoted by M_S^1.

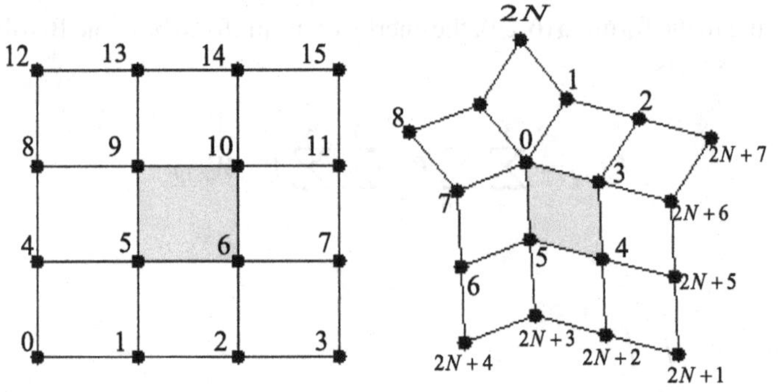

(a) Regular face (b) Extraordinary face

Fig. 6.13 1-ring meshes of faces. We can obtain limit patches by subdividing these 1-ring meshes. The limit patch is a uniform bi-cubic B-spline patch if the shaded face is a regular face

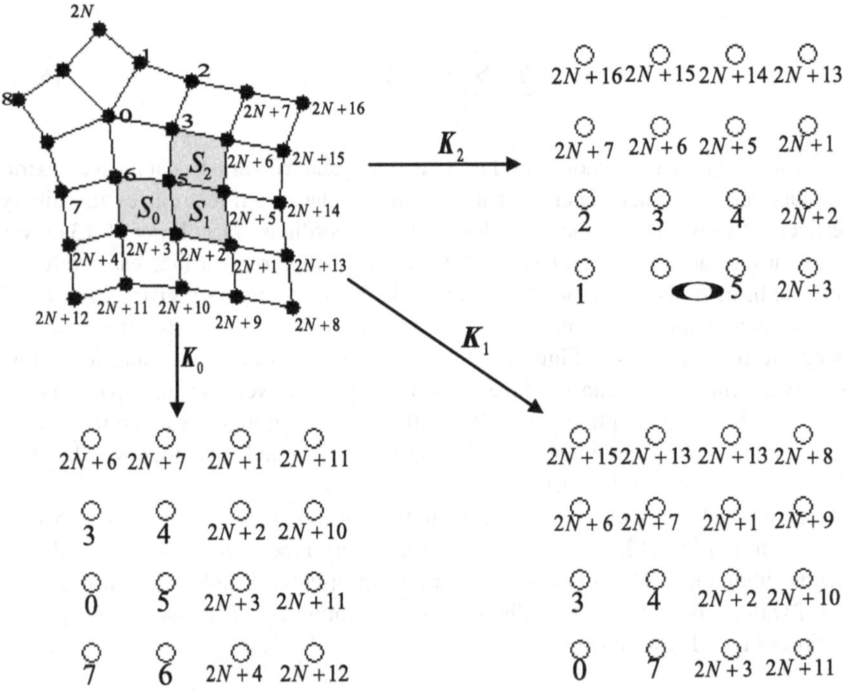

Fig. 6.14 A new mesh is produced by subdividing the 1-ring mesh of an extraordinary face. The new mesh contains three regular faces and an extraordinary face. Using three picking matrices, we obtain the 1-ring mesh of three regular faces. If the extraordinary vertex has valence 3, the indices of the first vertex of the 1-ring of the first regular face is 3, not 8, the new extraordinary face and its 1-ring mesh shall have the topology as same as the topology of the old ones

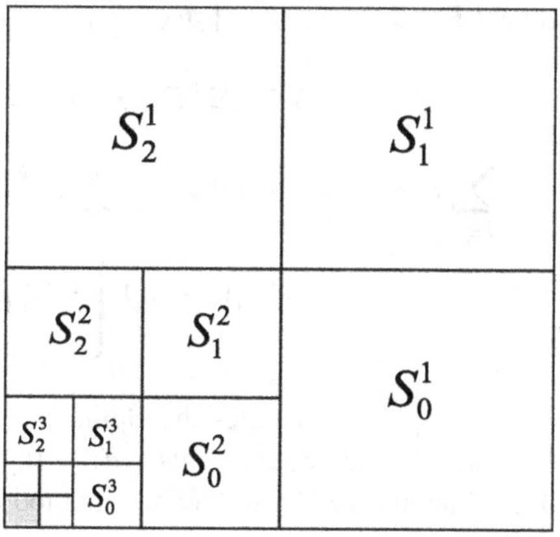

Fig. 6.15 Three regular faces and an extraordinary face appear when subdividing an extraordinary faces for the k time. The shaded face is an extraordinary face, and every regular face corresponds to a limit patch S_i^k

Arrange the vertices of these mesh as per the column vectors and denote them by: $\vec{V}_S$, $\vec{V}_{SS}^1$, $\vec{V}_{SS,R1}^1$, $\vec{V}_{SS,R2}^1$, $\vec{V}_{SS,R3}^1$, $\vec{V}_S^1$, and M_S, M_{SS} are the subdivision matrices satisfying with the following conditions:

$$\begin{aligned}\vec{V}_S^1 &= M_S\vec{V}_S\\ \vec{V}_{SS}^1 &= M_{SS}\vec{V}_S^1\end{aligned}\tag{6.34}$$

$K_i (i = 0, 1, 2)$ are the picking matrices satisfying with the following conditions:

$$\begin{aligned}\vec{V}_{SS,R1}^1 &= K_0\vec{V}_{SS}^1\\ \vec{V}_{SS,R2}^1 &= K_1\vec{V}_{SS}^1\\ \vec{V}_{SS,R3}^1 &= K_2\vec{V}_{SS}^1\end{aligned}\tag{6.35}$$

From the Formulas (6.34) and (6.35), we have

$$\begin{aligned}\vec{V}_{SS,R1}^k &= K_0 M_{SS}(M_S)^{k-1}\vec{V}_S\\ \vec{V}_{SS,R2}^k &= K_1 M_{SS}(M_S)^{k-1}\vec{V}_S\\ \vec{V}_{SS,R3}^k &= K_2 M_{SS}(M_S)^{k-1}\vec{V}_S\end{aligned}$$

where $k \geqslant 1$. Consequently,

$$E(S_i^k) = (K_i M_{SS}(M_S)^{k-1}\vec{V}_S)^{\mathrm{T}} E(K_i M_{SS}(M_S)^{k-1}\vec{V}_S)$$

Let $M_S = \Xi\Lambda\Xi^{-1}$ and $E_{\mathrm{PA}} = (K_i M_{SS})^{\mathrm{T}} E(K_i M_{SS})$, we have

$$E(S_i^k) = (\vec{V}_S)^{\mathrm{T}}(\Xi^{-1})^{\mathrm{T}}[\Lambda^{k-1}(\Xi^{\mathrm{T}} E_{\mathrm{PA}}\Xi)\Lambda^{k-1}]\Xi^{-1}\vec{V}_S$$

Let $\boldsymbol{\Xi}^{\mathrm{T}}\boldsymbol{E}_{\mathrm{PA}}\boldsymbol{\Xi} = \boldsymbol{E}_{\mathrm{PAV}} = \{(\boldsymbol{E}_{\mathrm{VAP}})_{i,j}\}$, we have

$$\boldsymbol{\Lambda}^{k-1}(\boldsymbol{\Xi}^{\mathrm{T}}\boldsymbol{E}_{\mathrm{PA}}\boldsymbol{\Xi})\boldsymbol{\Lambda}^{k-1} = \{(\boldsymbol{E}_{\mathrm{VAP}})_{i,j}\boldsymbol{\Lambda}_{i,i}^{k-1}\boldsymbol{\Lambda}_{j,j}^{k-1}\}$$

$$\begin{aligned}\sum_{k=1}^{\infty} E(S_i^k) &= (\vec{\boldsymbol{V}}_S)^{\mathrm{T}}(\boldsymbol{\Xi}^{-1})^{\mathrm{T}}\sum_{k=1}^{\infty}[\boldsymbol{\Lambda}^{k-1}(\boldsymbol{\Xi}^{\mathrm{T}}\boldsymbol{E}_{\mathrm{PA}}\boldsymbol{\Xi})\boldsymbol{\Lambda}^{k-1}]\boldsymbol{V}^{-1}\vec{\boldsymbol{V}}_S \\ &= (\vec{\boldsymbol{V}}_S)^{\mathrm{T}}(\boldsymbol{\Xi}^{-1})^{\mathrm{T}}\left\{\sum_{k=1}^{\infty}[(\boldsymbol{E}_{\mathrm{VAP}})_{i,j}(\boldsymbol{\Lambda}_{i,i}\boldsymbol{\Lambda}_{j,j})^{k-1}]\right\}\boldsymbol{\Xi}^{-1}\vec{\boldsymbol{V}}_S \end{aligned} \tag{6.36}$$

where $(\cdot)_{i,j}$ or $\cdot_{i,j}$ denotes the element (i, j) of the element matrix $\cdot$. $\{\cdot_{i,j}\}$ denotes the matrix formed by elements $\cdot_{i,j}(i = 0, \ldots, m-1;\ j = 0, \ldots, n-1)$. The matrix $\boldsymbol{A}_S$ is diagonalized by the Stam's method [40]:

$$\boldsymbol{A}_S = \boldsymbol{\Xi}\boldsymbol{\Lambda}\boldsymbol{\Xi}^{-1} \tag{6.37}$$

where

$$\boldsymbol{\Xi} = \begin{pmatrix} \boldsymbol{D}_0 & 0 \\ \boldsymbol{D}_1 & \boldsymbol{W}_1 \end{pmatrix}$$

We have discussed the method to compute the eigenstructure in Sect. 4.4. For the $\boldsymbol{\Lambda}$ in the Formula (6.37), we have

$$\boldsymbol{\Lambda}_{1,1} = 1, 0 < \boldsymbol{\Lambda}_{i,i} < 1, (i > 1)$$

Consequently,

$$\sum_{k=1}^{\infty}[(\boldsymbol{\Lambda}_{i,i}\boldsymbol{\Lambda}_{j,j})^{k-1}] = \begin{cases} \infty, i = 1, j = 1 \\ 1/(1 - \boldsymbol{\Lambda}_{i,i}\boldsymbol{\Lambda}_{j,j}), \text{other} \end{cases}$$

We find that

$$(\boldsymbol{E}_{\mathrm{VAP}})_{1,1} = 0 \tag{6.38}$$

We have theoretically proved the Formula (6.38) and have made numerical computations for vertices with valence $N = 3, \ldots, 50$; based on the formula (6.38), we define

$$\sum_{k=1}^{\infty}[(\boldsymbol{E}_{\mathrm{VAP}})_{i,j}(\boldsymbol{\Lambda}_{i,i}\boldsymbol{\Lambda}_{j,j})^{k-1}] = \begin{cases} 0, i = 1, j = 1 \\ (\boldsymbol{E}_{\mathrm{VAP}})_{i,j}\sum_{k=1}^{\infty}[(\boldsymbol{\Lambda}_{i,i}\boldsymbol{\Lambda}_{j,j})^{k-1}], \text{other} \end{cases}$$

Let $(\boldsymbol{V}^{-1})^{\mathrm{T}}\left\{\sum_{k=1}^{\infty}[(\boldsymbol{E}_{\mathrm{VAP}})_{i,j}(\boldsymbol{\Lambda}_{i,i}\boldsymbol{\Lambda}_{j,j})^{k-1}]\right\}\boldsymbol{V}^{-1} = \boldsymbol{E}_N$, the expression (6.36) becomes:

$$\sum_{k=1}^{\infty} E(S_i^k) = (\overrightarrow{V}_S)^{\mathrm{T}} \boldsymbol{E}_N \overrightarrow{V}_S \tag{6.39}$$

Based on the Expressions (6.32) and (6.37), we can calculate the energy norm of an extraordinary face by the Formula (6.40):

$$E(S) = \sum_{i=0}^{2} \left(\sum_{k=1}^{\infty} E(S_i^k) \right) \tag{6.40}$$

Note that the $\boldsymbol{E}_N$ is a matrix only related with valences of an extraordinary vertices. In practice, we can precompute these matrices to a certain maximum and store them in a file. In our experiments, we set 20 for the maximum valence.

Using the Formula (6.23) and the Formula (6.38), we can compute the energy norm of a mesh M. Assume that $S_{R,i} (i = 0, \ldots, m-1)$ are the limit patches of regular faces and $S_{Sj} (j = 1, \ldots, n)$ are limit patches of extraordinary faces. We can have the energy of M:

$$E(M) = \sum_{i=1}^{m} E(S_{R,i}) + \sum_{j=1}^{n} E(S_{S,j}) \tag{6.41}$$

If there is not other extraordinary vertex in the 2-ring mesh of any extraordinary vertex of M, we can abandon the energy of the extraordinary face in Formula (6.41), that is, we can compute the energy of M by using the Expression (6.42):

$$E(M) = \sum_{i=0}^{m-1} E(S_{R,i}) \tag{6.42}$$

From the Formula (6.23), we can evaluate the energy norm for M^k by using M^{k+r}. Consequently, we can evaluate the energy norm for M^k by using the Formula (6.42). There is no large difference between the results obtained from the Formula (6.41) and the Formula (6.42), respectively.

6.2.7 Examples

In Fig. 6.16, we give some specific example. We construct interpolatory surfaces by computing M^0, M^1 and M^2. It is easy to know that shapes more approach to the shape of the given mesh I^0 when the subdivision time k increases. The interpolatory surfaces seem to be plumper when the subdivision time k decreases

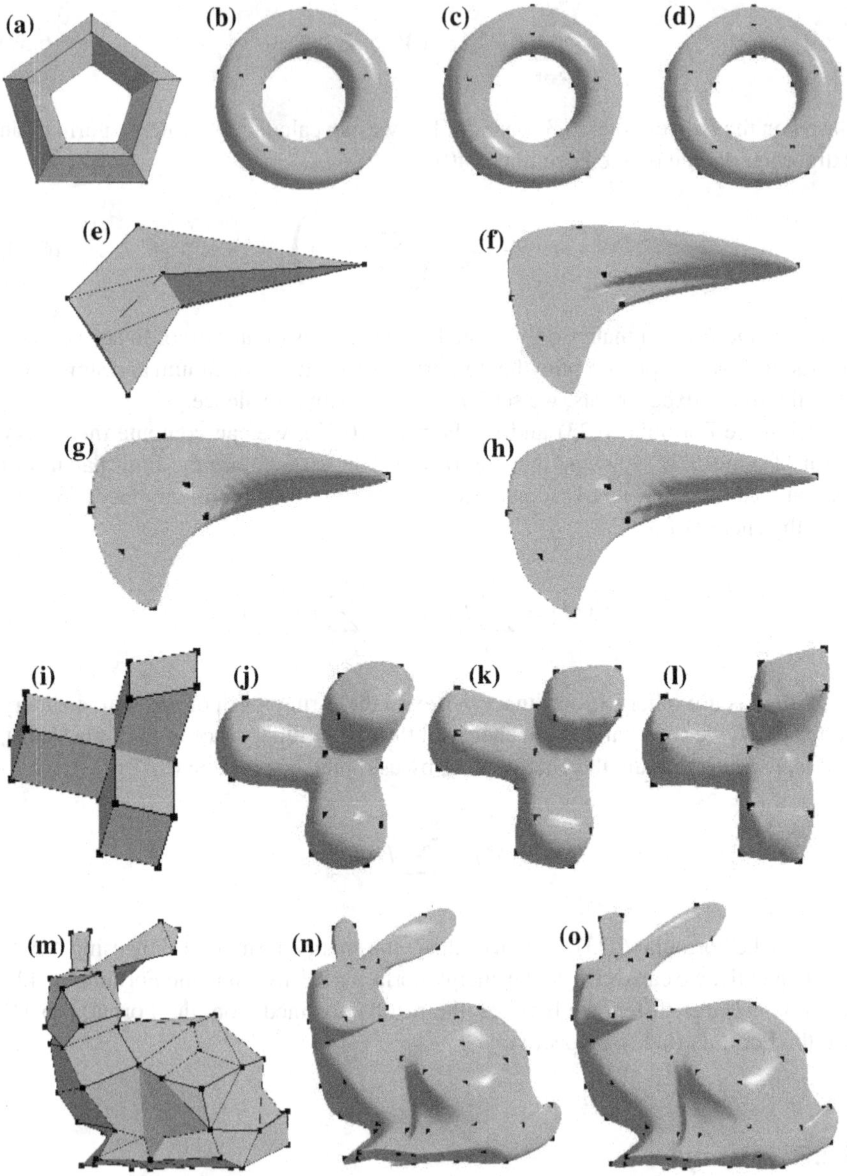

Fig. 6.16 Construct interpolatory surfaces by computing M^0, M^1, and M^2

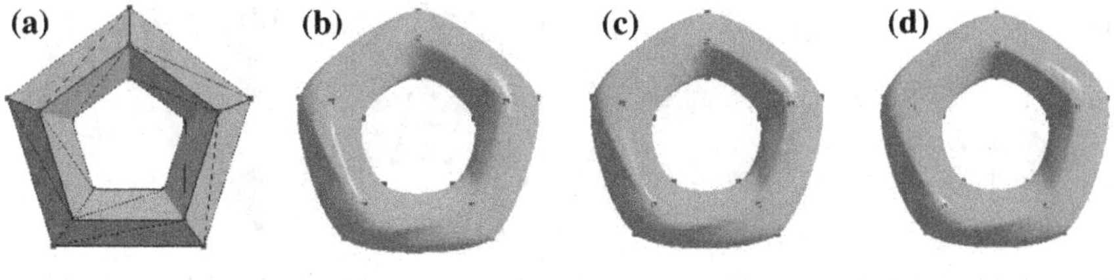

Fig. 6.17 Construct interpolatory surfaces by computing M^1. Several energy evaluation methods are used. **b**: Evaluate energy of M^1 using the energy of M^2. **c**: Evaluate energy of M^1 using the energy of M^3. **d**: Evaluate energy of M^1 using the energy of all regular faces of M^3

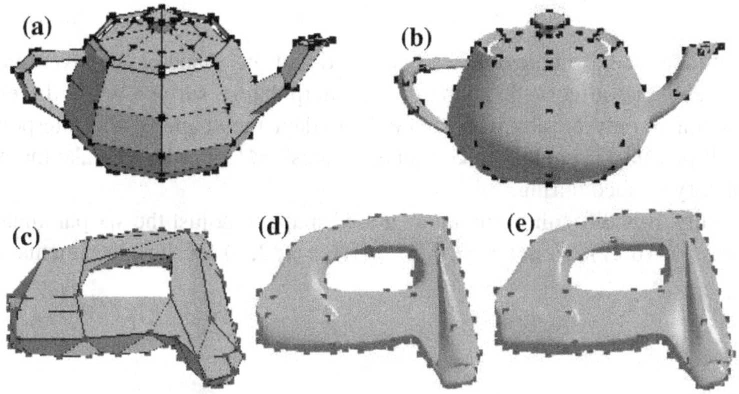

Fig. 6.18 Interpolatory surfaces construct upon the open meshes. **b** Surface obtained by computing M^0. **d** and **e** are the surfaces obtained by computing M^0 and M^1, respectively

When the subdivision time k increases, computation time will drastically increase. This is the reason why we only compute M^0, M^1 for the initial mesh (m) in Fig. 6.16m.

We can evaluate energy norms for a mesh after subdividing it several times. Figure 6.17 shows interpolatory surfaces constructed upon computation, where M^1 is evaluated the energy norms by several methods. From which it can be seen there is no large difference between the surfaces constructed, respectively, by abandoning and by using the energy of the extraordinary faces.

For the examples in Fig. 6.18, we compute the control meshes by computing the boundaries in advance, where their boundaries curves are the uniform cubic B-spline curves. For the examples in Fig. 6.19, we compute all control vertices via the global optimization model. We can find a drawback in Fig. 6.19, where in Fig. 6.19b, some depressed regions are near to the corner, but the whole control mesh is convex. Just as Shuhua Lai et al. [184] pointed out, it is a tricky task to determine the location of boundary vertices and it is not a good idea to determine the location of boundary vertices only by the global optimization model. Some skills are very necessary when the additional rings of vertices are added along the current boundaries.

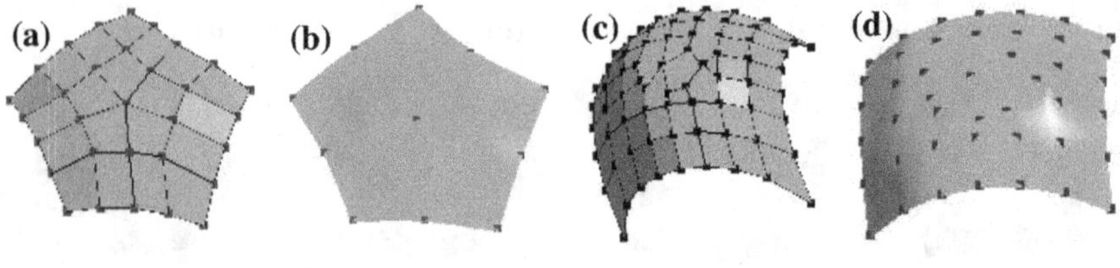

Fig. 6.19 Interpolatory surfaces constructed upon the open meshes via the global optimization model; where the control mesh of the interpolatory surfaces is M^0 and the boundaries are abandoned when M^0 is subdivided, i.e., the boundary vertices and the boundary edges do not produce new points in the process of subdivision

From the above examples, we can also find that the shape of the given mesh is a main factor that affects the shape of the interpolatory surface while the energy optimization is only an auxiliary method to adapt the shape of the interpolatory surface. It possibly is enough to compute the mesh M^1 in order to make the whole interpolatory surface fairing.

Generally, it is difficult to obtain a "good" shape to adjust the six parameters in the expression (6.27). We set $\alpha_i = 0$, $\beta_i = 1 (i = 1, 2, 3)$ for the experiments in this paper.

Remarks

This chapter constructs blending patches and interpolatory surfaces by using physical PDE. In view of the relationship between the optimization model and the control mesh, the former can be divided into the global optimization models and the local optimization models. In the global optimization model, all vertices of the considered control mesh are variables of the optimization model; while in the local optimization model, some vertices of the meshes are the variables of the optimization model. It is needed to solve multiple local optimization model in order to determine vertices of the control mesh; in the local optimization model, only some vertices of a considered control mesh are variables of the optimization model. We only have to solve many optimization models in order to determine vertices of the control mesh. Generally, the surfaces constructed by using the global optimization method have better fairness than that of the surfaces constructed by using the local optimization method. However, the local optimization method has higher computation speed than that of the global optimization method. This chapter uses the global optimization method. Though we only use the optimization models to construct surfaces, the optimization method is still the effective method to fair surfaces. In the last several chapters in this book, the optimization method is used for constructing surfaces. For example, the fairness of the subdivision surfaces interpolated the curve nets is improved by adopting the geometric PDE and the fairing deformation surfaces are constructed by using the physical PDE.

Exercises

1. Let $\boldsymbol{p}(u) = \sum_{i=0}^{8} P_i N_{i,3}(u)$, $\boldsymbol{p}(u) = \sum_{i=0}^{8} \boldsymbol{P}_i N_i$ is the uniform cubic B-spline curve. $\boldsymbol{P}_i (i = 0, 1, 2, 6, 7, 8)$ are given. Compute the vertices $\boldsymbol{P}_i (i = 3, 4, 5)$ using the optimization model (6.1).
2. Assume that there are three patches on a plane as shown in following figure. They have the same shapes and symmetric positions. These patches are uniform cubic B-spline surfaces and three matrix lattices are their control polygons. Now, we have to construct a Catmull–Clark subdivision surface to blend the three patches. The topology structure is shown in the following figure. Compute the black point coordinates according to the optimization model (6.11).
3. Assume that there is a Catmull–Clark subdivision surface interpolating all vertices of a unit cubic. The initial control mesh of the subdivision surface has the same topology as the unit cubic. Compute the initial control vertices of the subdivision surface.

Chapter 7
Interactive Shape Editing for Subdivision Surfaces

This chapter firstly provides two free-form deformation algorithms of subdivision surfaces by combining mesh deformation with shape editing under simple geometrical constraints or potential function constraints. Secondly, local editing algorithms for limit surfaces under geometrical constraints are described using local parameter representation of subdivision surfaces. The first two methods are generally consistent in the deformation process, i.e., the relevant constraints of points, lines, and faces are imposed on the current subdivision level mesh, and the mesh deforms under these constraints. The mapping relationships of constraints between successive subdivision levels are then deduced according to the subdivision rules. Finally, mesh deformation under simple geometrical or potential function constraints is re-conducted on the desired subdivision level, and the final results are obtained. By defining the position and normal vector of any point and isoparametric line in the regular mesh region on the subdivision surface, the last method sets up a constrained linear system and solves it with using least-squares method and energy optimization method. This allows the surface to be modified under different geometrical constraints.

7.1 Adjustment of Subdivision Meshes Under Simple Geometrical Constraints

Using the idea of Lanquetin et al. [194] as a basis, we first impose geometrical constraints on the current subdivision level mesh (e.g., point, line, and face constraints) and deform the subdivision mesh under these constraints. Secondly, we solve the transferring map between the current subdivision level mesh and its next subdivision level mesh, according to the subdivision rules. Thirdly, we set the next subdivision

W. Liao et al., *Subdivision Surface Modeling Technology*,
DOI 10.1007/978-981-10-3515-9_7

level as the current subdivision level. Lastly, we impose geometrical constraints on the current subdivision level mesh and deform the mesh under these imposed constraints.

7.1.1 Deformation under Simple Geometrical Constrains

Here, we take the points constraints as an example to explain the deformation method under simple geometrical constrains. The locations of constraint points and target points are first specified by the user in the input mesh, and the target point is a point that a constraint point should reach after deformation. The deformation area is then specified, and the corresponding deformation reference curve is chosen.

There are many forms of deformation reference curves, but it is not the case that any curve may be used as a deformation reference curve. For example, using a straight line as a deformation reference curve cannot produce a smooth deformation effect, especially in a boundary area. To overcome this limitation, a cubic Hermite curve is recommended. It is G^1 continuous in the boundaries of the deformation area, and it allows the user to control the shape of the deformation reference curve more conveniently. Figure 7.1 shows the corresponding Hermite deformation reference curve of any point $\boldsymbol{Q}$ in the mesh. The curve equation is:

$$\boldsymbol{Q}^*(t) = (2t^3 - 3t^2 + 1)\boldsymbol{Q}_c + (t^3 - 2t^2 + t)\boldsymbol{m}_0 + (t^3 - t^2)\boldsymbol{m}_1 + (-2t^3 + 3t^2)\boldsymbol{P}^*$$

where $\boldsymbol{Q}_c$ is the intersection point between the boundary curve of the deformation region and the plane S, determined by point $\boldsymbol{Q}$, mesh constraint point $\boldsymbol{P}$, and point $\boldsymbol{P}^*$. Point $\boldsymbol{P}^*$ is the target point of $\boldsymbol{P}$, and $\boldsymbol{m}_0$ and $\boldsymbol{m}_1$ are the tangent vectors of, respectively, $\boldsymbol{Q}_c$ and the point $\boldsymbol{P}^*$ in the plane Π. Position $\boldsymbol{Q}^*$ after the deformation of $\boldsymbol{Q}$ can be determined by the corresponding parameter:

$$t = |\boldsymbol{Q}_c\boldsymbol{Q}_p|/|\boldsymbol{Q}_c\boldsymbol{P}|,$$

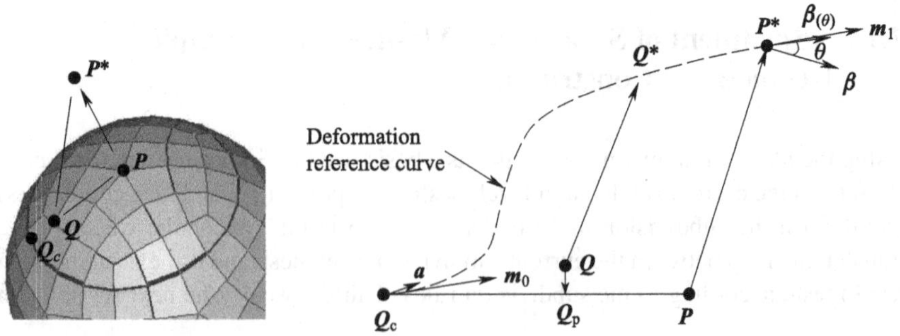

(a) Point constraint (b) Deformation reference curve(dashed line)

Fig. 7.1 Hermite deformation reference curve

Table 7.1 Examples of point-constrained deformations, using different adjustment parameters

Deformation mesh	$k_0 = 0.3$ $k_1 = 1.0$ $\theta = \pi/6$	$k_0 = 0.3$ $k_1 = 1.0$ $\theta = \pi/2$	$k_0 = 1.5$ $k_1 = 2.5$ $\theta = \pi/2$	$k_0 = 1.5$ $k_1 = 2.5$ $\theta = -\pi/6$	$k_0 = 5.0$ $k_1 = 4.0$ $\theta = \pi/2$	$k_0 = 1.5$ $k_1 = 2.5$ $\theta = 2\pi/3$
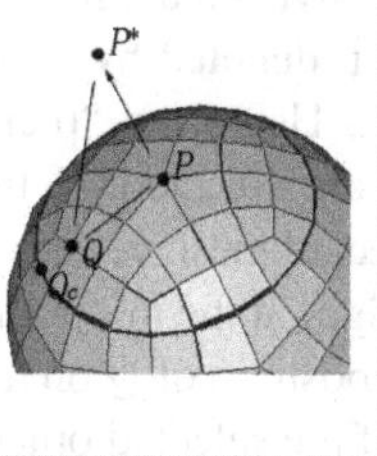	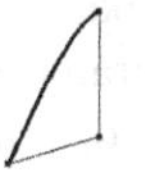					

where, $\boldsymbol{Q}_p$ is the projection point of $\boldsymbol{Q}$ in the line segment $\boldsymbol{Q}_c\boldsymbol{P}$. The tangent vectors $\boldsymbol{m}_0$ and $\boldsymbol{m}_1$ may be calculated by the formulae:

$$\boldsymbol{m}_0 = k_0|\boldsymbol{Q}_c\boldsymbol{P}|\alpha$$

and

$$\boldsymbol{m}_1 = k_1|\boldsymbol{P}^*\boldsymbol{P}|\beta_{(\theta)},$$

where α is the unit tangent vector of the mesh at point Q_c in the direction determined by the mesh and the plane S; $\beta_{(\theta)}$ is the vector obtained by rotating the unit vector β by an angle of θ around the point P^*; β is a unit vector in the plane S passing through the point P^* and vertical to PP^*; and k_0, k_1 and θ are the adjustment parameters for the deformation.

Table 7.1 gives examples of point-constrained deformations using different k_0, k_1, and θ values. The curve in this table is the Hermite deformation reference curve corresponding to the mesh vertex Q in the left-hand figure. From Table 7.1, it can be seen that when k_0 and k_1 take too large values, or θ takes an unreasonable value, the deformed meshes obtained are not ideal, and self-intersection may even occur (e.g., the last two columns shown in the table). To ensure that the mesh deformation does not produce too much distortion, the value ranges of k_0, k_1, and θ are limited: $k_0, k_1 \in [0, 3]$ and $\theta \in [-\pi/2, \pi/2]$.

7.1.2 Mapping the Relationship of Constraint Points Between Successive Subdivision Levels

Line constraints and face constraints may be implemented by point constraints. Therefore, a solution for mapping the relationship of constraint lines and constraint faces between successive subdivision levels may be converted into a solution for mapping the relationship of constraint points. Here, this process will be explained using the Catmull–Clark subdivision surface as an example.

(1) Determining the positions of constraint points in the mesh before subdivision

The elements of the mesh (vertices, edges, and faces) before subdivision are named using the "pick" mechanism of OpenGL, and a constraint point, denoted by Q, is produced by user interaction through a mouse-click operation. Using the "pick" operation for mesh elements, either vertices, edges, or faces are selected. If the selected object is a vertex, then the constraint point Q is the selected mesh vertex. If the selected object is an edge, the edge is divided into two subedges at the midpoint, and the subedge to which the constraint point Q belongs and the position of Q on the subedge are determined through use of intersection judgment. If the selected object is a face, a triangular segmentation of this face is made, and the division rules follow the corresponding subdivision rules for the original face. The centroid of this face is taken as the new F-point, the midpoint of each edge is taken as a new E-point, and the new V-point locates to the original vertex position. We then connect the new F-point with the new E-point, the new E-point with the new V-point, and the new F-point with the new V-point. An $\boldsymbol{n}$-sided face is thereby divided into $2n$ triangular subfaces, and the index arrangement of these subtriangles is based on the index arrangement of the vertices in the original face. After this triangulation, the subface to which the constraint point Q belongs, and the position of Qin the subface, may be determined by intersection with each subface. Figure 7.2 illustrates two cases of triangular segmentation, for a triangle and quad face, in a Catmull–Clark subdivision surface; other polygons may be segmented using a similar approach.

(2) The image of the constraint point at successive subdivision levels

The new V-points created in the process of subdivision correspond with vertices of the preceding subdivision level mesh, so they have the same vertex index. The new E-points and F-points created during subdivision are also placed in the vertex chain list in which the new V-points are placed. The constraint point in the mesh before subdivision may be a mesh vertex, or a point on the interior of an edge or face. If the constraint point is a mesh vertex, then its image in the next subdivision level is simply the corresponding new V-point; since the index of the new V-point is unchanged, the

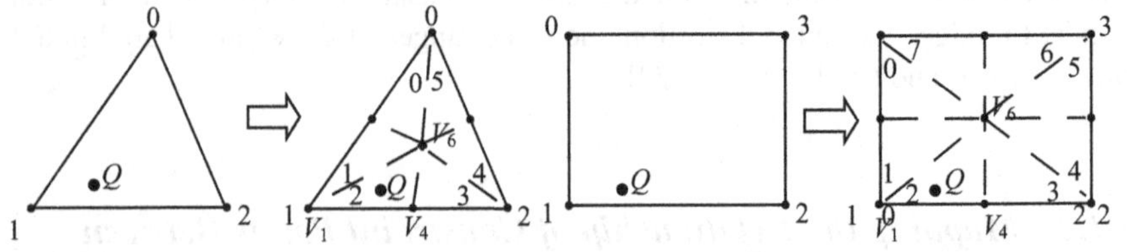

(a) Triangle mesh face segmentation (b) Quad mesh face segmentation

Fig. 7.2 Triangular segmentation of the mesh face (Catmull–Clark subdivision)

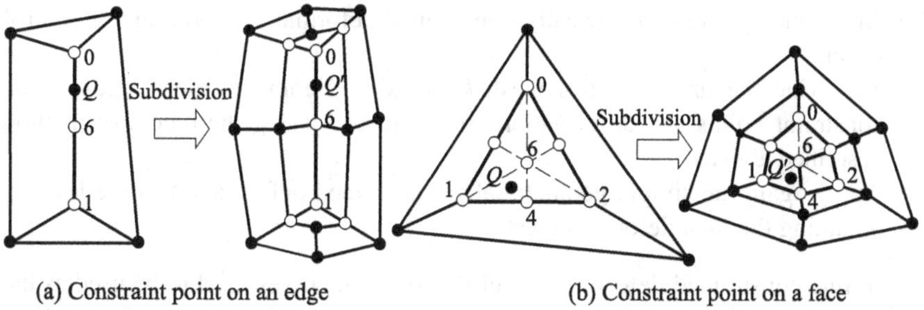
0
Q
Subdivision
6
1
0
Q'
6
1
0
6
Q
1
2
4
Subdivision
0
6
1
Q'
2
4
(a) Constraint point on an edge
(b) Constraint point on a face

Fig. 7.3 The image of the constraint point in the next Catmull–Clark subdivision level

image may be found easily in the index. If the constraint point is on the interior of some edge or face of the mesh before subdivision, the process of finding its image is as follows. In Fig. 7.3, the numbers represent the indices of corresponding vertices, and V_i and V_i' are vertices with index i before and after subdivision, respectively. Let us first consider the situation of a constraint point on the interior of an edge, as shown in Fig. 7.3a. If a constraint point Q before subdivision, located on the subedge V_0V_6 of an edge V_0V_1, can be expressed as $Q = uV_0 + vV_6$, its image Q' in the next subdivision level can be calculated by $Q' = uV_0' + vV_6'$, where V_0' and V_6' are the corresponding images of V_0 and V_6, respectively. The situation for a constraint point located on the interior of a face is similar to that for a constraint point on an edge. Let the constraint point Q be located on the subface $V_1V_4V_6$ of a mesh face $V_0V_1V_2$, as shown in Fig. 7.3b; this may be expressed as $Q = uV_1 + vV_4 + wV_6$. The image Q' of the vertex Q in the next subdivision level can then be expressed as $Q' = uV_1' + vV_4' + wV_6'$, where V_1', V_4' and V_6' are the corresponding images of V_1, V_4, and V_6, respectively. Note that the meshes need to be triangulated after Catmull–Clark subdivision.

The above process may be iterated on successive subdivision levels to follow any point on the mesh. From this, a simple geometrically constrained deformation can be imposed on any given level, and this deformation will remain on successive meshes obtained following further subdivision. The resulting shape will thus meet the necessary requirements.

7.1.3 The Deformation Algorithm

For a point-constrained deformation, the constraints are imposed on the initial control mesh, and the deformation mesh is computed after k subdivision steps. The specific deformation algorithm is obtained as follows:

(1) Set the constraint points, target points, deformation area, deformation reference curve, and adjustment parameters k_0, k_1, and θ for the initial control mesh.

(2) Apply the simple geometrically constrained deformation to the initial control mesh.
(3) Subdivide the initial control mesh k times, and compute the images of the constraint points and deformation area at level k using the recursive method described above.
(4) Apply again the simple geometrically constrained deformation to the level k, obtaining the final deformed mesh.

For line-constrained deformations, closed or open curves need to be used as the constraint curves for the deformation mesh, and the deformation region also requires defining. After specifying the target constraint curves after deformation, the mesh in the deformation region will deform. In contrast to point-constrained deformation, where only the constrained points in the deformation region are relevant to the vertices to be deformed, an important step in line-constrained deformation is determining which point in the constraint curve is relevant to the vertex to be deformed; this is known as the correlative constraint point of the vertex to be deformed. As shown in Fig. 7.4a, Delaunay triangulation is first established for all vertices on the constraint curves and boundary curves of the deformation region, and then the whole mesh in the deformation region is projected to a plane using harmonic mapping. Thus, the spatial triangles derived from the earlier triangulation are also mapped on to a plane.

There are two possibilities for the position of the harmonic mapping point in a triangle, for example, points Q_1' and Q_2' shown in Fig. 7.4. Q_1' is the harmonic mapping point of mesh point Q_1, which lies in a triangle determined by two vertices in the harmonic mapping constraint curve and one vertex in the boundary curve of the harmonic mapping deformation area. Q_1 is called type-I mesh point. In contrast, Q_2', the harmonic mapping point of Q_2, lies in a triangle determined by one vertex in the harmonic mapping constraint curve and two vertices in the boundary curve of the harmonic mapping deformation area. Q_2 is called a type-II mesh point. For type-I mesh point Q_1, join Q_1' with the vertex of the triangle, containing Q_1', that lies in the boundary curve of the harmonic mapping deformation area. The intersection point between this straight line and the harmonic mapping constraint curve may then be obtained, and this intersection point is the mesh point that corresponds to the correlative constraint point of point Q_1. For type-II mesh point Q_2, the correlative constraint point is the mesh point that corresponds to the vertex of the triangle, containing Q_2', that lies in the harmonic mapping constraint curve.

For surface-constrained deformation, the constraint surface in the deformation mesh is set, the target constraint surface after the deformation is specified, and the deformation area is delimited. A one-to-one correspondence relationship is set up between the vertices in the constraint surfaces and those in the target surfaces. The deformation method applied to the other vertices is consistent with that used for line-constrained deformation. As shown in Fig. 7.4b, a Delaunay triangulation is first set up between the boundary curves of the deformation regions and the boundary curves of the constraint surfaces. The mesh in the deformation area is then harmonically mapped to a plane, and the correlation between these vertices and the points on the

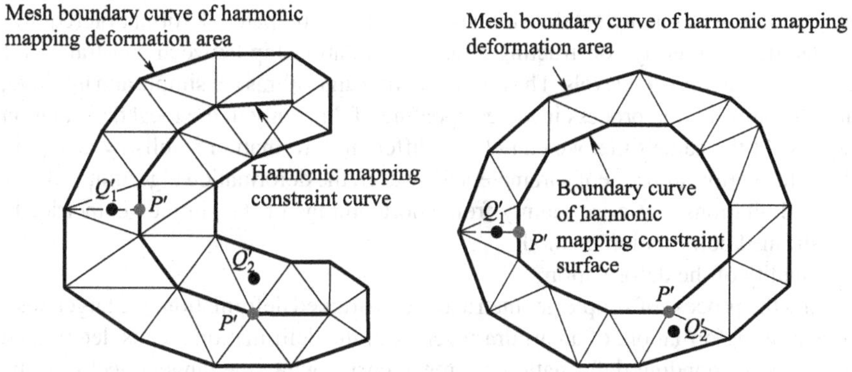

(a) Mesh points and constraint curves (b) Mesh points and constraint surfaces

Fig. 7.4 Correlation between mesh points, constraint curves, and constraint surfaces

Table 7.2 Examples of simple geometrically constrained deformations

Constraint type	Initial mesh model to be deformed	Adjustment parameters	Deformation results on the initial mesh	Deformation results on the meshafter two subdivisions	
				Edges visible	Edgesinvisible
Point constraint	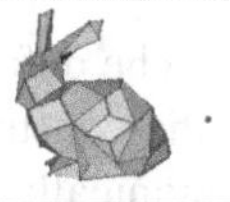	$k_0 = 1.5$ $k_1 = 1.0$ $\theta = \pi/2$	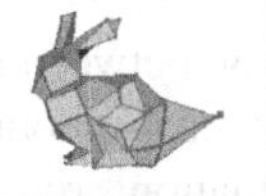		
Curve constraint	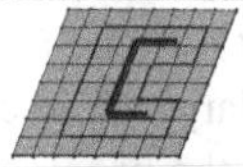	$k_0 = 2.5$ $k_1 = 2.5$ $\theta = \pi/2$	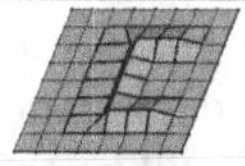		
Surface constraint	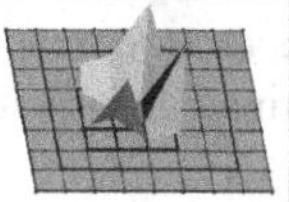	$k_0 = 1.5$ $k_1 = 1.5$ $\theta = \pi/6$	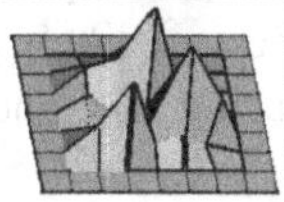		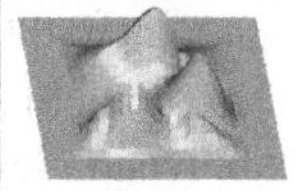

constraint surface boundary is established. According to this correlation, the vertices are deformed.

Table 7.2 lists three different examples of subdivision surface deformations under geometrical constraints.

7.2 Editing of Subdivision Level Meshes Using Potential Functions

The previous section has discussed simple geometrically constrained deformation algorithms for subdivision surfaces, where the constraint types are point, line, or face constraints. According to the requirements of the deformation, the constraints may be imposed on any subdivision level, and the anticipated deformation requirements may be met by constructing a mapping relationship between constraints for successive subdivision levels. These deformation algorithms are simple and intuitive, and the deformation process involves updating of data only in the local deformation region. At the same time, we can obtain different deformation results by changing the deformation adjustment parameters. However, the deformation algorithm still has some limitations that arise mainly from shortcomings of the simple geometrically constrained deformation method:

(1) Quality of the deformation

During the process of simple geometrically constrained deformation, the target position, after deformation, of an arbitrary vertex in the deformation area is determined by the corresponding deformation reference curve, while the tangent vector of the first point on the deformation reference curve is directly influenced by the boundary curves of the deformation area. Therefore, if the interactively specified boundary curves are not sufficiently smooth, this will have a serious impact on the quality of the final deformation. For example, if the deformation boundary curves are selected so improperly that the shapes of the deformation reference curves for two nearby vertices in the deformation area are completely different, a shear phenomenon may appear on the mesh model after deformation.

(2) Harmonic mapping

In order to obtain the correlation between any mesh vertex to be deformed and the points in the constraint curves or the boundary curves of the constraint surfaces, the whole mesh in the deformation area needs to be harmonically mapped to a plane. Harmonic mapping is a process in which a linear system is solved using the least-squares method. For a deformation mesh with a large number of vertices, the dimensions of the equations will be greater, and hence the process of harmonic mapping will have a higher time cost, likely causing an interaction–response barrier. This problem mainly occurs during line- and face-constrained deformations of high subdivision level meshes.

(3) Computational stability

Since the deformation reference curves for each of the mesh vertices are computed independently, and these curves do not share similar harmonics, it is difficult to estimate and control the quality of the resulting deformation surface.

Due to these problems of simple geometrically constrained deformations, mesh-constrained deformation under a potential function is a useful alternative; its application for subdivision surface shape editing is discussed below.

7.2.1 *Mesh-Constrained Deformations Under Potential Functions*

We first define the mesh-constrained deformation under the potential function. Taking point-constrained deformation as an example, the constraint point, the target point after deformation and an effective influence radius are specified by the user and expressed as P, P^*, and R, respectively. Let $r(Q, P)$ be the distance of any point Q in the mesh model to a designated constraint point P, and $f(r, R)$ be the potential function defined in the distance space r. Make the potential function value the deformation weight and define a deformation function *deform*(Q), which makes a point Q transform to *deform*(Q): *deform*$(Q) = Q + (P^* - Q) \times f(r, R)$.

The potential function $f(r, R)$ has many potential expressions [189], including Blinn's power function, Nishimura's piecewise quadratic polynomial, Murakami's quartic polynomial, and Wyvill's six-degree polynomial. Of these, Wyvill's polynomial function has the following useful properties: $f(0, R) = 1$, $f'(0, R) = 0$, $f(R, R) = 0$, $f'(R, R) = 0$ and $f(R/2, R) = 1/2$; it therefore has numerous applications in many situations. Here, we take Wyvill's polynomial as the potential function; it is expressed as follows:

$$f(r, R) = \begin{cases} -\frac{4}{9}\left(\frac{r}{R}\right)^6 + \frac{17}{9}\left(\frac{r}{R}\right)^4 - \frac{22}{9}\left(\frac{r}{R}\right)^2 + 1 & 0 \le r \le R \\ 0 & r > R \end{cases} \tag{7.1}$$

There are different definitions for the distance function $r(Q, P)$. Since Euclidean distance and Manhattan distance ignore the spacial structure of the manifold, we take the geodesic distance.

For mesh deformation under the constraints of line or face, since all vertices in the constraint curve or boundary curve of the constraint surface are constraint points, there are some mesh vertices that lie within the influence radii of numerous constraint points, requiring that all deformation influences are taken into account simultaneously. Here, in order to obtain the final deformation, we take a weighted average of the deformations of a vertex under the constraints of all the influencing points.

Taking line-constrained deformation as an example, let the target positions after deformation of the constraint points $P_i (i = 0, 1, 2, \ldots, n-1)$ in the constraint curve be $P_i^* (i = 0, 1, 2, \ldots, n-1)$, respectively. For any point Q on the mesh model, its new position Q^* may be obtained using the formula:

$$Q^* = \sum_{i=0}^{n-1} w_i Q_i^*,$$

in which Q_i^* is the position of Q after deformation under the constraint of P_i, and w_i meets the criteria: $\sum_{i=0}^{n-1} w_i = 1$ and $w_i \geqslant 0$. Therefore, the new location of Q after deformation, Q^*, may be expressed as a linear combination of all Q_i^*. There are many methods to set the weights of $w_i (i = 0, 1, 2, \ldots, n-1)$; here, we define w_i with the following formula to make its influence inversely proportional to the distance from P_i to Q:

$$w_i = 1/r(Q, P_i)/\sum_{i=0}^{n-1}\left(1/r(Q, P_i)\right). \tag{7.2}$$

Instead of resorting to deformation reference curves using the simple geometrically constrained deformation method, mesh-constrained deformation under a potential function takes advantage of both potential functions and deformation functions. By combining these techniques, triangulation and harmonic mapping operations required for simple geometrically constrained deformation methods are avoided.

7.2.2 *Updating of the Potential Function*

Solving the mapping relationship of the constraints and updating the potential function between successive subdivision levels are important steps in the deformation algorithm that applies the mesh-constrained deformation under the potential function to edit the shape of subdivision surfaces. Computation of the former may be achieved with the method described above. The latter mainly involves updating of the effective influence radius R when the mesh-constrained deformation under the potential function is re-imposed on the next subdivision level. Due to shrinkage of the approximating subdivision scheme, it is difficult to adjust the effective influence radius R directly; for this reason, another approach is required, which considers the value of r/R in the potential function expression.

We will first introduce the concept of uniform splitting, taking Catmull–Clark subdivision as an example. The topological rules of uniform splitting are the same as those for a Catmull–Clark subdivision scheme, while the geometrical rules are defined as follows:

(1) Generate a new V-point for each mesh vertex, whose location is the same as the corresponding original vertex.
(2) Generate a new E-point for each edge, whose location is at the midpoint of the corresponding edge.
(3) Generate a new F-point for each face, whose location is at the centroid of the corresponding face.

From the geometrical rules described above, it may be seen that uniform splitting is an interpolating subdivision. For the same number of splits or subdivisions, the split mesh and subdivided mesh are topologically homeomorphous and close in shape. Furthermore, in the deformation algorithm used here, it is on the initial mesh that we first apply the mesh-constrained deformation under the potential function and determine the effective influence radius R of the constraint.

According to the above analysis, calculation of the ratio of r/R for a standard subdivision mesh vertex can be converted into computation of r/R for the corresponding mesh vertex after uniform splitting, where r is the distance of the corresponding

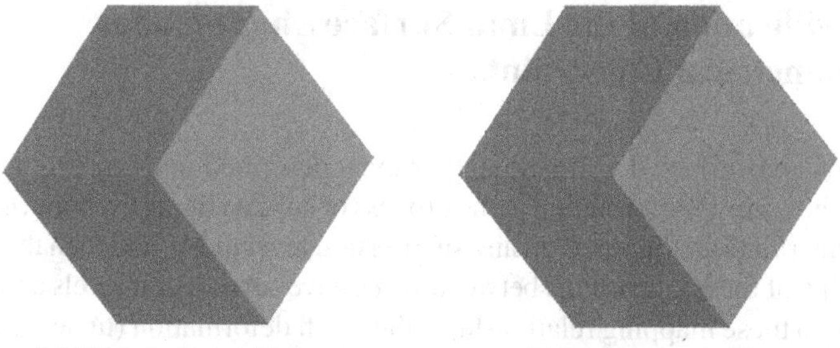

(a) Mesh before splitting (b) Mesh after uniform splitting twice

Fig. 7.5 An example of uniform splitting

Table 7.3 Examples of constrained deformations under potential functions

Type of constraint	Initial mesh to be deformed	Deformation under potential function imposed on initial mesh	Deformation under potential function imposed on the mesh subdivided once or twice	
			Edge visible	Edge invisible
Point constr-aint	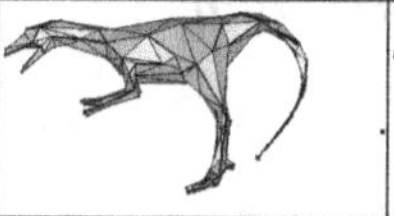	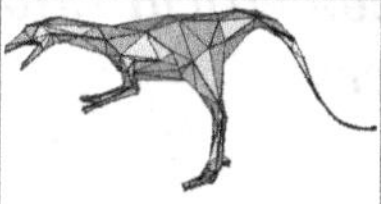	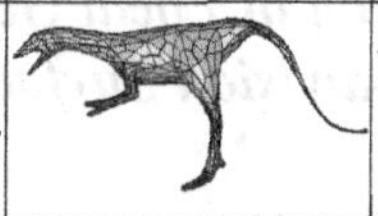	
Curve constr-aint	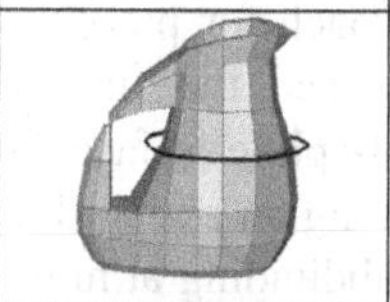	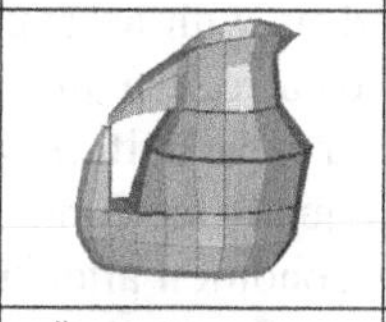	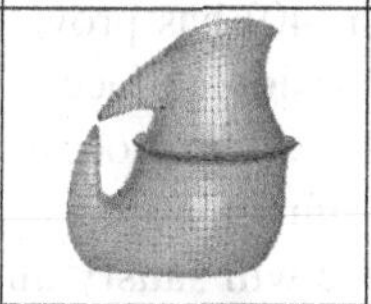	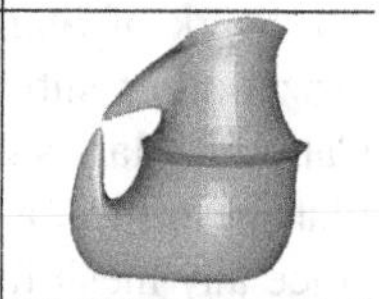
Surface constr-aint				

vertex to the constraint, and R is taken as having the same value as in the initial mesh. Figure 7.5 gives an example of uniform splitting.

Table 7.3 lists examples of subdivision surface deformations under potential functions, with three different types of constraints.

7.3 Modification of the Limit Surface Shape Under Geometrical Constraints

The subdivision surface deformation algorithms described in the previous two sections take the form of deformation in the process of subdivision, *i.e.*, point, curve, and surface constraints are imposed on any subdivision level mesh, and then the mapping relationships of these constraints between successive subdivision levels are deduced. According to these mapping relationships, the mesh deformation (under simple geometrical constraints or under a potential function) is applied again to subsequent subdivision levels. This deformation approach makes full use of the multiresolution property of subdivision surfaces, and the method is intuitive, simple to understand, and easy to realize. However, due to the shrinkage of approximating subdivision surfaces, these methods are not suitable for situations that demand high precision. This section will introduce a precise algorithm for modification of subdivision surfaces, with geometrical constraints imposed directly on the limit surface.

7.3.1 Definition of Local Geometrical Information for Subdivision Surfaces

The work of Stam [40] has provided us with an approach for piecewise parameterization of subdivision surfaces. For a regular mesh face, the limit surface of a Catmull–Clark subdivision process is a cubic uniform B-spline surface. For an irregular mesh face, assume that this is a quad face containing an extraordinary vertex, since any mesh face will satisfy this condition after subdividing at most twice; the corresponding limit surface can be approximated by infinite B-spline surfaces. The parameter field is shown in Fig. 7.6, where the extraordinary point corresponds to the origin, but cannot be expressed with a formula. Fortunately, Halstead [44] gives a method to evaluate the limit position and tangent vector at the extraordinary point.

Each quadrilateral mesh face of a Catmull–Clark subdivision surface corresponds to a B-spline patch, and the geometrical variables, such as the location and tangent vector, can be defined by setting up a local coordinate system on this patch. Figure 7.7 shows an example of a Catmull–Clark subdivision surface based on piecewise parameterization, in which the limit surface of the red mesh face is also shaded in red.

Since there is not a global parameter field for a subdivision surface, strictly speaking, there are no isoparametric lines for a subdivision surface. However, a Catmull–Clark subdivision surface can be expressed by cubic uniform B-spline surfaces in most regions, except for a few extraordinary points; since B-spline surfaces have the local support property, we can therefore designate local isoparametric curves in the regions without an extraordinary point.

Definition 7.1 A polygon is called a mesh curve if it is formed by a series of edges in order, no matter whether it is open or closed. A mesh curve is called a regular mesh

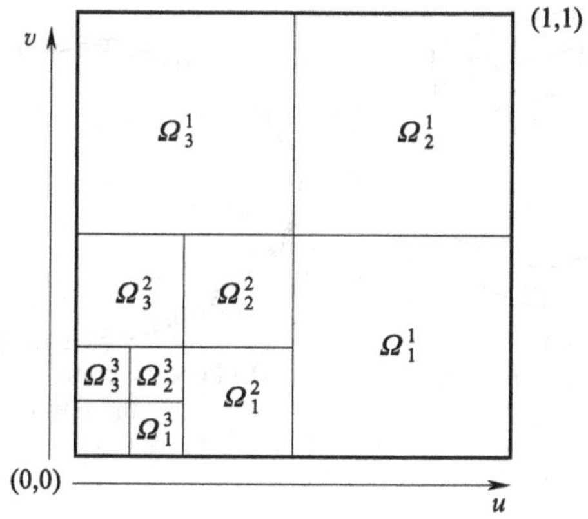

Fig. 7.6 Partition of the parameter field of an extraordinary face

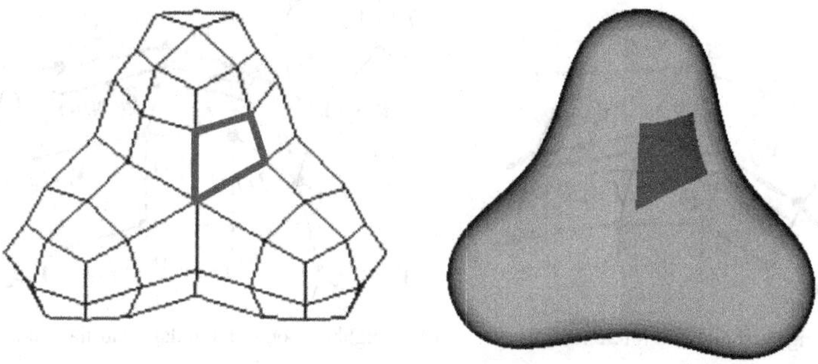

(a) Control mesh (b) Limit surface represented by a piecewise parameterization method

Fig. 7.7 Expression of a subdivision surface using piecewise parameterization

curve if any two neighboring edges of the mesh curve are not on the same face, and all the endpoints of the edges are regular points. The 1-neighborhood of a regular mesh curve forms a regular mesh belt.

A regular mesh belt corresponds to a polygon complex, as shown in Fig. 7.8a. Fig. 7.8b is the result of the complex in Fig. 7.8a after subdividing three times with the Catmull–Clark subdivision scheme. It may be seen that the regular mesh belt shrinks in the subdivision process and eventually converges to a continuous limit curve that is a cubic uniform B-spline curve.

Definition 7.2 The limit curve of a regular mesh belt is called the local isoparametric curve of the subdivision surface; for short, an isoparametric curve.

According to the theory of B-spline curves and tensor product surfaces, the ith control vertex $\boldsymbol{p}_i$ of an isoparametric curve may be calculated from three control vertices in the longitudinal direction of the corresponding regular mesh belt (see Fig. 7.9a):

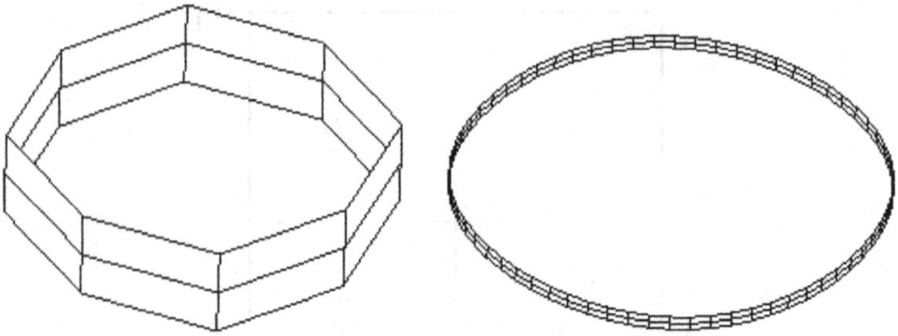

(a) Initial polygon complex (b) Polygon complex subdivided three times

Fig. 7.8 Regular mesh belts

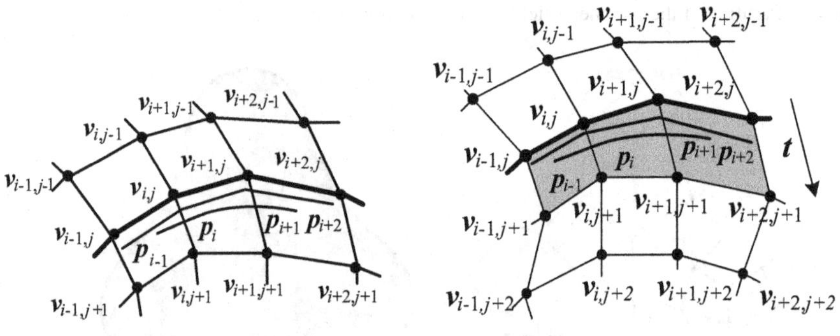

(a) 1-neighborhood of a reference curve (b) 1-neighborhood of a regular quadrilateralmesh belt

Fig. 7.9 A regular quadrilateral mesh belt and the isoparametric curve defined by it

$$\boldsymbol{p}_i = 1/6\boldsymbol{v}_{i,j-1} + 2/3\boldsymbol{v}_{i,j} + 1/6\boldsymbol{v}_{i,j+1}. \tag{7.3}$$

Definition 7.3 If two regular mesh belts have, and only have, two common regular mesh curves, they are called regular quadrilateral mesh belts.

The mesh belt shown in Fig. 7.9b is a regular quadrilateral mesh belt, and the Catmull–Clark subdivision surface defined by it is a cubic uniform B-spline surface. Taking an isoparametric line on the surface as a reference curve, we can obtain a group of isoparametric curves with the same direction, as shown in Fig. 7.9b. Let the isoparametric curve corresponding to the bold mesh curve be the reference curve which corresponds to the surface parameter $t = 0$, and let another regular mesh line of the regular quadrilateral mesh belt corresponds to the surface parameter $t = 1$; thus, any $t \in [0, 1]$ defines an isoparametric curve between these two curves, and the ith control vertex $\boldsymbol{p}_i$ can be calculated from four control vertices in the longitudinal direction:

$$\boldsymbol{p}_i = N_0(t)\boldsymbol{v}_{i,j-1} + N_1(t)\boldsymbol{v}_{i,j} + N_2(t)\boldsymbol{v}_{i,j+1} + N_3(t)\boldsymbol{v}_{i,j+2} \tag{7.4}$$

where $N_j(t)$, $(j = 0, 1, 2, 3)$ are cubic uniform B-spline basis functions.

7.3.2 Subdivision Surface Shape Modification Based on the Least-Squares Method

The basic idea of applying the least-squares method to surface shape modification is to minimize the total amount of disturbance to the values of the control vertices, under the demands of a given shape modification.

(1) Point constraints

First, specify the local coordinate (f_i, u_i, v_i) of the point $\boldsymbol{p}_i$ to be modified, and its position variation $\Delta\boldsymbol{p}_i$; the local coordinate is inputted by the user directly or calculated by the system according to the user's choice in the multiresolution approximating mesh surface. The corresponding control vertices $\boldsymbol{v}_{i,j}$, $j = 1, 2, \ldots, N_i$ are then determined by the coordinates of the points to be modified. Since our goal is to minimize the total amount of disturbance to the control vertex, we need only adjust the local control vertices that have effects on the designated deformation.

Let the local parameterization representation of $\boldsymbol{p}_i$ in the subdivision surface be $\boldsymbol{S}_i(u_i, v_i) = \sum_{j=1}^{N_i} a_{i,j}\boldsymbol{v}_{i,j} = \boldsymbol{p}_i$, the adjustment quantity of the control vertex $\boldsymbol{v}_{i,j}$ be $\Delta\boldsymbol{v}_{i,j}$, $\boldsymbol{X}_i^{\mathrm{T}} = (\Delta\boldsymbol{v}_{i,1}, \Delta\boldsymbol{v}_{i,2}, \ldots, \Delta\boldsymbol{v}_{i,N_i})$, $\boldsymbol{C}_i^{\mathrm{T}} = (\boldsymbol{v}_{i,0}, \boldsymbol{v}_{i,1}, \ldots, \boldsymbol{v}_{i,N_i})$, and $\boldsymbol{A}_i = (a_{i,1}, a_{i,2}, \ldots, a_{i,N_i})$. Therefore:

$$\boldsymbol{A}_i(\boldsymbol{C}_i + \boldsymbol{X}_i) = \boldsymbol{p}_i + \Delta\boldsymbol{p}_i.$$

We can then obtain the following constraint equation system:

$$\boldsymbol{A}_i\boldsymbol{X}_i = \Delta\boldsymbol{p}_i.$$

(2) Normal vector constraints

Sometimes, the user will impose the normal vector constraint on some point $\boldsymbol{p}_i$ in the surface. Let the modified normal vector of a vertex $\boldsymbol{p}_i$ be $\boldsymbol{n}_i$, and let $\boldsymbol{T}_{i,j}$, $j = 0, 1$ be two tangent vectors of the surface at $\boldsymbol{p}_i$ which are not co-linear; with $\boldsymbol{C}_i$ and $\boldsymbol{X}_i$ defined as above, then:

$$\boldsymbol{T}_{i,j}(\boldsymbol{X}_i + \boldsymbol{C}_i)\boldsymbol{n}_i^{\mathrm{T}} = 0, \quad j = 0, 1.$$

This can simplified to the following constraint equation:

$$\boldsymbol{T}_{i,j}\boldsymbol{X}_i\boldsymbol{n}_i^{\mathrm{T}} = -\boldsymbol{T}_{i,j}\boldsymbol{C}_i\boldsymbol{n}_i^{\mathrm{T}}, \quad j = 0, 1.$$

(3) Isoparametric line constraints

As mentioned above, if a user selects a regular mesh line in the control mesh and takes the corresponding isoparametric line as the reference curve, the system will give a reference direction $\boldsymbol{t}$, and the user can set a parameter $t (t \in [-1, 1])$ according to $\boldsymbol{t}$ in order to select a local isoparametric line. If the isoparametric line to be adjusted is the limit curve corresponding to the selected mesh line, set $t = 0$ and put the 1-neighborhood mesh vertices of this mesh line into a vertices array. If the isoparametric line to be adjusted lies in the same direction as $\boldsymbol{t}$, input a positive value of $t(t \in (0, 1))$, test the regularity of the mesh face belt on this side (as shown in the shaded faces of Fig. 7.9b), and save the control vertices of the regular quadrilateral mesh face belt. If the curve to be adjusted has an opposite direction to $\boldsymbol{t}$, a negative parameter value t is needed, and then a similar technique as above is applied to the other side of the quadrilateral mesh face belt. If the modification quantity of each control vertex on the isoparametric line is set as $\Delta \boldsymbol{d}_i$, the constraints of the isoparametric line can be converted to those of the control vertices:

$$\sum_j b_{i,j} \Delta \boldsymbol{v}_{i,j} = \Delta \boldsymbol{d}_i.$$

(4) Solution of the constraint equations

Lemma 7.1 [200] *Let $\boldsymbol{A}_i$, $\boldsymbol{B}_i (i = 1, 2, \ldots, N)$ and $\boldsymbol{C}$ be $m \times n$, $p \times q$ and $m \times q$ known matrices, respectively, and $\boldsymbol{X}$ be an $n \times p$ unknown matrix. The sufficient and necessary condition of $\boldsymbol{X}$ being the solution of the matrix function $\sum_{i=1}^{N} \boldsymbol{A}_i \boldsymbol{X} \boldsymbol{B}_i = \boldsymbol{C}$ is that $\boldsymbol{x} = vec(\boldsymbol{X})$ is the solution of the linear equation $\boldsymbol{G}\boldsymbol{x} = \boldsymbol{c}$, in which $\boldsymbol{G} = \sum_{i=1}^{N} \boldsymbol{B}_i^{\mathrm{T}} \otimes \boldsymbol{A}_i$, $\otimes$ is a Kronecker product of two matrices, and $\boldsymbol{c} = vec(\boldsymbol{C})$ is the column expansion of a matrix.*

Lemma 7.2 [200] *If the linear system $\boldsymbol{A}\boldsymbol{X} = \boldsymbol{b}$ is consistent, $\boldsymbol{X} = \boldsymbol{A}^{+}\boldsymbol{b}$ is its minimum normal solution; If $\boldsymbol{A}\boldsymbol{X} = \boldsymbol{b}$ is a contradictory linear system, $\boldsymbol{X} = \boldsymbol{A}^{+}\boldsymbol{b}$ is its minimum least-squares solution.*

The geometrical constraints given above are all defined in a local coordinate system and need to be converted into a global linear system about the disturbance quantity of the control vertices to be adjusted in the vertex array. Let the number of normal vector constraints be M_N. If $M_N \neq 0$, the linear system contains constraint equations with forms both $\boldsymbol{A}\boldsymbol{X}\boldsymbol{B} = \boldsymbol{C}$ and $\boldsymbol{D}\boldsymbol{X} = \boldsymbol{P}$. Let

$$\overline{\boldsymbol{A}} = \begin{pmatrix} \boldsymbol{B}^{\mathrm{T}} \otimes \boldsymbol{A} \\ \boldsymbol{E}_3 \otimes \boldsymbol{D} \end{pmatrix}, \quad \overline{\boldsymbol{B}} = \begin{pmatrix} \mathrm{vec}(\boldsymbol{C}) \\ \mathrm{vec}(\boldsymbol{P}) \end{pmatrix} \text{ and } \overline{\boldsymbol{X}} = vec(\boldsymbol{X}),$$

in which $\boldsymbol{E}_3$ is a unit matrix of order 3. We can then conclude from Lemma 7.1 that the above matrix equation can be converted into a linear system $\overline{\boldsymbol{A}}\,\overline{\boldsymbol{X}} = \overline{\boldsymbol{B}}$, and from Lemma 7.2 that the optimal solution based on the generalized inverse matrix is $\overline{\boldsymbol{X}} = \overline{\boldsymbol{A}}^{+}\overline{\boldsymbol{B}}$. If $M_N = 0$, the matrix equation can be solved directly by: $\boldsymbol{X} = \boldsymbol{D}^{-1}\boldsymbol{P}$.

7.3.3 Subdivision Surface Shape Modification with an Energy Optimization Method

The Catmull–Clark subdivision surface shape modification algorithm, based on the least-squares method described in the section above, is suitable for performing local adjustments. However, if the adjustment quantity is too large, the smoothness of the whole surface cannot be assured, because no constraints are imposed on the smoothness of the surface while solving the equations. Surface energy is an important concept that reflects the smoothness of a surface, and a variety of energy expressions are available [190]. Here, we take the most commonly used form, plate energy, as an example to discuss optimization of the modified algorithm for a Catmull–Clark subdivision surface.

For any mesh face $\boldsymbol{f}_i$ of a Catmull–Clark subdivision surface, let the corresponding patch be $\boldsymbol{S}_i(u, v) = (S_i^x(u, v), S_i^y(u, v), S_i^z(u, v))$; the plate energy of $\boldsymbol{S}_i(u, v)$ is then:

$$\begin{aligned} E(\boldsymbol{S}_i) &= \iint \left(\|\boldsymbol{S}_{iuu}\|^2 + 2\|\boldsymbol{S}_{iuv}\|^2 + \|\boldsymbol{S}_{ivv}\|^2\right) dudv \\ &= \sum_{g=x,y,z} \iint \left((S_{iuu}^g)^2 + 2(S_{iuv}^g)^2 + (S_{ivv}^g)^2\right) dudv \\ &= vec(\boldsymbol{C}_i)^T (\boldsymbol{E}_3 \otimes \boldsymbol{Q}_0) vec(\boldsymbol{C}_i), \end{aligned}$$

in which $\boldsymbol{Q}_0 = \int_0^1 \int_0^1 \left(\boldsymbol{b}_{u,u}\boldsymbol{b}_{u,u}^{\mathrm{T}} + 2\boldsymbol{b}_{u,v}\boldsymbol{b}_{u,v}^{\mathrm{T}} + \boldsymbol{b}_{v,v}\boldsymbol{b}_{v,v}^{\mathrm{T}}\right) dudv$ is a 16×16 order energy matrix.

The energy matrix $\boldsymbol{Q}_i$ is relative to the local coordinate system and hence needs to be converted into a global matrix about all the vertices to be adjusted. Add all the adjusted energy matrices together to obtain a final energy matrix $\boldsymbol{Q}$. Let all the position disturbances of the adjusted vertices be $\boldsymbol{X}$, and let $\overline{\boldsymbol{X}} = vec(\boldsymbol{X})$, and $\overline{\boldsymbol{Q}} = \boldsymbol{E}_3 \otimes \boldsymbol{Q}$. Setting up the global constraint equation $\overline{\boldsymbol{A}}\,\overline{\boldsymbol{X}} = \overline{\boldsymbol{B}}$, and using a similar method to that described in the section above, we can obtain the following optimization model:

$$\begin{aligned} &\text{Min} \quad \frac{1}{2}\overline{\boldsymbol{X}}^{\mathrm{T}}\overline{\boldsymbol{Q}}\,\overline{\boldsymbol{X}} \\ &\text{s.t.} \quad \overline{\boldsymbol{A}}\,\overline{\boldsymbol{X}} = \overline{\boldsymbol{B}}. \end{aligned}$$

Solve the model with the penalty function method, and let:

$$f(\overline{\boldsymbol{X}}) = \overline{\boldsymbol{X}}^{\mathrm{T}}\overline{\boldsymbol{Q}}\,\overline{\boldsymbol{X}} + \lambda(\overline{\boldsymbol{A}}\,\overline{\boldsymbol{X}} - \overline{\boldsymbol{B}})^{\mathrm{T}}(\overline{\boldsymbol{A}}\,\overline{\boldsymbol{X}} - \overline{\boldsymbol{B}}).$$

Calculate the derivative about $\overline{X}$, and let it be zero to obtain a linear system:

$$(\overline{Q} + \lambda\overline{A}^{\mathrm{T}}\overline{A})\overline{X} = \lambda\overline{A}^{\mathrm{T}}\overline{B}$$

where, the penalty factor λ usually takes a large positive number. Finally, from Lemma 7.2, we can obtain the explicit solution:

$$\overline{X} = \lambda(\overline{Q} + \lambda\overline{A}^{\mathrm{T}}\overline{A}) + \overline{A}^{\mathrm{T}}\overline{B}.$$

7.3.4 Examples and Analysis

The least-squares method only needs to modify the control vertices affecting the designated deformation region, and thus it has the advantages of high calculation speed and local properties. Using the local property of the least-squares method, adjusting there solution control mesh can produce different modification effects with the same constraint conditions. The energy optimization method minimizes the plate energy of the disturbance surface, so the surface will off set when the geometrical constraint conditions are too small and the shape modification request cannot be satisfied. Other geometric constraint conditions can be imposed to prevent the whole surface from offsetting, such as fixing the boundaries of the open surface or specifying other auxiliary constraint conditions.

As these two methods are converted to linear operations on an invertible matrix, and since the linear transformation has useful properties such as reversibility, exchangeability, and associativity, both methods also possess these excellent properties: There is an inverse operation in each modification to revoke the last operation; the modification has nothing to do with the order of operations; and the shape modification has a stackable property, which means that the modified result is related only to the final modification quality, no matter what the intermediate process is.

Table 7.4 gives some examples of surface deformations, under three different geometrical constraints.

Remarks

Geometrical modeling is usually a repeated process in which interactive shape modification is inevitable. Therefore, shape editing is an important research field in surface modeling. T Shape editing of free-form surfaces is an important topic in surface modeling. NURBS surface is a classical free-form surface, and there are many publications on shape editing of NURBS surfaces [233–236]. Methods in these literatures modify shapes of NURBS surfaces by using knot vectors, control vertices, and weights of NURBS surfaces. Wang [235] present a method to modify

Table 7.4 Examples of surface editing of the limit surface under geometrical constraints

Constraint type	Original surface & constraint conditions	Deformation surfaces with least-squares method	Deformation surfaces with energy optimization method	
			Without boundary constraints	With boundary constraints
Point constraint	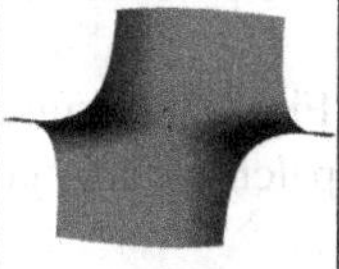	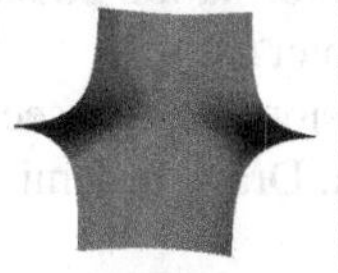	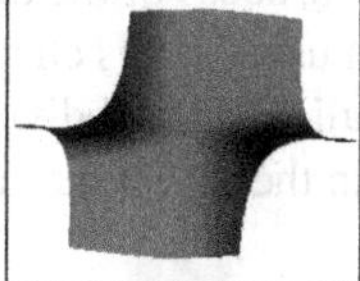	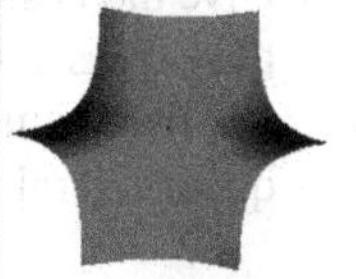
Point & normal vector constraint		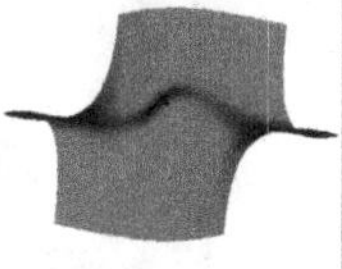	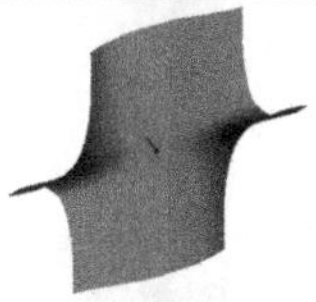	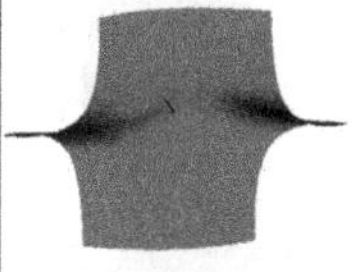
Isoparametric line constraint	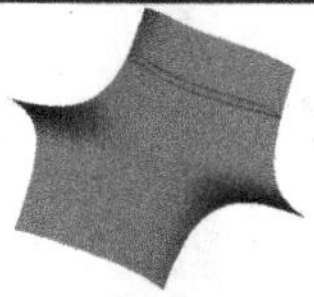			
Mixed constraints	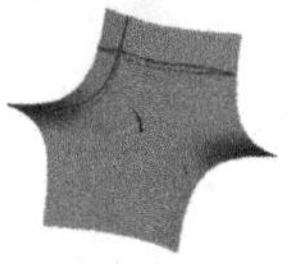		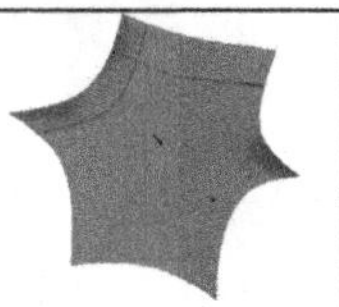	

shapes of B-spline surfaces by using geometric constraints, such as points, normals, curves, and faces. In shape modification algorithms presented in this chapter, we also use these geometric constraints to modify shapes of subdivision surfaces. Since multiresolution is an important property of subdivision surfaces, many shape modification methods [199, 238] use the property. So do methods presented in this chapter. Surface shape modification is also the surface deformation. The FFD method is an extensively applied method. It can be classified into two types: the medium mapping deformation [239–247] and the direct constraint deformation [248–252]. The former insert the model to be deformed into a parameter volume space (e.g., a Bézier volume with 3 parameters). By adjusting the shape of the volume (i.e., medium), the shape of the model is adjust. The latter deforms surfaces by solving linear systems or optimization models that are constructed based on given geometric constraints. These methods given in this chapter belong to the direct constraint deformation.

Exercises

(1) What is the DRC deformation? For a point in the deformation region of a surface, how does the DRC deformation define its new position?
(2) Give the basic process of the deformation under the potential function. What is the expression used in this chapter?
(3) For the Catmull–Clark subdivision surface, see Fig. 7.7. Assume that there is a quadrilateral in the initial mesh. Draw the limit patch of the quadrilateral.

Chapter 8
Intersection and Trimming of Subdivision Surfaces

Based on the multiresolution approximation and piecewise parameterization representation of subdivision surfaces, this chapter introduces precise intersection and trimming algorithms. The control meshes at certain subdivision levels, which satisfy certain error requirements, are discretized and intersected first, and then an intersection mesh belt is constructed. Following this, local coordinate systems are set up at the mesh elements in the intersection mesh belt, and the subdivision surface is divided into a series of small patches, with each vertex evaluated as a parameter in the local coordinate system to which it belongs. The meshes are subdivided with a revised skirt-removed approach, and intersecting vertices and corresponding parameters are determined. The final precise solution is obtained with an iterative method. According to the trimming region designated by the user, the direction of each intersect line is reset. The control mesh faces of the trimmed surface are classified into three types: reserved faces, trimmed faces, and discarded faces. The trimming domain of each trimmed face is set, and then the whole subdivision surface is trimmed.

8.1 Related Work

Most intersection algorithms are based on multiresolution properties of subdivision surfaces, and make an approximate solution using the concept of discretization and partition. Nasri [48] studied an intersection algorithm for Doo–Sabin subdivision surfaces, based on a boundary-generating method and an axis-directioned bounding box checking technique; this was a typical "divide-conquer" method. Lanquetin [194] introduced a discretization method based on a bidirectional graph, which first searched intersection faces, and then set up a relationship graph between them; subsequently, it was checked whether intersections were between related surfaces, and calculations were made of the intersection points between mesh faces subdivided to a

W. Liao et al., *Subdivision Surface Modeling Technology*,
DOI 10.1007/978-981-10-3515-9_8

certain level. Although the above methods provide discrete intersections for the control mesh, it is time-consuming to obtain intersections with high precision. Although the introduction of association faces reduces the scope of the comparison, the iterative subdivision process still requires significant time. In a different approach, Zhu et al. [195] used a Marching method to intersect by virtue of an exact evaluation formula and an active affine frame, and this greatly improved the precision of intersecting.

For the trimming of subdivision surfaces, Habib [196] split a subdivision surface into many patches by inserting vertices and edges and revising subdivision rules; however, this method could only trim the original surface along parameter directions. Biermann [197] proposed a multiresolution representation and a feature-modeling method that trimmed the surface along feature boundaries. Litke [198] trimmed subdivision surfaces with a combined subdivision method that could interpolate intersection curves by adjusting the control mesh near the trimming curves. Since the surface shape changed near the intersection curves, this method only produced approximate trimmed surfaces. The method of Zhou et al. [70] was even simpler: The intersection points of subdivided triangular meshes were taken as those of subdivision surfaces, and only the topological information was changed while trimming, i.e., splitting along the intersection points; as a result, the trimmed surface was actually only the discretization of the original surface and included a large amount of data, which was not convenient for succedent surface operations.

8.2 Initial Discrete Intersection of Subdivision Surfaces

8.2.1 Basic Idea of Discrete Intersection

Subdivision surfaces are the limit surfaces of subdivided polygon meshes, so a natural approach to intersection operations is to intersect the subdivision meshes at some subdivision level. Intersection faces and related faces may be designated for each face, according to the discrete intersection principle based on a bidirectional graph as reported in the literature [194], and an intersection check made only between associated faces. In view of the rapid increment of control vertices in the subdivision process, here a revised skirt-removing approach is taken, ensuring that only two-ring neighborhood vertices of the intersection faces are retained, and all others are discarded in order to improve efficiency.

An intersection test is carried out first for the bounding boxes of the two corresponding subdivision surfaces. If the boxes are not intersected at all, the two surfaces have no intersection points and the algorithm ends; otherwise, further tests are made. Since the control mesh of the approximate subdivision surfaces contracts during subdivision, the coarseness of the control mesh should be checked before the intersection test. If the control mesh is coarse, it should be subdivided until it satisfies a certain approximation precision. The value of the initial relative error threshold, ε_0, should be chosen according to efficiency and stability. If ε_0 is too small, the number of initial subdivisions will be too large, leading to too many control points and a negative impact on the efficiency of the algorithm. It is preferable to depress the initial precision request as much as possible, provided no intersection points are missed. According to experiments, we can obtain ideal results if we take the value of ε_0 as about 0.1. If the control mesh has never been subdivided, another test for regularity should be made, which means that the mesh should be subdivided once if there exists an irregular face with two or more irregular vertices. If the control meshes satisfy the initial error threshold and regularity, an intersection test is made between them. The intersection faces are found, and their one-neighbor faces are set as related faces. In order to ensure a one-neighbor mesh of intersection faces, related faces may be obtained with the revised skirt-removed approach. The two-neighborhood vertices of the intersection faces are preserved as the control vertices of the next subdivision, other mesh faces are deleted, and the intersection test is applied only between related faces. The remaining mesh is subdivided by the revised skirt-removed approach, and the related faces of the subdivided faces are inherited from their parent faces; the intersection test and construction of new related faces then proceed between the related faces of the subdivided faces. The process is repeated until the error between the subdivision surfaces and the control meshes meets the requirements. Eventually, the intersection points of the discrete faces provide the approximate solutions.

8.2.2 Intersection Test Between Quad Meshes

There are two main approaches used for intersection tests: space decomposition methods and hierarchical bounding volumes methods. The former splits the space into small iso-volume elements, and the intersection test occurs only between objects in the same or neighboring elements. Space decomposition methods based on octree and binary space partition (BSP) tree are typical. The core idea of the latter is to wrap complex objects with larger boxes that are simple geometric shapes, and make the intersection test between these boxes first. Further intersecting computation between the wrapped objects continues only when the boxes intersect. If a tree structure is introduced in the construction of bounding boxes, the elements that do not intersect can be ruled out, reducing substantial unnecessary computation and improving test efficiency. Typical bounding boxes include axis aligned bounding boxes (AABB), sphere bounding boxes, and oriented bounding boxes (OBB).

This chapter uses AABBs and a hierarchical bounding volume method to judge whether the control meshes of two subdivision surfaces are intersected. The features of AABB determine that the subdivision level should not be too deep to impact on the efficiency of the algorithm. Figure 8.1 depicts the structure of hierarchical bounding boxes, where the upperscript in $B^k_{i,j}$ indicates the split level of the bounding box.

To obtain all intersection quad mesh face pairs for the two control meshes, M^0_0 and M^0_1, we have to search all intersected quads in each pair of intersected bounding

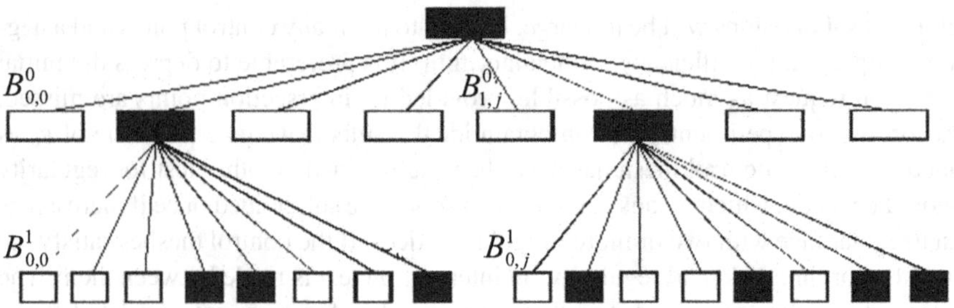

Fig. 8.1 Structure of hierarchical bounding boxes

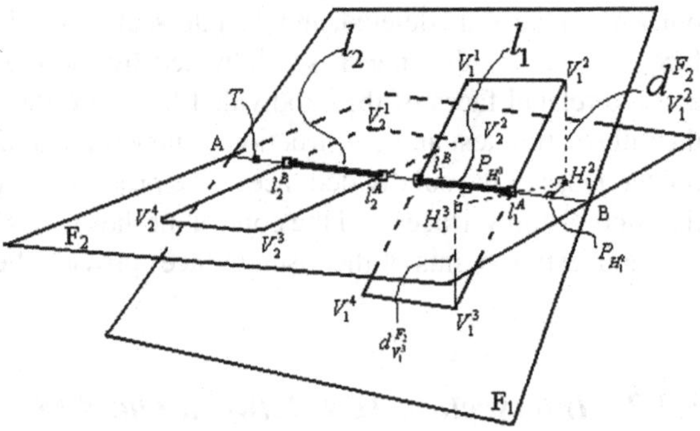

Fig. 8.2 Judgment of intersection between quads

boxes. Since the number of quads in the two intersection boxes may be large, the following acceleration method is recommended:

Let $\boldsymbol{V}_1^1$, $\boldsymbol{V}_1^2$, $\boldsymbol{V}_1^3$, $\boldsymbol{V}_1^4$ and $\boldsymbol{V}_2^1$, $\boldsymbol{V}_2^2$, $\boldsymbol{V}_2^3$, $\boldsymbol{V}_2^4$ be the four vertices of quads Q_1 and Q_2, F_1 and F_2 be the planes determined by Q_1 and Q_2, and AB be the intersection line of the planes, as shown in Fig. 8.2. Judgement of whether the two quads are intersected is made by the determination of whether the intersection line AB has a common part on the two quads.

Let the equation of the plane F_2 be

$$\boldsymbol{n}_2(\boldsymbol{X} - \boldsymbol{V}_2^4) = 0,$$

$\boldsymbol{X}$ is an arbitrary point in F_2, and

$$\boldsymbol{n}_2 = (\boldsymbol{V}_2^1 - \boldsymbol{V}_2^3) \times (\boldsymbol{V}_2^2 - \boldsymbol{V}_2^4)$$

is a unit normal of F_2. The distance of each vertex $\boldsymbol{V}_1^i$ to F_2 is

$$d_{V_1^i}^{F_2} = \boldsymbol{n}_2 \boldsymbol{V}_1^i - \boldsymbol{n}_2 \boldsymbol{V}_2^4 .$$

The intersection property of two quads Q_1 and Q_2 can be judged by the sign of $d_{V_1^i}^{F_2}$:

①If $d_{V_1^i}^{F_2} = 0 (i = 1, 2, 3, 4)$, the quads Q_1 and Q_2 are co-planar, and the intersection problem converts to finding the overlapping parts in a 2D plane; if this exists, set overlapping flag and return; otherwise, there are no intersection points.

②If $d_{V_1^i}^{F_2} \neq 0 (i = 1, 2, 3, 4)$, none of the four vertices of Q_1 are in the plane F_2. If all $d_{V_1^i}^{F_2} (i = 1, 2, 3, 4)$ have the same signs, Q_1 and Q_2 are not intersected; otherwise, there must be an intersection line between Q_1 and F_2, and an intersection line between Q_2 and F_1. Respectively denote the two intersection lines as L_1 and L_2, shown as the bold lines in Fig. 8.2. If L_1 and L_2 have common parts, we say that Q_1 and Q_2 are intersected: Set intersection flag and return.

When all intersected quad pairs between the two initial meshes, M_0^0 and M_1^0, are found with the above method, the corresponding vertices are stored in the mesh nodes link list, and all the flagged quads form a quad mesh belt.

8.2.3 Subdivision of the Intersected Mesh Belt

The skirt-removed approach was first proposed by Li [223], as a new subdivision surface boundary-generating method in the construction of n-sided blending patches. The basic idea of this technique is to first generate the blending mesh with the original subdivision rules, and then delete the skirt of the mesh. It may be seen that the process of generating and deleting boundaries is redundant in our intersection operations on discrete subdivision surfaces, for we are concerned only with the surface near the intersection lines. To improve efficiency, we have to make a revision to the original algorithm.

Before we introduce the skirt-removed approach, we will discuss some basic concepts. The boundary vertex and boundary edge generated by removing the skirt are called the new boundary vertex and new boundary edge, respectively. The face without the new boundary vertex and new boundary edge is called the inner face; that with just one new boundary vertex is called the inner corner face; that with just one new boundary edge is called the new boundary face; that with two new boundary vertices and one new boundary edge is called the pseudo-boundary face; that with two new neighboring boundary edges is called the outer corner face; and that with three new neighboring boundary edges is called the byland face. Figure 8.3 is a sketch map illustrating all these faces. Suppose that the face with flag I is a new boundary

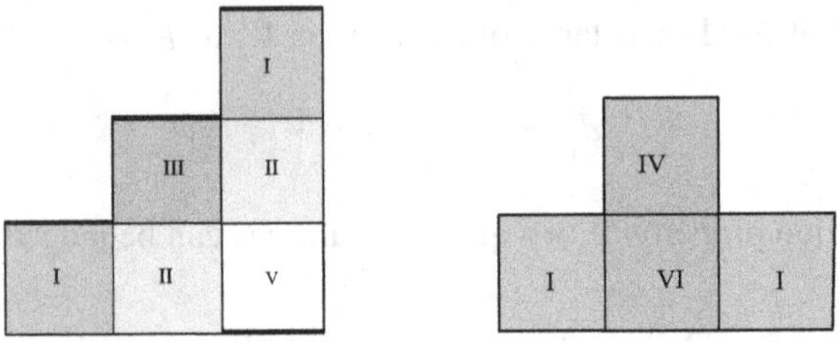

Fig. 8.3 Definitions of the different types of face

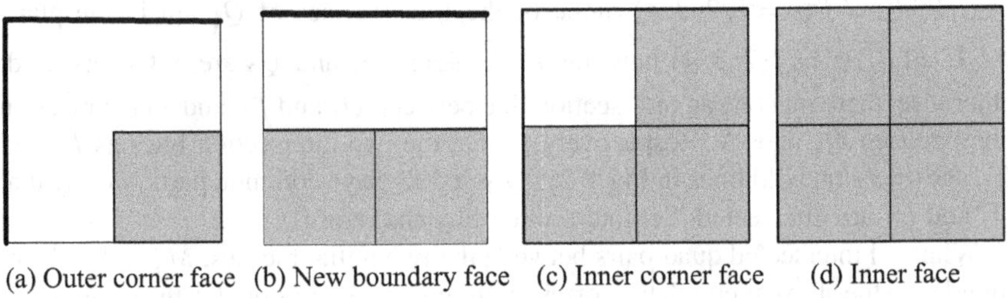
(a) Outer corner face
(b) New boundary face
(c) Inner corner face
(d) Inner face

Fig. 8.4 Topology of four types of face, subdivided once

face, the faces with flags II-VI would then correspond to the inner corner face, outer corner face, byland face, inner face, and pseudo-boundary face, respectively.

In the subdivision process, the byland face is deleted directly, and the pseudo-boundary face with an ordinary boundary edge becomes a new boundary face. The mesh topology after one subdivision of the other four types of face is shown in Fig. 8.4, in which the bold edges are new boundary edges, the gray rectangles are the faces to be retained, and all others are deleted. New boundary edges and new boundary vertices no longer generate new edge points and new vertex points, and other new face points, new edge points, and new vertex points are calculated with the original subdivision rules explained in Chap. 3 of this book. Thus, the generation and deletion work on the skirt mesh are ruled out, while the shape of the original surface is unchanged.

The characteristics of the various types of face are listed in Table 8.1.

8.2.4 *Error Control for Approximate Intersection*

The precision of initial discrete intersection can be measured by the approximate precision of the control mesh of the subdivision surface. In general, 2-4 times the final error threshold may be used as the upper boundary of the initial intersection, and further elevation of intersection precision is left to the later iteration process in order to improve efficiency. The error upper boundaries given in [86, 87] can be

Table 8.1 Characteristics of the different types of face in the revised skirt-removed method

Face type	Number of new boundary edges	Number of new boundary boundaries	Number of new faces subdivided once
Inner face	0	0	4
Inner corner face	0	1	3
New boundary face	1	2	2
Outer corner face	2	3	1
Byland face	3	4	0
Pseudo-boundary face	0	2	2

used as a transcendental estimation method, but these formulae are somewhat coarse, and the subdivision time calculated by these techniques is usually longer than that actually needed. Since the number of control vertices increases exponentially in the subdivision process, it is expensive to subdivide a mesh many times. So here, we use a posterior error evaluation method to control the precision. According to the limit formula (2.1), we can get:

$$e(\boldsymbol{v}_0) = |\boldsymbol{v}_0^{\infty} - \boldsymbol{v}_0| = \left| \frac{4\sum_{i=0}^{n-1} e_i - \sum_{i=0}^{n-1} \boldsymbol{f}_i - 5n\boldsymbol{v}_0}{n(n+5)} \right|.$$

Once we find a vertex whose error exceeds the threshold, we will end the computation and continue subdividing the related faces and testing their intersection properties. If all vertices satisfy error requirements, we stop the subdivision process and calculate the intersection points between the intersecting faces.

8.2.5 Construction of Intersection Lines

Let the normal vector of the plane F_1 determined by the quad Q_1 be $\boldsymbol{n}_1$; according to the analysis in Sect. 8.2.2, the equation of the intersection line AB of the two planes is:

$$\boldsymbol{L} = \boldsymbol{T} + t\boldsymbol{c}_L, \tag{8.1}$$

where $\boldsymbol{T}$ is an arbitrary vertex in the line AB, and

$$\boldsymbol{c}_L = \frac{\boldsymbol{n}_0 \times \boldsymbol{n}_1}{|\boldsymbol{n}_0 \times \boldsymbol{n}_1|}$$

We can see from Fig. 8.2 that if the quads Q_1 and Q_2 are truly intersected, the line segments L_1 and L_2 must have an overlapping part in the line AB, and the intersection line of Q_1 and Q_2 is in fact the common part.

The process for resolving the ends $\boldsymbol{L}_1^A$ and $\boldsymbol{L}_1^B$ of the line segment L_1 is as follows: Referring again to Fig. 8.2, let the projection point of the vertex $\boldsymbol{V}_1^2$ of the quad Q_1 to the plane F_2 be $\boldsymbol{H}_1^2$, and the projection point of $\boldsymbol{H}_1^2$ to the intersection line AB be $\boldsymbol{P}_{H_1^2}$; according to analytic geometry, it can then be deduced that:

$$\boldsymbol{P}_{H_1^2} = \boldsymbol{c}_L(\boldsymbol{V}_1^2 - \boldsymbol{T}).$$

The endpoint $\boldsymbol{L}_1^A$ of the line segment L_1 is the intersection point of the straight line $\boldsymbol{V}_1^2\boldsymbol{V}_1^3$ with AB, and the parameter $t_{L_1^A}$ of $\boldsymbol{L}_1^A$ in Eq. (8.1) is:

$$t_{L_1^A} = \boldsymbol{P}_{H_1^2} + \frac{(\boldsymbol{P}_{H_1^3} - \boldsymbol{P}_{H_1^2})\, d_{V_1^2}^{F_2}}{d_{V_1^2}^{F_2} - d_{V_1^3}^{F_2}}.$$

With the same method, we can obtain the parameter $t_{L_1^B}$ of $\boldsymbol{L}_1^B$:

$$t_{L_1^B} = \boldsymbol{P}_{H_1^1} + \frac{(\boldsymbol{P}_{H_1^4} - \boldsymbol{P}_{H_1^1})\, d_{V_1^1}^{F_2}}{d_{V_1^1}^{F_2} - d_{V_1^4}^{F_2}}.$$

So the parametric equation of L_1 is:

$$\boldsymbol{L} = \boldsymbol{T} + t\boldsymbol{c}_L \quad t \in [t_{L_1^A}, t_{L_1^B}]$$

and the parametric equation of L_2 can be constructed analogously:

$$\boldsymbol{L} = \boldsymbol{T} + t\boldsymbol{c}_L \quad t \in [t_{L_2^A}, t_{L_2^B}].$$

The intersection line of the quads Q_1 and Q_2 is the part of the straight line AB when $t \in [t_{L_1^A}, t_{L_2^B}]$ in Eq. (8.1).

When all the intersection points that satisfy the initial precision request are calculated, they are arranged to obtain the intersection (discrete) line according to the topological relationships between the original data link list of quads containing the intersection points. In order to facilitate the subsequent precision improvement and trimming operation, the parameters of the intersection points need to be calculated. The local parameter coordinates of the control vertices of different subdivision levels are set according to the method in Sect. 7.3.1, and the intersection point parameters are calculated with convex combination of the vertex parameters of the face to which they belong.

8.3 Calculating High Precision Solutions with an Iterative Method

Since the control mesh is only an approximation of the surface limit, a discrete method of subdivision can only produce an approximate solution. As the number of subdivision levels increases, the time required becomes ever larger. Hence, using only a discrete subdivision method cannot meet the requirement of high precision. The Marching method [202] overcomes the drawback of a discrete subdivision method, but it also has a fatal limitation: how to obtain a precise initial intersect point for each intersection line. In fact, subdivision surfaces have similar locality and continuity properties as spline surfaces, and a method for their local parameter representation has been discussed in Sect. 7.3, so we will improve intersection precision with an iterative method.

There are two main approaches to the iterative method: three-parameter iteration and four-parameter iteration. Here, we give the basic principles of three-parameter iteration. Let $s_1(u, v)$ and $s_2(s, t)$ be two surfaces to be intersected, and the projections of the initial intersection point $\boldsymbol{p}_0$ on the two surfaces be $\boldsymbol{p}_1(u_0, v_0)$ and $\boldsymbol{p}_2(s_0, t_0)$, respectively. Since $\boldsymbol{p}_0$ is not precise, $\boldsymbol{p}_1$ and $\boldsymbol{p}_2$ are not generally coincident (if $\boldsymbol{p}_1$ and $\boldsymbol{p}_2$ happen to be coincident, the iteration is unnecessary and $\boldsymbol{p}_1$ or $\boldsymbol{p}_2$ is the solution). By fixing the parameter u_0, the problem is converted into a nonlinear system about the parameters v, s, and t:

$$\boldsymbol{R}(v, s, t) = \boldsymbol{s}_1(u_0, v) - \boldsymbol{s}_2(s, t),$$

which can be solved with a Newton–Raphson method:

$$\boldsymbol{J}(v_{k+1}, s_{k+1}, t_{k+1}) \begin{bmatrix} \Delta v \\ \Delta s \\ \Delta t \end{bmatrix} = - \begin{bmatrix} \boldsymbol{R}(v_k, s_k, t_k)[x] \\ \boldsymbol{R}(v_k, s_k, t_k)[y] \\ \boldsymbol{R}(v_k, s_k, t_k)[z], \end{bmatrix} \tag{8.2}$$

where

$$\boldsymbol{J}(v_k, s_k, t_k) = \begin{bmatrix} \dfrac{\partial \boldsymbol{R}[x]}{\partial v} & \dfrac{\partial \boldsymbol{R}[x]}{\partial s} & \dfrac{\partial \boldsymbol{R}[x]}{\partial t} \\ \dfrac{\partial \boldsymbol{R}[y]}{\partial v} & \dfrac{\partial \boldsymbol{R}[y]}{\partial s} & \dfrac{\partial \boldsymbol{R}[y]}{\partial t} \\ \dfrac{\partial \boldsymbol{R}[z]}{\partial v} & \dfrac{\partial \boldsymbol{R}[z]}{\partial s} & \dfrac{\partial \boldsymbol{R}[z]}{\partial t} \end{bmatrix}_{(v,s,t)=(v_k,s_k,t_k)},$$

i.e., the Jacobian matrix. Solving the linear system (8.2), we can obtain the iteration increments Δv, Δs, and Δt. We revise the corresponding parameters v, s, and t and continue iterating until the following termination condition is satisfied:

$$\|\boldsymbol{R}(v_{k+1}, s_{k+1}, t_{k+1})\| \leq \varepsilon, \tag{8.3}$$

where ε is the given error upper boundary.

The four-parameter iteration method is based on geometry analysis, and its basic principle is shown in Fig. 8.5. The projection points of the approximate intersection point $\boldsymbol{p}$ are $\boldsymbol{p}_1 = s_1(u_0, v_0)$ and $\boldsymbol{p}_2 = s_2(s_0, t_0)$, and these are in-coincident. A new $\boldsymbol{p}_0$ position of $\boldsymbol{p}$ is calculated such that the distance of its projection points on the two faces is smaller than $|\boldsymbol{p}_1\boldsymbol{p}_2|$.

The method for calculating $\boldsymbol{p}_0$ is as follows: Construct two tangent planes to the two subdivision surfaces s_1 and s_2 at $\boldsymbol{p}_1$ and $\boldsymbol{p}_2$, and set the projection points of $\boldsymbol{p}_1$ and $\boldsymbol{p}_2$ at the intersection line of the two tangent planes to be $\boldsymbol{q}_1$ and $\boldsymbol{q}_2$. It can then be inferred from the geometrical relationship that:

$$\boldsymbol{p}_0 = \frac{1}{2}(\boldsymbol{p}_1 + \boldsymbol{p}_2) + \frac{1}{2}\frac{1}{(\boldsymbol{m}\cdot\boldsymbol{n})^2 - 1} \times[(\Delta\boldsymbol{p}\cdot\boldsymbol{n})(\boldsymbol{m}\cdot\boldsymbol{n})\boldsymbol{m} - \boldsymbol{n}] - (\Delta\boldsymbol{p}\cdot\boldsymbol{m})((\boldsymbol{m}\cdot\boldsymbol{n})\boldsymbol{n} - \boldsymbol{m}), \tag{8.4}$$

where $\Delta\boldsymbol{p} = \boldsymbol{p}_2 - \boldsymbol{p}_1$, and $\boldsymbol{m}$ and $\boldsymbol{n}$ are unit normal vectors of the tangent planes at $\boldsymbol{p}_1$ and $\boldsymbol{p}_2$, respectively. Take $\boldsymbol{p}_0$ as the initial value in the next iteration until $\|\Delta\boldsymbol{P}\| \le \varepsilon$.

The calculation of the projection point of a vertex onto a surface is in fact calculating the nearest point on the plane, which can be resolved with a Newton iteration method:

$$\begin{cases} [\boldsymbol{p}_i - (\boldsymbol{s}(\boldsymbol{t}_i) + \nabla\boldsymbol{s}(\boldsymbol{t}_i))]\cdot\boldsymbol{s}_u(\boldsymbol{t}_i) = 0 \\ [\boldsymbol{p}_i - (\boldsymbol{s}(\boldsymbol{t}_i) + \nabla\boldsymbol{s}(\boldsymbol{t}_i))]\cdot\boldsymbol{s}_v(\boldsymbol{t}_i) = 0, \end{cases} \tag{8.5}$$

where $\boldsymbol{t}_i = (u_i, v_i)$, $\Delta\boldsymbol{t}_i = (\Delta u_i, \Delta v_i)$, $\nabla\boldsymbol{s} = (\partial\boldsymbol{s}/\partial u, \partial\boldsymbol{s}/\partial v)$ is the Jacobian matrix, and the approximation intersection vertex parameters are the initial values for iteration. Parameter $\boldsymbol{t}_i$ is updated with $\Delta\boldsymbol{t}_i$ until $\|\Delta\boldsymbol{t}_i\| \le \delta$, where δ is the given error threshold.

There are three free parameters in the three-parameter iteration method, so this is prone to force the intersection point to some iso-parametric line. The four-parameter iteration method allows the four parameters to change freely, so it is easy to improve precision. We use both parameterization methods, chosen on the basis of the patch-

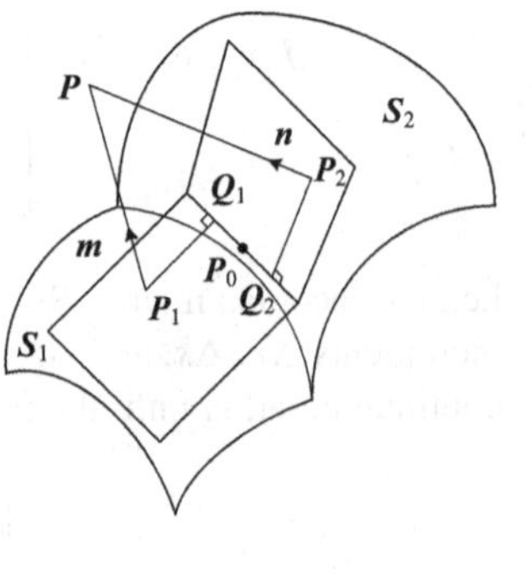

Fig. 8.5 Sketch map illustrating the four-parameter iteration method

wise properties and locally defined parameterization of subdivision surfaces, and the later trimming operation. For the convenience of the discussion, denote one subdivision surface as S_A, which will be trimmed in the subsequent trimming operation, and the other as S_B. For each initial intersection point, set its parameters corresponding to surfaces S_A and S_B as $(f_i^{\mathrm{A}}, u_i^{\mathrm{A}}, v_i^{\mathrm{A}})$ and $(f_j^{\mathrm{B}}, u_j^{\mathrm{B}}, v_j^{\mathrm{B}})$, respectively. If u_i^{A} or v_i^{A} is 0 or 1, the point must be on a parameter boundary of the f_i^A-th local coordinate system of the surface S_A, and so the three-parameter iteration method is preferred to make the point move in the boundary only. If both u_i^{A} and v_i^{A} are in the interval (0,1), the four-parameter iteration method is suitable. If u_i^{A} or v_i^{A} is smaller than 0 or larger than 1, set the value to be 0 or 1, respectively, and adopt the three-parameter method to continue the iteration. With regard to parameters u_j^{B} and v_j^{B}, if one of these is smaller than 0 or larger than 1, turn to the corresponding neighboring local coordinate system, take the coordinate in the new system, and continue the iteration.

8.4 Trimming Algorithm for Subdivision Surfaces

8.4.1 Determination of a Valid Parameter Domain for Trimmed Subdivision Surfaces

The trimming region is designated in an interactive manner. The initial control mesh faces are classified into three types: retained faces, trimmed faces, and deleted faces. The trimmed face is a central face to which there is intersection points in the local coordinate system. The retained face and deleted face are central mesh faces for which the patch should be entirely retained and deleted, respectively. According to the selection of the user, retained faces and deleted faces may be identified with a region growing method. As with trimming of B spline surfaces, inner and outer boundary loops of the trimmed surface need to be set. The direction of the intersection curves should be adjusted first. To do this, first set the direction of boundary loops for the trimmed surface: The positive direction of the inner loops is clockwise, while the positive direction of the outer loops is counter-clockwise. With the directions of the loops set, the directions of the intersection lines may be easily determined.

The valid parameter domain of the trimmed subdivision surface is composed of the retained face parameter domain and the valid parameter domain of the trimmed face. Since the retained faces have been determined, the key problem is to obtain the valid parameter domain of the trimmed face, which may be determined from the directions of the intersection lines and the stipulation of positive loop direction. Similar to the whole valid parameter domain, the outer boundary loop runs clockwise while the inner one runs counter-clockwise. The areas shaded gray in Fig. 8.6 represent the valid parameter domains of trimmed faces in different situations.

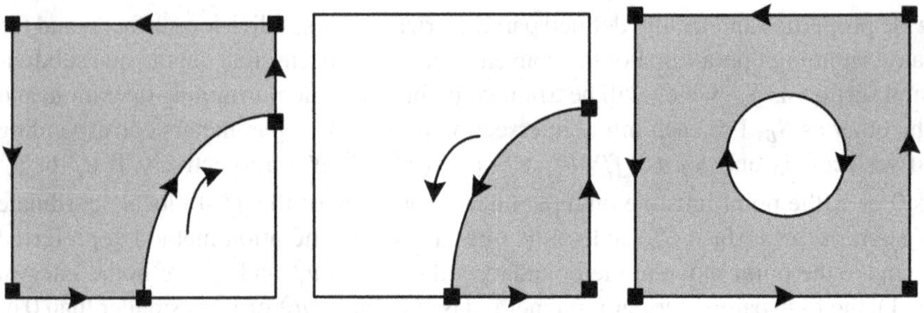

Fig. 8.6 Valid parameter domains of trimmed faces

8.4.2 Display of Trimmed Subdivision Surfaces

The boundaries of the valid parameter domain for each trimmed face are split by scanning with iso-parametric lines. As shown in Fig. 8.7, the bold lines denote the outer loop of the valid parameter domain, and the arrows on them indicate their directions. The real iso-parametric lines are in the valid region and therefore should be displayed; the dashed ones are the discarded parts. Let $v = v_0$ and $v = v_1$ be two neighboring u-lines, and $u = u_0$ and $u = u_1$ be two neighboring v-lines; according to the number of intersection parameters (u_i, v_j) $(i, j = 0, 1)$ in the valid region, there are five situations that need to be considered for the rectangle parameter domain:

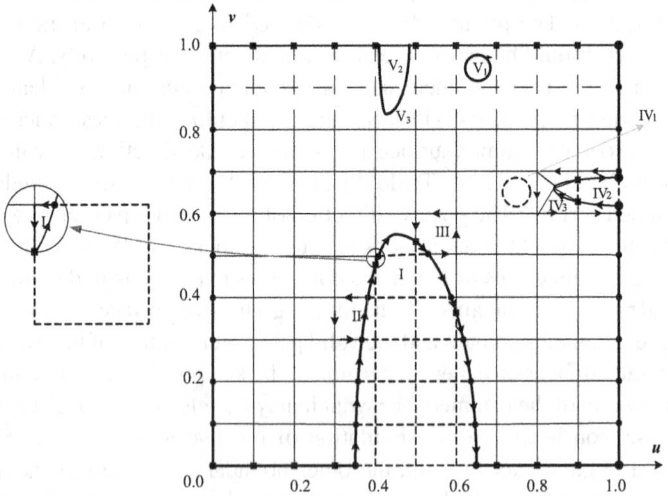

Fig. 8.7 Boundary splitting of the valid parameter domains for the trimmed face

(1) If there is only one pair of (u_i, v_j) in the valid domain, for example, region I in Fig. 8.7, the boundaries are composed of three parts: the trimming curve and two u-lines, $v = v_0$ and $v = v_1$.
(2) If there are two pairs of (u_i, v_j) in the valid domain, as in region II shown in Fig. 8.7, then: If the two pairs have a common u_i, the trimming curve and three iso-parametric lines $u = u_i$, $v = v_0$, and $v = v_1$ form the valid region; if the two pairs have a common v_j, the trimming curve and three iso-parametric lines $v = v_j$, $u = u_0$, and $u = u_1$ form the valid region.
(3) If there are three pairs of (u_i, v_j) in the valid domain, as in region III shown in Fig. 8.7, the trimming curve and four iso-parametric lines $u = u_0$, $u = u_1$, $v = v_0$, and $v = v_1$ form the valid region.
(4) If the four pairs of (u_i, v_j) are all in the valid domain, there are three possibilities:

(i) If there are small inner loops in the rectangular region, as in region IV_1 shown in Fig. 8.7, we should split the rectangular region into smaller regions, or increase the resolution to avoid this case. However, a situation in which the inner loop is smaller than the display resolution rarely occurs.
(ii) If the trimmed curve is split into two segments by the rectangular domain, as in region IV_2 shown in Fig. 8.7, divide the rectangular domain into two patches, and display each patch with the method described for situation II. Increasing the resolution can also avoid such a case.
(iii) If there is an iso-parametric line that is split into two by the trimming curve, divide the trimming curve into two at the position that is the farthest from the iso-parametric line, and add two lines linking the split point to the endpoints of the opposite iso-parametric line; the valid parameter domain is split into three parts, as in region IV_3 shown in Fig. 8.7.

(5) If none of the four pairs of (u_i, v_j) are in the valid domain, there are also three possibilities:

(i) If the outer loop of the valid region is completely within the rectangular region, as in region V_1 shown in Fig. 8.7, we should increase the display resolution to avoid such a situation.
(ii) If the trimming curve is split into two segments by the rectangular domain, as in region V_2 shown in Fig. 8.7, the two trimming curves and iso-parametric lines form the quad valid region.
(iii) If there is an iso-parametric line that is split into two by the trimming curve, as in region V_3 shown in Fig. 8.7, the iso-parametric line and the split trimming curves form a triangular valid region.

8.5 Examples

Figure 8.8 gives an example of two C-C subdivision surfaces intersecting at two open curves, where (a) shows the two subdivision surfaces to be intersected, (b) represents the two-neighbor mesh of intersection faces obtained by the revised skirt-removed

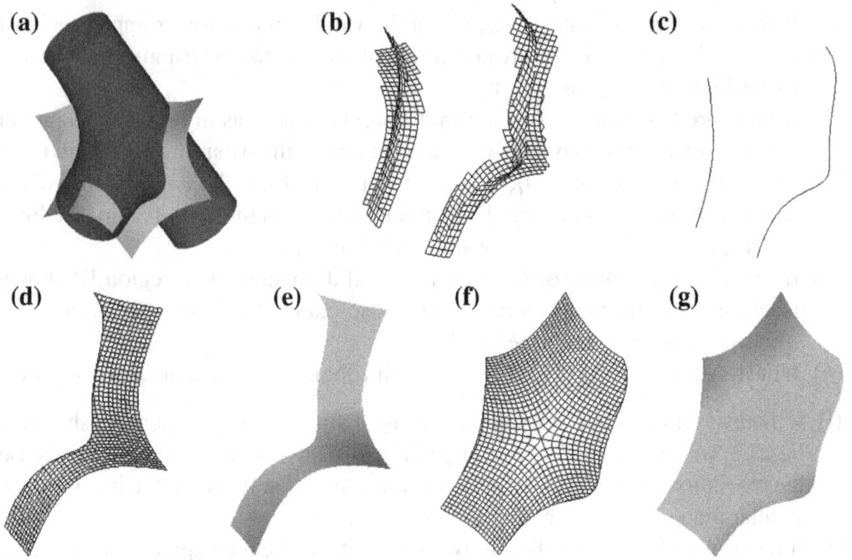

Fig. 8.8 Intersection and trimming of two subdivision surfaces I

method, (c) shows the intersection lines, and (d)–(g) illustrate the frame and shading graphs of the trimmed surfaces, respectively.

Figure 8.9 gives another example of two C-C subdivision surfaces intersecting at a closed curve, where (a) shows the two subdivision surfaces, (b) represents the two-neighbor mesh of intersection faces subdivided three times, (c) is the closed intersection line, and (d)–(e) illustrate the frame and shading graphs of the trimmed surfaces, respectively.

Remarks

Intersection and trimming are the most fundamental and important operations in CAD/CAM, and are widely applied in surface offsetting, interference inspection, and Boolean operations in solid modeling. The quality of the intersection and trimming thus impacts significantly on the robustness and applicability of the whole system. This chapter introduced two algorithms for intersection and trimming of subdivision surfaces: approximation representation with control meshes at certain subdivision levels, and piecewise parametric representation. The region of Boolean operations is closely related to intersection and trimming. Consequently, in addition to the work discussed in this chapter, there are other publications referring to intersection and trimming. For example, Litke [54] presents a trimming method for subdivision surfaces; Biermann [68] discusses Boolean operations on solids that are represented by

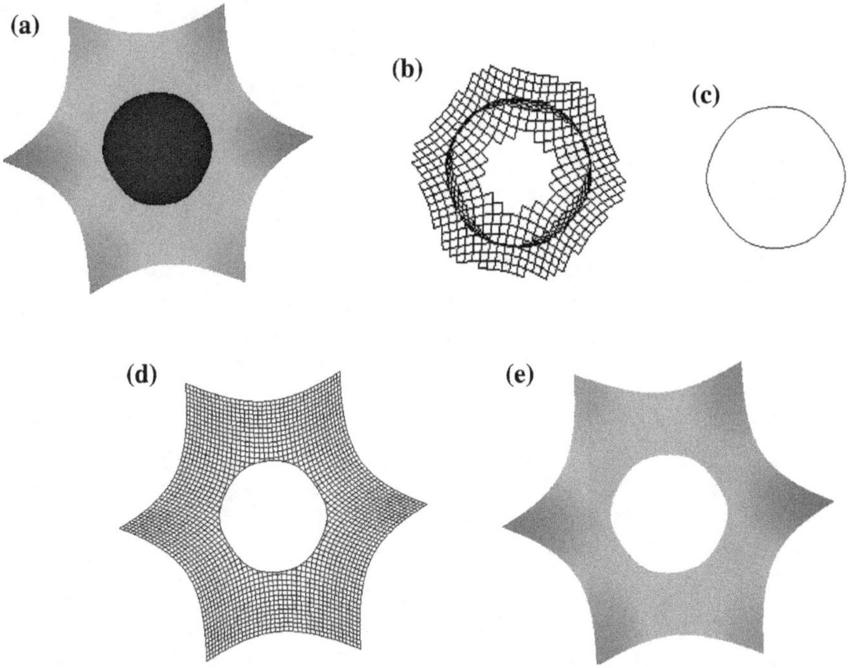

Fig. 8.9 Intersection and trimming of two subdivision surfaces II

subdivision surfaces; and Zhou [70] provides methods for intersection and trimming. The discussions in these publications are based on triangular subdivisions. Hui [69] describes a smooth blending method for C-C subdivision surfaces; blending of two surfaces can be regarded as a union operation.

Exercises

Give a concise process for the intersection operation of two subdivision surfaces according to the discussion of this chapter.

Chapter 9
Subdivision Surfaces and Curve Networks

Curve network interpolation is an important method for free-form surface modeling. This chapter first discusses the construction of a curve network of arbitrary topology and then introduces a technique for generation of a combined subdivision surface interpolating curve network. Finally, an improved approach to fairing is provided, and two shape modification methods for combined subdivision surfaces are described. The quality of the surface that interpolates the curve network is improved by modifying the second difference operator and the correction operator of the combined subdivision scheme. The first approach is based on updating the modified surface after local editing of the curve network, using the overlapping principle of the combined subdivision scheme. The second method combines vertex constraint and shape optimization with discrete PDE (partial differential equations), moving the locus of the vertex in its normal direction to make the discrete curvature distribution more ideal. Both techniques affect different regions of the combined subdivision surface during the process of shape modification.

9.1 Construction of a Curve Network

9.1.1 Basic Concepts

The definition of a curve network mainly consists of geometrical and topological information, as is the case for a polygonal mesh. The geometrical information includes the geometrical characteristics of all the curves, while the topological information describes the connecting relationships between them. In most previous research studies, curve networks have been limited to a rectangular topological structure, severely restricting the scope for their application [203–205]. In order to remove this restriction, the curve network is allowed to be of arbitrary topology, meaning

W. Liao et al., *Subdivision Surface Modeling Technology*,
DOI 10.1007/978-981-10-3515-9_9

that two or more curves may intersect at one point. The curve network may also be made up of a group of unconnected curves, but these must be connected by newly added curves to form a whole network. In order to describe a curve network more conveniently, some basic concepts and notations are explained briefly below.

1. Nodes

A point $\boldsymbol{p}$ is called a node if two or more curves $L_i (i = 0, 1, 2, \ldots)$ intersect at it; the number of curves passing through $\boldsymbol{p}$ is denoted by $N(\boldsymbol{p})$. The expression of L_i may be written as $\boldsymbol{L}_i(u)$. Curves in a curve network are united as a whole system, so that if a curve is moved, the others connected to it will move at the same time. Several nodes can be generated between two curves, while several curves may form only one node. The ends of open curves are commonly nodes. As Fig. 9.1a shows, curves L_1, L_2, and L_3 intersect and form a node p_0, which can be expressed as follows:

$$\boldsymbol{p}_0 = L_1 \cap L_2 \cap L_3 = \boldsymbol{L}_1(t_{p_0}) = \boldsymbol{L}_2(u_{p_0}) = \boldsymbol{L}_3(w_{p_0})$$

A curve network may also be composed of a group of unconnected curves, such as that shown in Fig. 9.1b, which is made up of curves L_4, L_5, and L_6.

2. Continuity

The continuity of a curve network is determined mainly by the curves it includes. At all nodes, the continuity is decided by the curves intersecting at them, while outside the nodes, continuity is the same as that of the corresponding curves.

For this reason, only the continuity of nodes needs attention. The continuity at nodes is at least G^0, on the basis of their definition. A curve network is said to be G^1 continuous at a node if the curves passing through it have a common tangent plane. If the number of curves intersecting at a node is more than two, more constraints are usually required to obtain G^1 continuity. For example, in Fig. 9.2 there are three

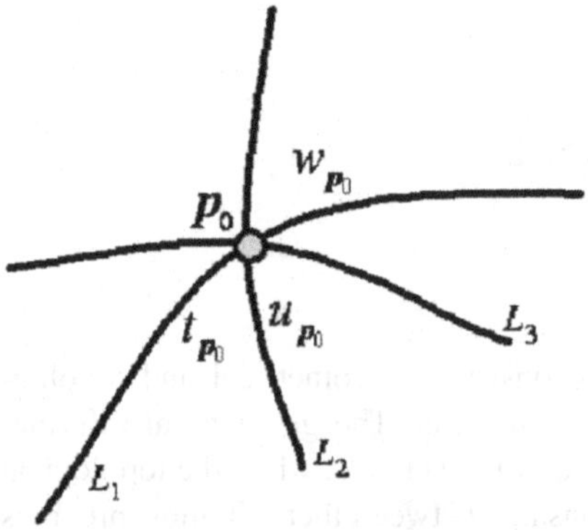

(a) A node created by three curves

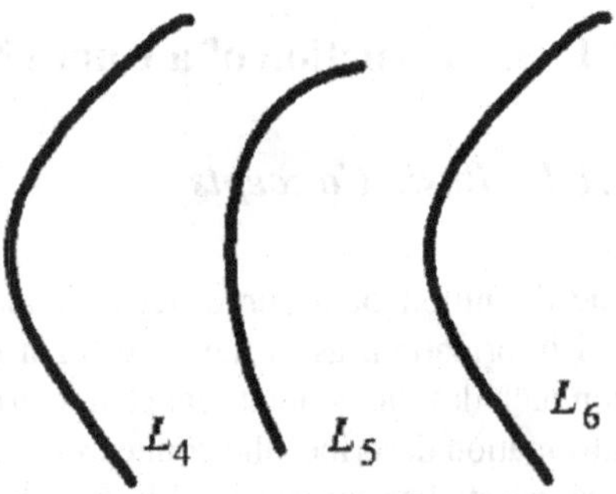

(b) A curve network without a node

Fig. 9.1 Topological structure of curve networks

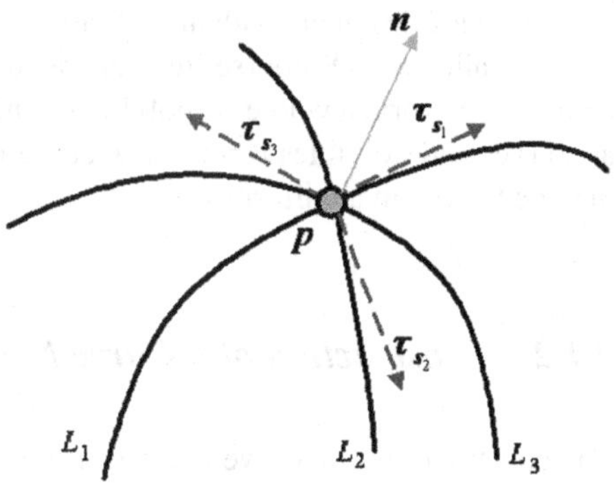

Fig. 9.2 Continuity of a curve network at a node

curves, L_1, L_2, and L_3, intersecting at a node $\boldsymbol{p}$; the G^1 continuity condition for the curve network at $\boldsymbol{p}$ is:

$$(\boldsymbol{\tau}_{L_1} \times \boldsymbol{\tau}_{L_2}) \cdot \boldsymbol{\tau}_{L_3} = 0$$

3. Topological structure

The topological structure of a curve network provides important information that describes the relationship between the curves. A curve network has a wireframe structure, similar to that of a polygonal mesh; hence, its topological structure may be represented in an analogous way. Therefore, we can represent a curve network using curve segments and curve loops.

A curve segment is a part of a curve and is defined by the parameters of its two endpoints on the curve. A curve loop is formed by curve segments, whose endpoints connect in sequence and enclose a loop. Nodes, curve segments, and curve loops together generate the complete topological information of a curve network. As shown in Fig. 9.3, the curve between $\boldsymbol{p}_1$ and $\boldsymbol{p}_2$ forms a curve segment, and the curve segments between any two of nodes $\boldsymbol{p}_0$, $\boldsymbol{p}_1$, and $\boldsymbol{p}_2$ form a curve loop. The topological correctness of a curve network may be tested by checking whether the mesh obtained by substitution of curve segments with straight-line segments satisfies the requirements of a manifold mesh.

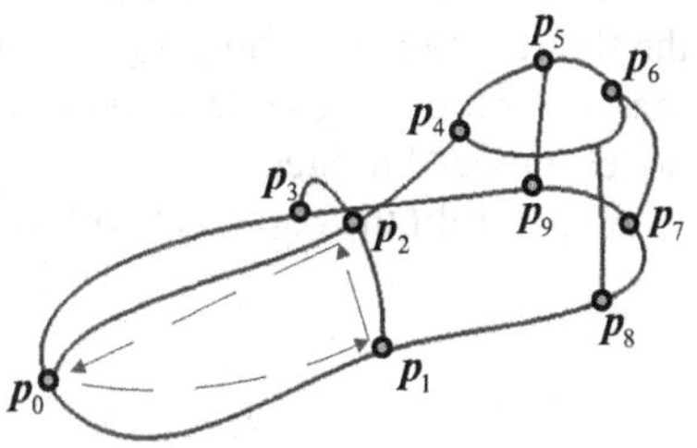

Fig. 9.3 Curve segments and curve loops in a curve network

To provide designers with more freedom, the existence of curve networks without nodes is allowed. Of course, the separated curves in the curve network should be connected to form a whole by polylines generated interactively at a subsequent step; all curves and polylines form a new curve network, with a complete topology, that may be treated in a uniform way.

9.1.2 Construction of a Curve Network

The construction of a curve network is a relatively independent process, compared with that for an interpolated surface, and forms the basis of combined subdivision surface modeling. The quality of a curve network directly influences the quality of the reconstructed combined subdivision surface and hence is particularly important to the whole curve network modeling system. There have been few recent research reports on curve network construction, and the approaches used may be classified into one of three techniques [205]: a method based on sketch, a method based on a polygonal mesh, and a method based on a point cloud. All three approaches can complete the construction of a curve network. The first has the advantages of flexibility and high efficiency, but is hard to realize due to the complexity required; the second is easy to carry out, but a mesh can provide relatively little information, and so is limited in its representation; the third is mainly used in reverse engineering. On the basis of these three techniques, a simple and efficient method for the construction of a curve network is described here, mainly using the following steps:

Step 1: Design the shapes and topological structures of the feature curves, boundary curves, and contour lines to express the shape.

Step 2: Select a current drawing plane Σ and input some points on the screen. Then, project the input points onto the plane Σ to obtain the corresponding foot points. Interpolate all the foot points to obtain a curve L_{new}, and put this into the curve network $\aleph$. This process continues until all curves are inputted, before proceeding to Step 3.

Step 3: Select two or more curves in $\aleph$ and calculate their intersection point as a node. If a node created by more than two curves does not meet a G^1 condition, the curves passing through the node need modification to satisfy the requirement of continuity. Once the defining of all nodes is completed, proceed to Step 4.

Step 4: Find all possible curve loops passing through a selected node $\boldsymbol{p}_{sel}$, using the shortest path-searching algorithm in graph theory, and interactively select the one that meets the design intention most closely. When this process is completed for all nodes, proceed to Step 5.

Step 5: Edit the curve network $\aleph$ to obtain a satisfactory shape.

9.2 A Combined Subdivision Scheme and Its Improvement

The combined subdivision scheme proposed by Levin [203, 204] is based on a classical subdivision scheme and can infinitely interpolate curves by modifying local subdivision rules. Properties such as continuity and approximation order near the interpolated curves have been analyzed, and C^2 continuity near the cubic interpolation curves has been proven theoretically by Levin [204]. The combined subdivision scheme possesses most of the advantages of a classical subdivision scheme and can conveniently and effectively interpolate a curve network; for this reason, it has been studied by numerous researchers. Kuragano [206] used a combined subdivision method to realize the automatic generation of a product geometric shape after inputting its feature curves, and Litke [207] applied a combined subdivision method to trim subdivision surfaces. The combined subdivision approach has the following advantages:

(1) There is no restriction on the representation form of the interpolated curves. Various types of curves may be used, such as Bezier, NURBS, or other parametric curves.
(2) The combined subdivision method has no restriction on topological structure and can precisely interpolate curve networks with complex topological structures.
(3) The limit surface is smooth and is at least G^1 continuous.
(4) Other classical subdivision methods, such as Catmull–Clark subdivision and Loop subdivision, can be applied to most parts of the mesh, except for the regions near the curve network. The interpolated surface thus has most of the advantages of the original subdivision method.
(5) The subdivision algorithm is simple and effective.

9.2.1 Basic Principles

A combined subdivision surface is defined by geometrical information and subdivision rules. The geometrical information includes the meshes and the interpolated curves, and the subdivision rules include ordinary subdivision operators and inverse operators. An inverse operator corresponds to a positive operator and is the key operator of a combined subdivision scheme. A positive operator is a regular subdividing operator, such as a Catmull–Clark subdivision operator or a Loop subdivision operator. If the positive operator is denoted by $\boldsymbol{S}$ and the inverse operator by $\boldsymbol{S}$, then $\boldsymbol{S}$ is the inverse of $\boldsymbol{S}^{\infty}$. The function of an inverse operator is to map a point on a curve to the curve vertex in a mesh. Let L denote the interpolated curve. If L is a cubic curve, the inverse operator in a combined subdivision method may be defined as follows [203]:

$$\overline{\boldsymbol{S}}(\boldsymbol{L}) = \boldsymbol{L} - \frac{1}{6}\left(\frac{\partial^2 \boldsymbol{L}}{\partial u^2} + \frac{\partial^2 \boldsymbol{L}}{\partial v^2}\right) \tag{9.1}$$

9.2.2 Control Meshes of Combined Subdivision Surfaces

It is important to distinguish between curve edges and common edges in the control meshes of combined subdivision surfaces, because these use different subdivision rules. *Curve edges* are the edges corresponding to the interpolated curves in the control mesh of a combined subdivision surface. Let the interpolated curve be denoted by $\boldsymbol{L}(u)$; a curve edge then corresponds to a parameter interval $[u_a, u_b]$ of $\boldsymbol{L}(u)$. The set of curve edges is denoted by Π_L^k where k is the number of subdivision steps. The two vertices that form a curve edge are called *curve vertices*, and the set of curve vertices is denoted by V_L^k. The points $\boldsymbol{p}_j^k$ on the curve that corresponds to the curve vertices are called *sample points*, and the set of sample points are denoted by P^k. As displayed in Fig. 9.4a, curve edge e_1 is formed by vertices $\boldsymbol{v}_0$ and $\boldsymbol{v}_2$, and their corresponding sample points are $\boldsymbol{p}_0$ and $\boldsymbol{p}_2$, respectively. Edges excluding curve edges are termed *ordinary edges*, and vertices excluding curve vertices are known

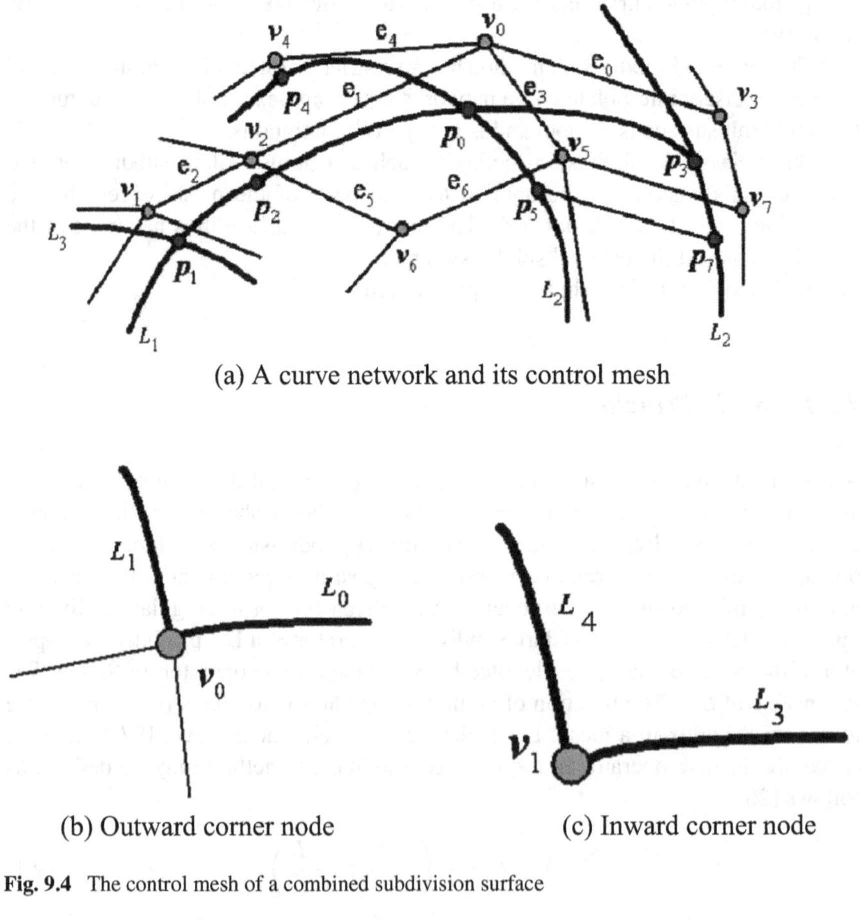

(a) A curve network and its control mesh

(b) Outward corner node (c) Inward corner node

Fig. 9.4 The control mesh of a combined subdivision surface as *ordinary vertices*. Examples, including the ordinary edge e_5 and the ordinary vertex $\boldsymbol{v}_6$, are shown in Fig. 9.4b. The sets of ordinary edges and ordinary vertices are denoted by Π_g^k and V_g^k, respectively, and the sets of all vertices and all edges are denoted by V^k and Π^k, respectively. All boundary edges in the control mesh are curve edges, and curve vertices are limited to the following six types [203]:

(1) **Internal intersection curve vertex**. There are four curve edges that emanate from this type of vertex. Of these, two opposite curve edges correspond to the same curve, and the other two opposite curve edges correspond to the other curve. Vertex $\boldsymbol{v}_0$, shown in Fig. 9.4a, is such a vertex.
(2) **Regular internal curve vertex**. There are two opposite curve edges that are associated with the same curve passing through the curve vertex, while the other opposite edges passing through this vertex are ordinary edges. Vertex $\boldsymbol{v}_2$, shown in Fig. 9.4a, is such a vertex.
(3) **Regular boundary curve vertex**. There are three edges that emanate from this type of vertex, two of which are opposite edges associated with the same boundary curve, while the other is an ordinary edge. An example, $\boldsymbol{v}_7$, is shown in Fig. 9.4a.
(4) **Boundary intersection curve vertex**. There are three curve edges that pass through this type of vertex, two of which are opposite curve edges associated with the same curve. Vertex $\boldsymbol{v}_3$, shown in Fig. 9.4a, belongs to this type of vertex.
(5) **Outward corner curve vertex**. Only two edges, corresponding to two different curves, pass through this type of vertex. Vertex $\boldsymbol{v}_0$, in Fig. 9.4b, is one such vertex.
(6) **Inward corner curve vertex**. Four edges pass through this type of vertex, and the opposite edges in each direction are associated with different curves. Vertex $\boldsymbol{v}_1$, in Fig. 9.4c, is an example.

9.2.3 Basic Operators

1. Subdivision operator
When $\boldsymbol{v} \in V_g^k$, the modified Catmull–Clark subdivision operator proposed by Sabin [204] is applied to ordinary vertices. This introduces a group of weightings into the subdivision rules that are related to the degrees of the corresponding vertices, in order to make surface regions near extraordinary points have limited curvature.
2. Second-order difference and second-order cross partial derivative
When $\boldsymbol{v} \in V_c^k$ is related to a curve L_i, there are two vectors associated with $\boldsymbol{v}$: the second-order difference $\boldsymbol{D}_{L_i}(\boldsymbol{v})$ and the second-order cross partial derivative $\boldsymbol{d}_{L_i}(\boldsymbol{v})$. If $\boldsymbol{v}$ corresponds only to a curve, the two operators may be simplified as $\boldsymbol{D}(\boldsymbol{v})$ and $\boldsymbol{d}(\boldsymbol{v})$. The second-order difference of the vertex $\boldsymbol{v}_0$ in Fig. 9.4a may be defined as:

$$\boldsymbol{D}_{L_i}(\boldsymbol{v}_0) = (\boldsymbol{p}_2 - \boldsymbol{p}_0) - (\boldsymbol{p}_0 - \boldsymbol{p}_3) \tag{9.2}$$

For the node $\boldsymbol{p}_3$ shown in Fig. 9.4a, the second-order difference of the related curve vertex $\boldsymbol{v}_3$ is defined as:

$$\boldsymbol{D}_{L_i}(\boldsymbol{v}_3) = 4\boldsymbol{p}_3 - 8\boldsymbol{L}\left(\frac{u_{p_3} + u_{p_0}}{2}\right) + 4\boldsymbol{p}_0 \quad (9.3)$$

When $\boldsymbol{v}$ is a curve vertex corresponding to an intersection point of two curves $L_i(i = 1, 2)$, the second-order cross partial derivative $\boldsymbol{d}_{L_i}(\boldsymbol{v})$ may be defined as:

$$\boldsymbol{d}_{L_i}(\boldsymbol{v}) = \boldsymbol{D}_{L_i}(\boldsymbol{v}_i), \quad i = 1, 2 \quad (9.4)$$

The second-order cross partial derivatives of other curve vertices are calculated according to the rules of combined subdivisions, after their initial values are obtained.

3. Inverse operator
The inverse operator $\overline{\boldsymbol{S}}$ is a particular operator of the combined subdivision scheme and is used mainly to map some sample point $\boldsymbol{p}$ on the interpolated curve to the corresponding vertex $\boldsymbol{v}$ on the control mesh. Different subdivision schemes have different inverse operators. When $\boldsymbol{p}$ is an intersection point of two curves, $L_i(i = 1, 2)$, $\overline{\boldsymbol{S}}$ may be defined as:

$$\boldsymbol{v} = \overline{\boldsymbol{S}}(\boldsymbol{p}) = \boldsymbol{p} - \frac{1}{6}(\boldsymbol{D}_{L_1}(\boldsymbol{v}) + \boldsymbol{D}_{L_2}(\boldsymbol{v})) \quad (9.5)$$

When $\boldsymbol{v}$ is not the intersection point of some curves, $\overline{\boldsymbol{S}}$ is defined as:

$$\boldsymbol{v} = \overline{\boldsymbol{S}}(p) = \boldsymbol{p} - \frac{1}{6}(\boldsymbol{D}(\boldsymbol{v}) + \boldsymbol{d}(\boldsymbol{v})) \quad (9.6)$$

4. Correction operator
In order to maintain the continuity of an interpolated surface near the interpolated curves, the loci where ordinary vertices connect directly with curve vertices need to be adjusted. As shown in Fig. 9.5, the correction operator is defined as [203]:

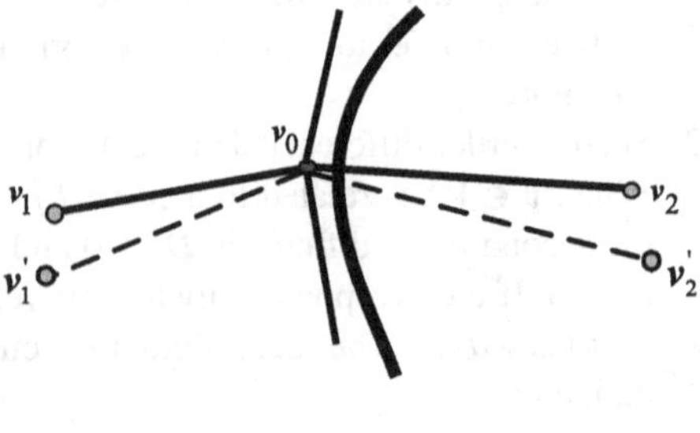

Fig. 9.5 Local correction of a vertex

$$\boldsymbol{v}_1' = \boldsymbol{C}(\boldsymbol{v}_1) = \boldsymbol{v}_0 + \frac{\boldsymbol{d}(\boldsymbol{v}_0)}{2} + \frac{\boldsymbol{v}_1 - \boldsymbol{v}_2}{2} \tag{9.7a}$$

$$\boldsymbol{v}_2' = \boldsymbol{C}(\boldsymbol{v}_2) = \boldsymbol{v}_0 + \frac{\boldsymbol{d}(\boldsymbol{v}_0)}{2} + \frac{\boldsymbol{v}_2 - \boldsymbol{v}_1}{2} \tag{9.7b}$$

where $\boldsymbol{v}_0 \in V_c^k$, and $\boldsymbol{v}_1$ and $\boldsymbol{v}_2$ are the ordinary vertices connected directly with $\boldsymbol{v}_0$. $\boldsymbol{v}_1'$ and $\boldsymbol{v}_2'$ are the corresponding new vertices of $\boldsymbol{v}_1$ and $\boldsymbol{v}_2$ after separately applying the correction operator.

9.2.4 Non-uniform Combined Subdivision Schemes

When the shape of the interpolated curves is complicated or the distribution of the sample points uneven, an area of fluctuation or depression will appear in the limit surface of a combined subdivision surface, as shown in Fig. 9.6.

Through analysis of the reasons for unfairness, studies [205] have made modifications to the second-order difference operator and the correction operator and proposed an improved combined subdivision scheme termed the non-uniform combined subdivision scheme.

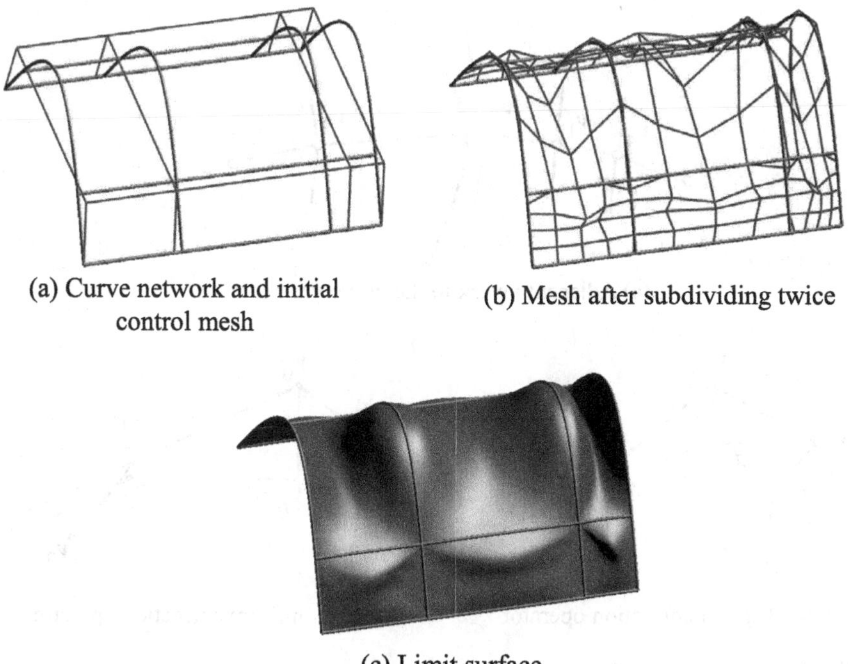

(a) Curve network and initial control mesh

(b) Mesh after subdividing twice

(c) Limit surface

Fig. 9.6 A combined subdivision surface with poor fairness

1. Non-uniform second-order difference operator
   The improved second-order difference operator is known as the non-uniform second-order difference operator and is represented by $\widehat{\boldsymbol{D}}$. It is defined by:

$$\widehat{\boldsymbol{D}}(\boldsymbol{v}) = \widehat{\boldsymbol{D}}(\boldsymbol{p}) = \frac{2d_1}{d_1 + d_2}(\boldsymbol{p}_2 - \boldsymbol{p}) - \frac{2d_2}{d_1 + d_2}(\boldsymbol{p} - \boldsymbol{p}_2) \tag{9.8}$$

   where d_1 and d_2 are the distances from $\boldsymbol{p}_1$ and $\boldsymbol{p}_2$ to $\boldsymbol{p}$, respectively. Although the original second-order difference operator is replaced by a non-uniform second-order difference operator, the method for calculating the position of a curve vertex, and the updating rule for the boundary cross partial derivative, remain unchanged. As subdivision proceeds, curve vertices become progressively more dense, and $\widehat{\boldsymbol{D}}(\boldsymbol{v})$ and $\boldsymbol{d}(\boldsymbol{v})$ converge to zero at a convergence rate of 1/4. Therefore, the curve vertices gradually converge to points on the curves, allowing interpolation to the target curve to be realized.
2. Non-uniform correction operator
   The function of the correction operator is to guarantee the continuity of the constructed surface near the interpolated curves. The definition of a correction operator is given in Eq. (9.9). Let Π be a plane determined by vertex $\boldsymbol{v}_0$ and parallel vectors $\boldsymbol{d}(\boldsymbol{v}_0)$ and $(\boldsymbol{v}_1 - \boldsymbol{v}_2)$, and the control polygon $\boldsymbol{v}_5\boldsymbol{v}_4\boldsymbol{v}_1\boldsymbol{v}_0\boldsymbol{v}_2\boldsymbol{v}_3$ be represented by $\boldsymbol{\Gamma}$, as shown in Fig. 9.7a. Figure 9.7b is a sketch map showing the projection

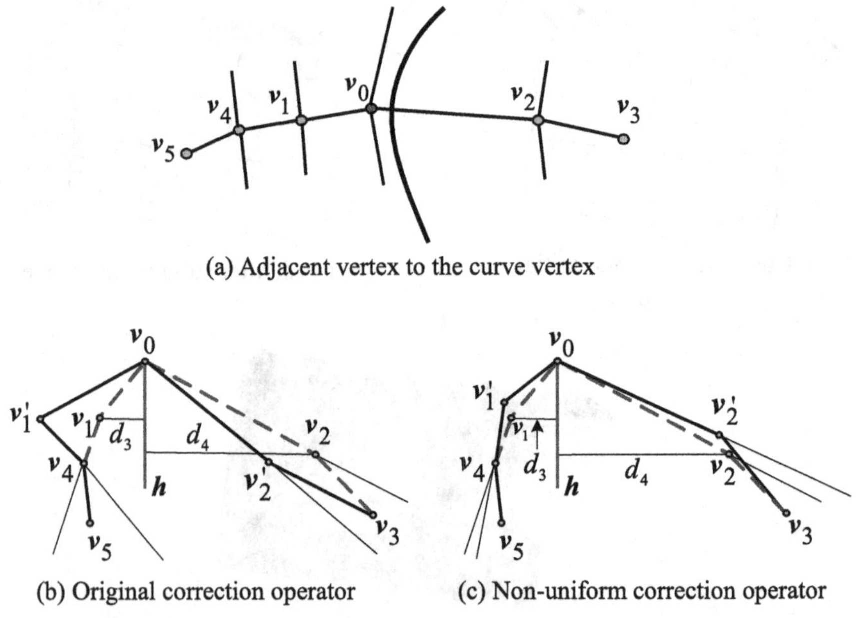

Fig. 9.7 Improvement of the correction operator

of $\boldsymbol{\Gamma}$ onto Σ where $\boldsymbol{v}_0\boldsymbol{h}$ parallels vector $\boldsymbol{d}(\boldsymbol{v}_0)$, and $\boldsymbol{d}_3$ and $\boldsymbol{d}_4$ are the distances from $\boldsymbol{v}_1$ and $\boldsymbol{v}_2$ to $\boldsymbol{v}_0\boldsymbol{h}$, respectively. $\boldsymbol{v}_3$, $\boldsymbol{v}_4$ and $\boldsymbol{v}_5$ are ordinary vertices. Let $\boldsymbol{\Gamma}$ be a convex polygon where $\boldsymbol{v}_5$ and $\boldsymbol{v}_3$ fall in the shaded region of Fig. 9.8b; polygon $\boldsymbol{v}_5\boldsymbol{v}_4\boldsymbol{v}_1'\boldsymbol{v}_0\boldsymbol{v}_2'\boldsymbol{v}_3$ will become concave after applying operator $\boldsymbol{C}$. As the difference between $\boldsymbol{d}_3$ and $\boldsymbol{d}_4$ gets larger, the probability that $\boldsymbol{\Gamma}$ turns from convex to concave becomes greater. The next subdivision may decrease the degree of depression near $\boldsymbol{v}_2'$, but it cannot completely avoid the depression near $\boldsymbol{v}_4$. The limit curve of $\boldsymbol{\Gamma}$ is thus unfair, and there are many unnecessary inflection points on its projection curve on Σ. This unfairness is due to the correction operator $\boldsymbol{C}$, especially when the difference between $\boldsymbol{d}_3$ and $\boldsymbol{d}_4$ becomes too large. For this reason, a coefficient

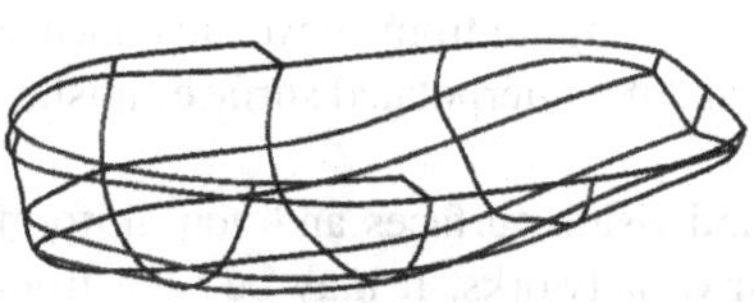

(a) Curve network

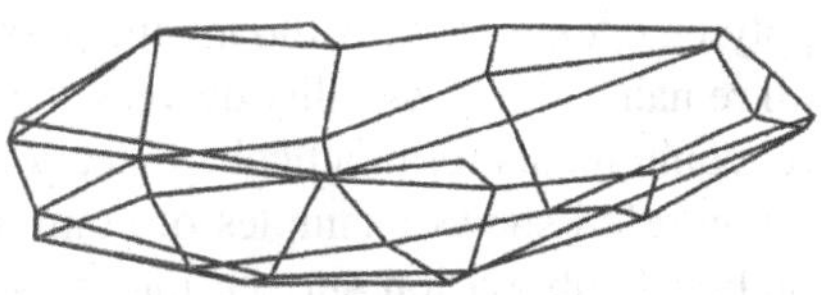

(b) Initial control mesh

(c) Mesh after combined subdivision twice

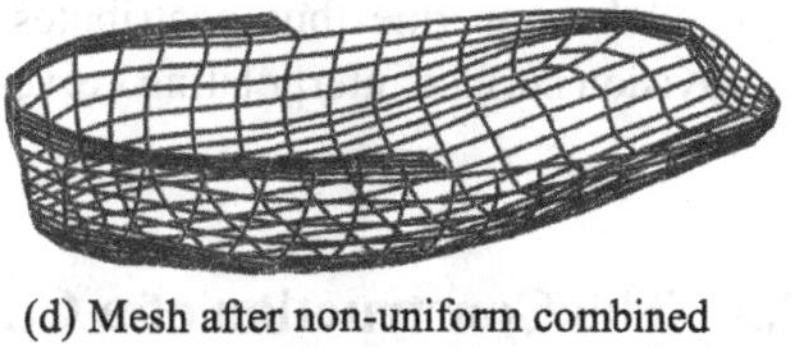

(d) Mesh after non-uniform combined subdivision twice

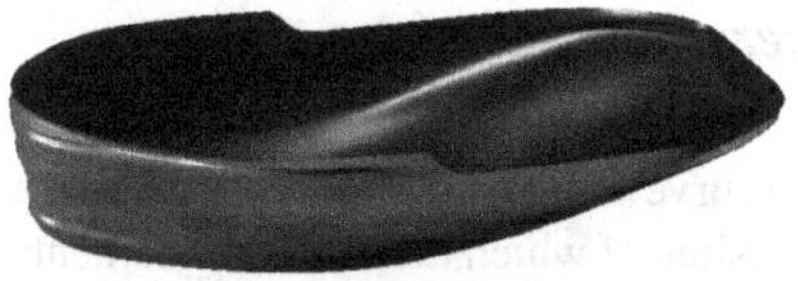

(e) Limit surface of the combined subdivision

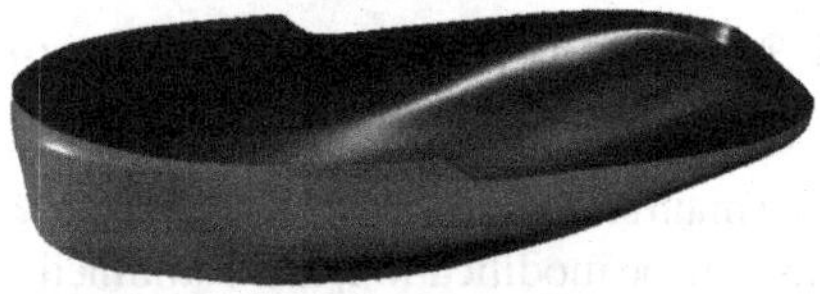

(f) Limit surface of the non-uniform combined subdivision

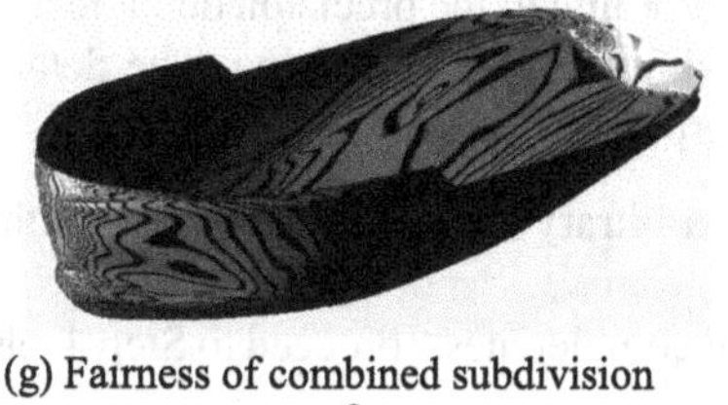

(g) Fairness of combined subdivision surface

(h) Fairness of non-uniform combined subdivision surface

Fig. 9.8 Comparison of combined subdivision with non-uniform combined subdivision

α is introduced to keep the ratio of $\boldsymbol{d}_3$ to $\boldsymbol{d}_4$ unchanged after applying the correction operator, and the correction operator $\boldsymbol{C}$ is modified to reduce the perturbation of vertices connected with curve vertices. Figure 9.7c illustrates the revised correction operator, called the non-uniform correction operator and represented as $\boldsymbol{C}_m$. It is defined as:

$$\boldsymbol{C}_m(\boldsymbol{v}_1) = \boldsymbol{v}_0 + \frac{\boldsymbol{d}(\boldsymbol{v}_0)}{2} + \alpha(\boldsymbol{v}_1)(\boldsymbol{v}_1 - \boldsymbol{v}_2) \tag{9.9}$$

where $\alpha(\boldsymbol{v}') = \dfrac{d_1}{d_1 + d_2}$.

It may be found from Fig. 9.7c that polygon $\boldsymbol{v}_5 v_4 \boldsymbol{v}'_1 \boldsymbol{v}_0 \boldsymbol{v}'_2 \boldsymbol{v}_3$ remains convex after applying operator $\boldsymbol{C}_m$. Although the possibility of changing from convex to concave still remains, this possibility decreases greatly, and the interpolated surface satisfies the requirements for practical applications.

Figure 9.8 shows examples of combined subdivision surfaces and non-uniform combined subdivision surface interpolating curve networks. It may be seen from Fig. 9.8 that the mesh to which operators $\boldsymbol{C}_m$ and $\widehat{\boldsymbol{D}}$ are applied distributes more uniformly, and its limit surface is fairer than that of the original limit surface. The new subdivision rule thus contributes to improving the quality of the combined subdivision surface interpolating curve network.

9.3 The Construction of a Curve Network Interpolation Surface

9.3.1 Steps for Constructing an Interpolated Surface

The main steps for surface modeling based on a curve network are initial mesh generation, shape modification, and combined subdivision, of which the first is fundamental before application of a combined subdivision. When a curve network is composed of unconnected curves, the curves need to be connected to obtain a complete mesh, and the cross second derivatives of the curve vertices require calculating. The subdivision surface can be subdivided to a certain depth according to the precision demands, and the interpolated surface is eventually obtained at the given precision. The detailed steps for construction of a curve network-interpolated surface are as follows:

Step 1: Construct curve network $\aleph$ with arbitrary topology, using the method described in Sect. 9.1.

Step 2: If the curve network has topological completeness, proceed to Step 4; else, go to Step 3.

Step 3: Select points on each curve, and connect the new sample points with polylines to generate a common edge in the mesh until a completed curve network is obtained.

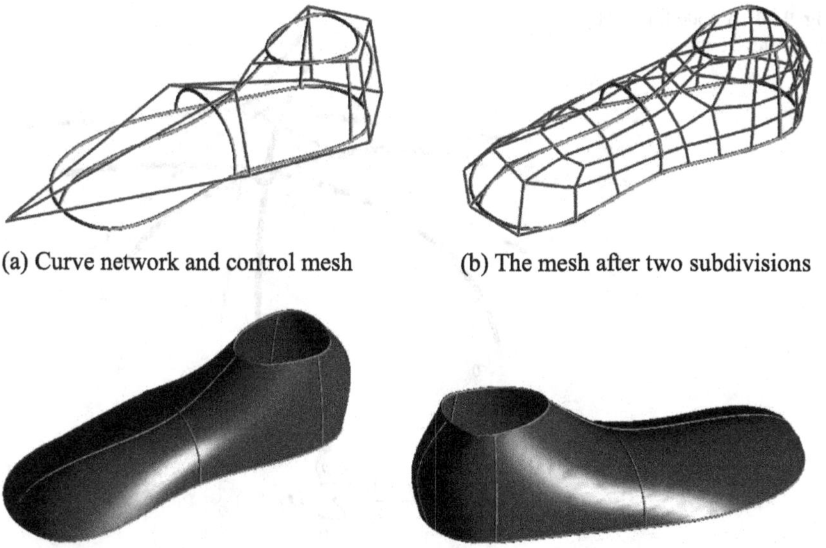

(a) Curve network and control mesh (b) The mesh after two subdivisions

(c) The final interpolated surface (d) Another view of the interpolated surface

Fig. 9.9 A curve network interpolation subdivision surface

Step 4: Based on the nodes and the curve face loops of the curve network $\aleph$, generate the topological structure of the initial mesh M^0.

Step 5: Compute the second cross derivative and second difference of the curve vertex v_j^0 of mesh M^0 and obtain the position of v_j^0 using the inverse operator $\overline{S}$.

Step 6: Subdivide the combined subdivision surface defined by $\aleph$ and M^k to a certain depth, until a certain precision is satisfied.

Figure 9.9 illustrates the process of generating a model of a shoe, using a subdivision surface interpolating curve network.

9.3.2 Extension of the Combined Subdivision

In order to ensure the C^2 continuity of the curve network interpolation surface, the combined subdivision surface proposed by Levin [203] does not allow two or more curves to intersect at one point; this restricts the scope of the curve network for representing shapes. Whereas C^2 continuity is difficult to achieve, C^1 continuity can satisfy most of the common demands of practical engineering design; on this basis, the combined subdivision scheme is extended so that it can interpolate a curve network with more than two curves intersected at a node, enhancing its ability to represent shapes. The definition of an inverse operator on point $\boldsymbol{p}$, shown in Fig. 9.10, is given by:

Fig. 9.10 A node formed by three intersecting curves

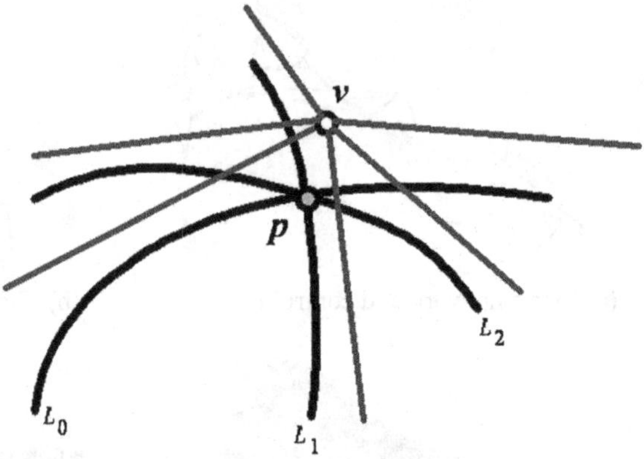

$$v = \overline{S}(p) = p - \frac{1}{3k}\sum_{i=0}^{N(p)} D_{L_i}(v) \qquad (9.10)$$

When the curve network is constructed, the nodes formed by more than two curves are required to be G^1 continuous, ensuring the G^1 continuity of the limit surface.

9.3.3 Generation of the Initial Mesh

Generation of the initial mesh is an important step in the construction of a curve network. The curve network is so general that it may contain unconnected curves and curve face loops composed of more than four curves; it therefore needs preprocessing in order to obtain the initial control mesh for the combined subdivision surface. The generation of the initial mesh includes the following steps:

Step 1: Add sample points p_{ij} at curve L_i according to the topological structure of curve network Ω, and connect the sample points on different curves to get ordinary edges linking the unconnected curves.

Step 2: Optimize the topological structure of the curve network $\aleph$ to reduce the number of curve segments in a single curve face loop and the degree of each node.

Step 3: Compute the second derivative $d(v_j^k)$, and make some adjustments to obtain a more satisfactory shape for interpolated surface.

Step 4: Apply the inverse operator $\overline{S}$ to obtain the positions of the vertices in the initial mesh M^0 of the combined subdivision surface.

Once a curve network with a completely connected relationship is set up, the initial mesh for the combined subdivision surface is constructed, the initial second derivatives of the curve vertices are calculated using Levin's method [203], and the regular combined subdivision is applied to obtain the final limit surface.

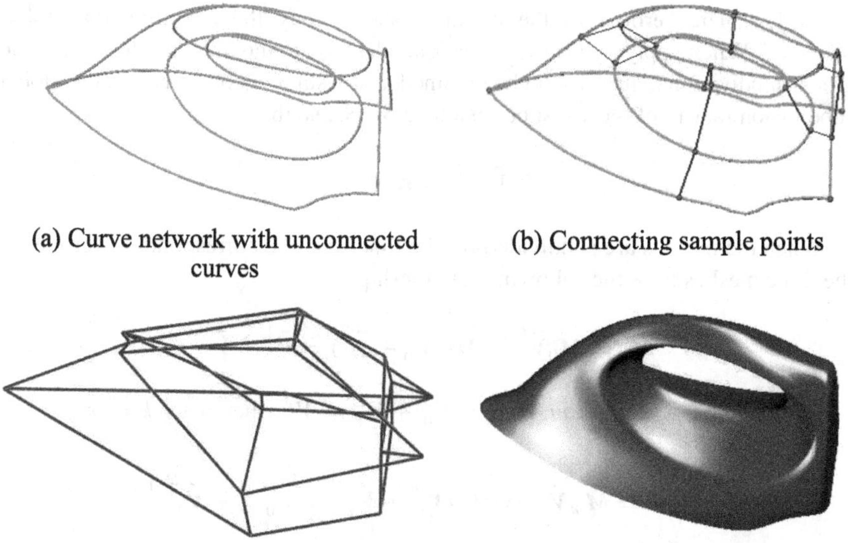

(a) Curve network with unconnected curves

(b) Connecting sample points

(c) Initial mesh

(d) Limit surface

Fig. 9.11 Construction of an interpolated curve network and a limit surface

Figure 9.11 shows an example of the construction of a control mesh from unconnected curves in a curve network for a model of an iron, and of generation of the final interpolated surface. In this example, a combined subdivision modeling method with feature is used.

9.4 Shape Modification Based on Curve Network Editing

Although an interpolating curve network is an effective approach for the design of surfaces, it is difficult to obtain a satisfactory geometric shape by interpolating the feature curves or boundary curves. In addition, surface design is a repeating process that needs to modify the shape many times to achieve a perfect result. For these reasons, shape editing functions are very important for methods using curve network modeling.

9.4.1 *Overlapping Subdivision Method for a Catmull–Clark Subdivision Surface*

Let M_0, M_1, and M_2 be three meshes with the same topological structure. After subdividing k times, their vertices are denoted by $\vec{V}_i^k = [v_{i0}^k, v_{i1}^k, \ldots, v_{in}^k]$ where

$i = 0, 1, 2$. The vertices of the initial meshes satisfy the condition that $V_2^0 = V_0^0 + V_1^0$. When applying Catmull–Clark subdivision to the three meshes, the same global subdivision matrix would be obtained at any subdivision level. Let the global subdivision matrix of the kth subdivision be $M_k S_k$, so that:

$$\vec{V}_i^{k+1} = M_k \vec{V}_i^k.$$

When the meshes are subdivided for the first time, the corresponding vertices in the three meshes have the following relationship:

$$\vec{V}_2^1 = M_0 \vec{V}_2^0 = M_0(\vec{V}_0^0 + \vec{V}_1^0) = \vec{V}_0^1 + \vec{V}_1^1$$

Provided that when $k = m$, there is $V_0^m + V_1^m = V_2^m$, then when $k = m + 1$, we obtain:

$$\vec{V}_2^{m+1} = M_m \vec{V}_2^m = M_m(\vec{V}_0^m + \vec{V}_1^m) = \vec{V}_0^{m+1} + \vec{V}_1^{m+1}$$

From the formula above, we can conclude that the vertex position vectors in mesh M_2 at any subdivision level can be obtained by adding the position vectors of the corresponding vertices in meshes M_0 and M_1 at the same subdivision level.

9.4.2 Overlapping Subdivision Method for Combined Subdivision Surfaces

Definition 9.1 Let $\aleph_0$ and $\aleph_1$ be two curve networks with the same topological structure. Their curves and nodes have one-to-one correspondence. $\aleph_0$ and $\aleph_1$ are called isomorphic curve networks when $\aleph_0$ and $\aleph_1$ satisfy the following conditions:

(1) the node vectors of curve $L_{0i}(u)$ in $\aleph_0$ are the same as those of the corresponding curve $L_{1i}(u)$ in $\aleph_1$.
(2) the node p_{0i} in $\aleph_0$ corresponds to the node p_{1i} in $\aleph_1$, and curves $L_{0j}(u)$ and $L_{1j}(u)$ separately pass p_{0i} and p_{1i} at the same parameter value.

The following analysis applies provided that $\aleph_0$, $\aleph_1$, and $\aleph_2$ are isomorphic curve networks, and their corresponding curves satisfy the following relation:

$$L_{2j}(u) = L_{0j}(u) + L_{1j}(u) \tag{9.11}$$

As shown in Fig. 9.12, let v_{i0}, v_{i1}, and v_{i2} be related to points p_{i0}, p_{i1}, and p_{i2}. Their parameters in curve $L_{ij}(u)$ are u_0, u_1, and u_2, respectively. The relationships of the second differences and cross-boundary partial derivatives in $\aleph_0$, $\aleph_1$, and $\aleph_2$ to the inverse operator and modified operator are analyzed below.

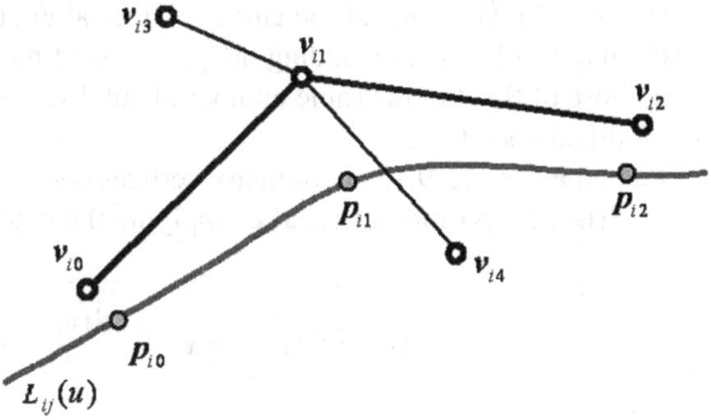

Fig. 9.12 Mesh overlapping of a combined subdivision surface

1. Second difference and cross second derivative
   According to the definition of the second difference, $\boldsymbol{D}(\boldsymbol{v}_{i1})$ is given by:

$$\boldsymbol{D}(\boldsymbol{v}_{i1}) = \boldsymbol{p}_{i0} + \boldsymbol{p}_{i2} - 2\boldsymbol{p}_{i1} \quad i = 0, 1, 2$$

   Adding the second-order difference of the vertex $\boldsymbol{v}_{01}$ to that of $\boldsymbol{v}_{11}$, we obtain:

$$\boldsymbol{D}(\boldsymbol{v}_{01}) + \boldsymbol{D}(\boldsymbol{v}_{11}) = (\boldsymbol{p}_{00} + \boldsymbol{p}_{10}) + (\boldsymbol{p}_{02} + \boldsymbol{p}_{12}) - 2(\boldsymbol{p}_{01} + \boldsymbol{p}_{11}) \tag{9.12}$$

   From Eq. (9.11), the following relation may be found:

$$\boldsymbol{p}_{0m} + \boldsymbol{p}_{1m} = \boldsymbol{p}_{2m}, \quad m = 0, 1, 2$$

   Putting the above formula into Eq. (9.12), we have:

$$\boldsymbol{D}(\boldsymbol{v}_{01}) + \boldsymbol{D}(\boldsymbol{v}_{11}) = \boldsymbol{p}_{20} + \boldsymbol{p}_{22} - 2\boldsymbol{p}_{21} = \boldsymbol{D}(\boldsymbol{v}_{21}) \tag{9.13}$$

   It is easy to prove that the cross second derivatives of the corresponding vertices in curve networks $\aleph_0$, $\aleph_1$, and $\aleph_2$ with the same topological structure satisfy the principle of similar overlap.
2. Inverse operator
   From the definition of inverse operator, the position of $\boldsymbol{v}_{i1}$ is:

$$\boldsymbol{v}_{i1} = \overline{\boldsymbol{S}}(\boldsymbol{p}_{i1}) = \boldsymbol{p}_{i1} - \frac{1}{6}(\boldsymbol{D}(\boldsymbol{v}_{i1}) + d(\boldsymbol{v}_{i1}))$$

   By combination with Eqs. (9.12) and (9.13), we obtain:

$$\boldsymbol{v}_{01} + \boldsymbol{v}_{11} = \boldsymbol{v}_{21} \tag{9.14}$$

   The parameter of a new edge vertex is the average of those of the curve edge endpoints, so the parameters of the vertices of the corresponding curve in the three curve networks have the same distribution during the subdivision of $\aleph_0$, $\aleph_1$, and $\aleph_2$. The positions of the curve vertices after subdividing the curve network $\aleph_2$ may be obtained by adding the positions of the corresponding vertices in $\aleph_0$ to those of $\aleph_1$ after the same number of subdivisions.

3. Modified operator
$\boldsymbol{v}_{i3}$ and $\boldsymbol{v}_{i4}$ in Fig. 9.12 are ordinary vertices connected directly with curve vertex $\boldsymbol{v}_{i1}$. The new position of $\boldsymbol{v}_{i3}$ after applying the modified operator is calculated as follows:

$$\boldsymbol{v}'_{i3} = \boldsymbol{C}(\boldsymbol{v}_{i3}) = \boldsymbol{v}_{i1} + \frac{\boldsymbol{d}(\boldsymbol{v}_{i1})}{2} + \frac{\boldsymbol{v}_{i3} - \boldsymbol{v}_{i4}}{2}$$

Supposing $\boldsymbol{v}_{03} + \boldsymbol{v}_{13} = \boldsymbol{v}_{23}$ and $\boldsymbol{v}_{04+\boldsymbol{v}_{14}} = \boldsymbol{v}_{24}$, then from Eq. (9.14) we obtain:

$$\boldsymbol{v}_{03} + \boldsymbol{v}_{13} = \boldsymbol{v}_{23} \tag{9.15}$$

The positions of ordinary vertices are obtained gradually by subdivision and in a combined subdivision surface are determined by the curve vertices, Catmull–Clark subdivision rules, and the modified operator. After analyzing the relationship of $\boldsymbol{v}_{23}$ to $\boldsymbol{v}_{03}$ and $\boldsymbol{v}_{13}$, it may be concluded that Catmull–Clark subdivision satisfies the principle of a mesh overlapping subdivision. Essentially, the vertex positions subdivided from curve network $\aleph_2$ at any subdivision level may be obtained by adding the positions of the corresponding vertices subdivided from curve networks $\aleph_0$ and $\aleph_1$ at the same subdivision level.

9.4.3 Local Updating of the Combined Subdivision Surface

Only local editing of the curve network will be considered in this section. A new curve network is constructed from the difference curves created by subtracting the original ones from the edited ones. By using the method proposed in the previous section, the subdivided mesh of the edited curve network can be obtained by overlapping the subdivision mesh of the original curve network with the subdivision mesh of the difference curve network. The detailed steps are as follows:

Step 1: Subdivide the original curve network $\aleph_0$ k times using the combined subdivision method and obtain mesh M_0^k.

Step 2: Edit curve network $\aleph_0$ locally and obtain curve network $\aleph_2$.

Step 3: Subtract the curves in $\aleph_0$ from the corresponding curves in $\aleph_2$ to obtain curve network $\aleph_1$.

Step 4: Subdivide $\aleph_1$ k times using the combined subdivision method, and obtain mesh M_1^k.

Step 5: The vertex positions in mesh M_2^k, generated by subdividing mesh $\aleph_2$ k times, may be obtained by adding the positions of the corresponding vertices in mesh M_0^k to those of mesh M_1^k.

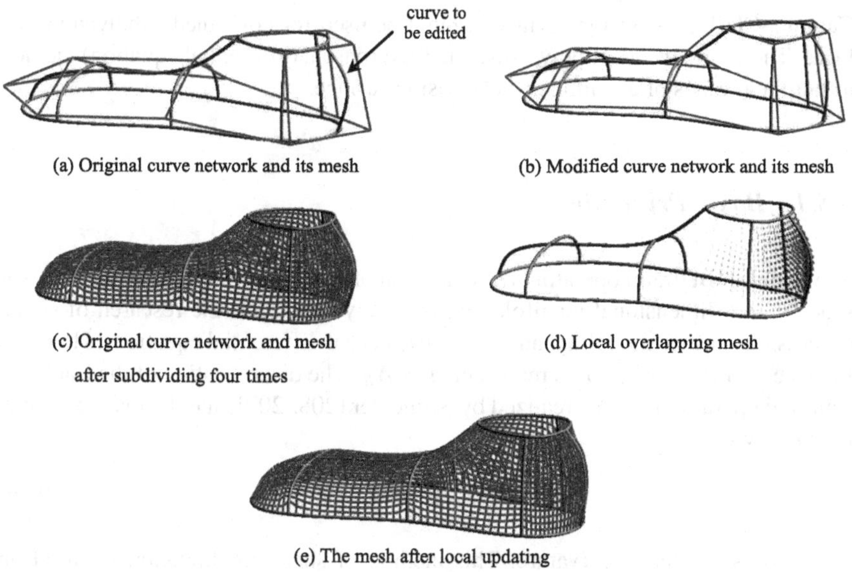

Fig. 9.13 Example of the modification of a combined subdivision interpolation surface, based on local editing of a curve network

As the curve network is edited locally, with only the curve segment between two end nodes edited, and the positions and tangential vectors of the two end nodes unchanged, all other curve segments in $\aleph_0$ and $\aleph_2$ remain the same. Most curve segments in $\aleph_1$ are "zero," except those associated with the edited curves, so only the submeshes associated with the edited curve loops are subdivided locally.

Figure 9.13 gives an example of the modification of a combined subdivision interpolation surface, based on the editing of a local curve network.

9.5 Shape Fairing of a Combined Subdivision Surface Based On Discrete PDE

After applying the method for combined subdivision surface modification, based on curve network editing, we can generally obtain a satisfactory curve network. However, in some cases the combined subdivision surface will still have flaws, creating a surface with low quality. This can make designers discontented with the local shape of the interpolation surface and can limit the scope for use of combined subdivisions. This section focuses on a fairing method for modification of combined subdivision surfaces. Due to the special subdivision rules near curves in a curve network, it is difficult to realize exact parameterization and computation of energy. For this reason, a fairing method based on energy optimization that is commonly used for

Catmull–Clark subdivision surfaces cannot be used for combined subdivision surfaces. Therefore, we will apply a discrete PDE (partial differential equation) method to fair the meshes of a combined subdivision surface.

9.5.1 Basic Principle

The Laplace–Beltrami operator is an extension of the Laplace operator from a planar space to a 2-dimensional manifold and is widely applied in the research of signal processing, surface modeling, and geometric partial differential equations. Here, the Laplace–Beltrami operator is represented as $\boldsymbol{\Delta}_B$. The discretization of the Laplace–Beltrami operator has been realized by Schneider [208, 209], and the main principle is as follows:

$$\boldsymbol{\Delta}_B H = 0 \tag{9.16}$$

where H is the mean curvature. This method of shape modification is based on geometrical invariance and can be derived from the Euler–Lagrange equation that minimizes the energy function $\int_S (k_1^2 + k_2^2) dS$. It can be used as an independent shape optimization method and has advantages over ordinary energy minimization methods. For example, this method can create surfaces such as spherical and cylindrical surfaces, whose mean curvatures are constant with boundary conditions of G^1 continuity, and for which extreme values [208] of mean curvature on internal surfaces do not exist, similar to a planar cloth-like spline.

9.5.2 A Discrete PDE Fairing Method

The combined subdivision surface can approximate the limit surface during the process of subdivision. After several subdivisions, the control mesh can be used as the approximation of the limit surface. Theoretically, the more subdivisions there are, the better the mesh approximates its limit surface; we can therefore fair the combined subdivision surface by fairing its mesh. In order to ensure the interpolation of a curve network, the curve edges must remain unchanged during the process of fairing, which differs from common surface fairing methods. If we divide the mesh of a combined subdivision surface by the curve edges and set its boundary conditions by the neighboring submesh, we only need solve the linear system of the submesh when applying the discrete PDE fairing method, avoiding integration and optimization steps and improving fairing efficiency.

1. Preprocessing the control mesh of the combined subdivision surface

The fairing algorithm in Schneider's research [208, 209] is designed for triangle meshes with boundary conditions, while the combined subdivision surface here is based on the quadrilateral mesh of a Catmull–Clark subdivision surface. To apply the fairing algorithm of Schneider, we apply the following regularization and preprocessing techniques to the mesh of the combined subdivision:

(1) Subdivide the control mesh once or twice to make all faces quadrangular, and denote the subdivided mesh by M_0.
(2) Since all boundaries are curve edges, the control meshes of the combined subdivision surface M_k are divided into a series of small meshes by these curve edges. Suppose that all these small meshes sew together without overlap to just form the completed mesh M_k and that there is no curve edge on the interior of the small meshes. Such small meshes are called the submeshes of the original meshes and are denoted by $\boldsymbol{B}_i^k$. When fairing a submesh, its adjacent submeshes form its boundaries. Fairing the submeshes at different subdivision levels can realize the global and local fairing of the original combined subdivision surface.
(3) The neighborhood of a control vertex $\boldsymbol{v}$ is a set, denoted by $N(\boldsymbol{v})$, that consists of the vertices that belong to the same face with vertex $\boldsymbol{v}$. The number of vertices in $N(\boldsymbol{v})$ is called the degree of $N(\boldsymbol{v})$ and is denoted by $n(\boldsymbol{v})$.
(4) The normal of vertex $\boldsymbol{v}_i$ is calculated as follows:

$$\boldsymbol{n}_{\boldsymbol{v}_i} = \frac{\sum\limits_j a_j \boldsymbol{n}_j}{\sum\limits_j a_j}$$

where a_j and $\boldsymbol{n}_j$ are the region and the normal vector of the ith face passing vertex $\boldsymbol{v}_i$, respectively.

2. Discrete curvature of control vertices

In order to ensure an effective solution for the discrete PDE, a uniform discrete curvature calculation method should be adopted for all vertices of the mesh, no matter whether they are general control points or vertices defining the interpolated curves. In addition, the discrete curvature should be a continuous function of the vertices in the mesh. Fortunately, the discrete curvature computing method proposed by Moreton [210] satisfies these two demands: the basic principle is that the discrete curvature can be calculated by solving the equations determined by the relationships between normal curvature, principle curvature, and principle direction.

Let the set of vertices that share the same face with vertex $\boldsymbol{v}_i$ be $\mathrm{Ne}(\boldsymbol{v}_i) = \{\boldsymbol{q}_j | j = 0, \ldots, 2n(\boldsymbol{v}_i) - 1\}$. As shown in Fig. 9.14, suppose $\boldsymbol{t}_j$ to be a unit vector projected by the vector $\overrightarrow{\boldsymbol{v}_i \boldsymbol{q}_j}$ in the mesh tangent plane defined by normal $\boldsymbol{n}_{\boldsymbol{v}_i}$, and make a circle crossing $\boldsymbol{q}_j$ and $\boldsymbol{v}_i$, with the tangent at $\boldsymbol{v}_i$ being $\boldsymbol{t}_i$. Let the curvature $\widetilde{\kappa}_j$ of the circle be the approximation of normal curvature, that is:

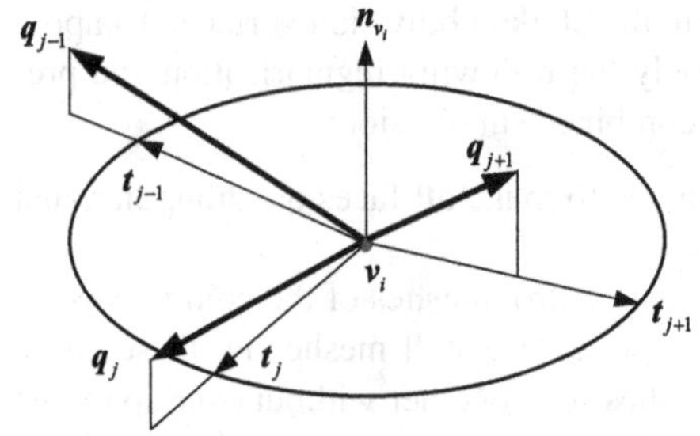

(a) Projection of the edge on tangent planes

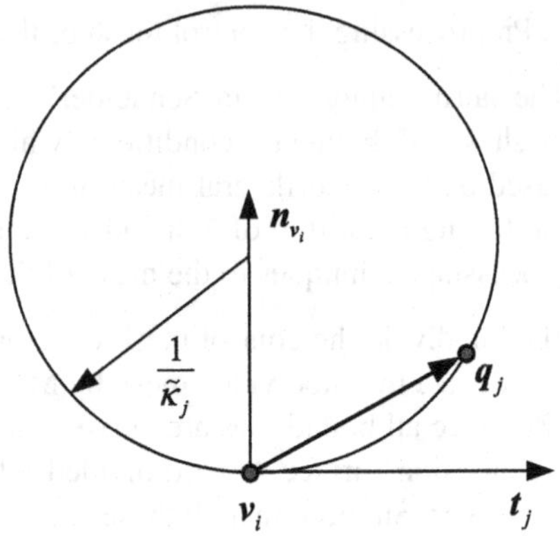

(b) Normal curvature

Fig. 9.14 Computation of discrete curvature

$$\widetilde{k}_j = \frac{2(\boldsymbol{q}_j - \boldsymbol{v}_i)\boldsymbol{n}_{\boldsymbol{v}_i}}{(\boldsymbol{q}_j - \boldsymbol{v}_i)(\boldsymbol{q}_j - \boldsymbol{v}_i)} \tag{9.17}$$

Let $\boldsymbol{b}_x$ and $\boldsymbol{b}_y$ be a group of orthogonal units based on $\boldsymbol{v}_i$ on the tangent plane, the coordinate of the vector $\boldsymbol{t}_j$ be $(t_{j,x}, t_{j,y})$, and the coordinate of the main direction at position $\boldsymbol{v}_i$ be (e_x, e_y). According to Euler's theorem, the normal curvature k_j of $\boldsymbol{v}_i$ can be represented as:

$$\widetilde{k}_j = [t_{j,x}, t_{j,y}]\boldsymbol{K}\begin{bmatrix} t_{j,x} \\ t_{j,y} \end{bmatrix}$$

where $K = \begin{bmatrix} e_x & e_y \\ -e_y & e_x \end{bmatrix}\begin{bmatrix} k_1 & 0 \\ 0 & k_2 \end{bmatrix}\begin{bmatrix} e_x & e_y \\ -e_y & e_x \end{bmatrix}^{-1}$.

Let $j = 0, \ldots, m$ and $m = n(\boldsymbol{v}_i) - 1$; the following linear system may be obtained as $\boldsymbol{Ax} = \boldsymbol{d}$ where:

$$\boldsymbol{A} = \begin{bmatrix} t_{0,x}^2 & t_{0,x}t_{0,y} & t_{0,y}^2 \\ t_{1,x}^2 & t_{1,x}t_{1,y} & t_{1,y}^2 \\ \vdots & \vdots & \vdots \\ t_{m,x}^2 & t_{m,x}t_{m,y} & t_{m,y}^2 \end{bmatrix}, \quad \boldsymbol{d} = \begin{bmatrix} \widetilde{k}_0 \\ \widetilde{k}_1 \\ \vdots \\ \widetilde{k}_m \end{bmatrix},$$

$$\boldsymbol{x} = \begin{bmatrix} x_0 \\ x_1 \\ x_2 \end{bmatrix} = \begin{bmatrix} e_x^2 k_1 + e_y^2 k_2 \\ 2e_x e_y (k_1 - k_2) \\ e_x^2 k_2 + e_y^2 k_1 \end{bmatrix}$$

The discrete mean curvature of the point $\boldsymbol{v}_i$ may be calculated by:

$$H = \frac{1}{2}(x_0 + x_2) \tag{9.18}$$

Since $m + 1 > 3$, this equation system is overdetermined and may be solved with a generalized inverse matrix method:

$$\boldsymbol{x} = (\boldsymbol{A}^t\boldsymbol{A})^{-1}\boldsymbol{A}^t\boldsymbol{d} \tag{9.19}$$

3. Discrete PDE based on the Laplace–Beltrami operator

There are several discrete modes for the Laplace–Beltrami operator [208–211], deduced from the relationship between the Laplace–Beltrami operator and the mean normal curvature of the surface. The discrete curvature calculating method [211] proposed by Desbrun is suitable for any mesh because it is simple and effective. As shown in Fig. 9.15, Eq. 9.16 can be discretized with this method as:

$$\sum_{\boldsymbol{v}_j \in \mathrm{Ne}(\boldsymbol{v}_i)} (\cos\alpha_j + \cos\beta_j)(H_j - H_i) = 0 \tag{9.20}$$

where $Ne(\boldsymbol{v}_i)$ is the neighborhood of the vertex $\boldsymbol{v}_i$, H_i is the mean curvature of vertex $\boldsymbol{v}_i$, and α_j and β_j are the angles as shown in Fig. 9.15.

When all the vertices in the interior of the submesh satisfy the above equations, the PDE will be discretized into sparse linear equations, in which the mean curvatures of the vertices in V_I are the variables to be solved while those for the vertices in V_B and V_C are known where V_I is the vertex set of the investigated mesh, V_B are the boundary vertex set, and V_C are the constrained set. The coefficients of the equations are:

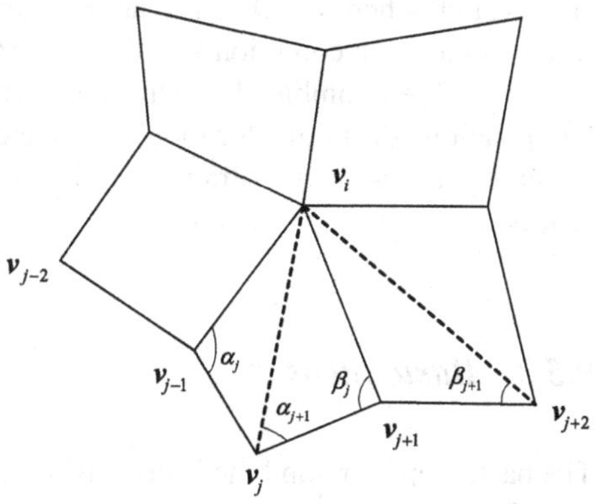

Fig. 9.15 Sketch map of the Laplace–Beltrami operator in quad meshes

$$c_{i,j} = \begin{cases} \sum(\cos\alpha_j + \cos\beta_j) & i = i \\ -(\cos\alpha_j + \cos\beta_j) & \boldsymbol{v}_j \in Ne(\boldsymbol{v}_i) \\ 0 & \text{other} \end{cases}$$

These equations can be solved using a complete pivoting Jordan elimination method, and their solutions are the mean curvatures of the vertices in V_I.

4. Updating the positions of the control vertices

The ideal mean curvature $\widetilde{H}_i$ of the inner vertex $\boldsymbol{v}_i^k$ in V_I can be obtained by solving the discrete PDE equations. The positions of the vertices have to be adjusted in the normal direction to make $H(\boldsymbol{v}_i^{k+1}) = \widetilde{H}_i$ where $\boldsymbol{v}_i^{k+1}$ is the vertex after adjusting $\boldsymbol{v}_i^k$. Let the deviation quantity in the normal direction of $\boldsymbol{v}_i^k$ be t; then, $\boldsymbol{v}_i^{k+1} = \boldsymbol{v}_i^k + t\boldsymbol{n}_{\boldsymbol{v}_i^k}$. Since the coefficient matrix A in Eq. (9.19) remains unchanged, it is clear from Eq. (9.17) that the discrete mean curvatures are only related to $\widetilde{k}_i$, and so $\widetilde{k}_i$ after adjustment can be represented as follows:

$$\widetilde{k}_j = \frac{2(\boldsymbol{q}_j - \boldsymbol{v}_i - t\boldsymbol{n}_{v_i^k})\boldsymbol{n}_i}{(\boldsymbol{q}_j - \boldsymbol{v}_i - t\boldsymbol{n}_{v_i^k})(\boldsymbol{q}_j - \boldsymbol{v}_i - t\boldsymbol{n}_{v_i^k})}$$

Suppose that the normal adjustment quantity of $\boldsymbol{q}_j$ is so small that the variation of $\boldsymbol{q}_j$ can be neglected relative to the value of $\|\boldsymbol{q}_j - \boldsymbol{v}_i\|$; then, from the formula above, we obtain:

$$\widetilde{k}_j \approx \frac{2(\boldsymbol{q}_j - \boldsymbol{v}_i)\boldsymbol{n}_i}{(\boldsymbol{q}_j - \boldsymbol{v}_i)(\boldsymbol{q}_j - \boldsymbol{v}_i)} - \frac{2t}{(\boldsymbol{q}_j - \boldsymbol{v}_i)(\boldsymbol{q}_j - \boldsymbol{v}_i)} \tag{9.21}$$

This equation together with Eqs. (9.18) and (9.19) forms a linear system, and the values of t_i may be obtained by solving this system. In the practicalities of the adjustment process, the error of Eq. (9.12) will expand if t_i is too large. In order to avoid this situation, multiple steps can be adopted in the adjustment strategy, *i.e.*, let $\boldsymbol{v}_i^{k+1} = \boldsymbol{v}_i^k + \lambda t_i \boldsymbol{n}_{v_i^k}$ where $0 < \lambda \le 1$. A superior result was obtained in our research when $\lambda = 0.8$. The adjustment of the position of the vertices will end when all meet the condition that $\Delta H_i = |H_i - \widetilde{H}_i| < \varepsilon$. The curvature distribution of the modified combined subdivision surface will then be more reasonable. The interpolation constraints for the vertices and curves are assured, since the positions of curve vertices and constrained vertices remain unchanged during the adjustment process.

9.5.3 Basic Steps

The basic steps for combined subdivision surface shape modification, based on discrete PDE, are as follows:

Step 1: Mesh M^k is obtained by subdividing the curve network k times, using the combined subdivision scheme.

Step 2: The constraint vertex set V_c is obtained by imposing vertex constraint conditions. The boundary vertex set V_B and the inner vertex set V_I may also be obtained by searching the mesh M^k from the neighborhood of constrained vertices.

Step 3: Set up discrete PDE equations $\Delta_B \widetilde{H}_i = 0$. The ideal curvature values ($\widetilde{H}_i$) of the vertices (v_i) in V_I are obtained by solving these equations.

Step 4: Repeatedly adjust the positions of ordinary vertices in V_I, according to the ideal curvature value in Step 3. The adjustment will terminate if the difference $\Delta H_i = |H_i - \widetilde{H}_i|$ between the actual discrete curvature and the ideal discrete curvature is smaller than the given error ε.

Step 5: If the result is satisfactory, proceed to step 6; otherwise, return to step 2 and modify the surface again.

Step 6: Continue subdividing the modified mesh to obtain the approximation of the limit surface of the interpolated surface.

9.5.4 Example

Figure 9.16 uses a model of a car as an example to illustrate shape modification of a curve network-interpolated surface. The combined subdivision surface between the interpolated curves blends uniformly after fairing, and the quality of the interpolated surface is clearly improved.

Remarks

Curve network interpolation is an important free-form surface modeling method. The interpolated curves may be feature curves, section curves, or contours of the surface, whose layout and quality are key factors that influence the quality and efficiency of modeling. Coons surfaces, Gordon surfaces, and B-spline skinning methods are all based on infinite interpolation of curves and are widely used in the fields of CAD, geometric modeling, and computer graphics. However, surfaces constructed by these methods are all restricted to rectangular parameter domains, and in general it is difficult to construct a whole interpolated surface for complex models. A curve network usually has to be divided into a few small ones with rectangular topology, and the split networks are interpolated independently and stitched into a completed interpolated surface, greatly reducing the modeling efficiency. For some time, researchers have wanted to break the constraints of curve network topology and model more freely [203–205]. Subdivision surface modeling based on curve network constraints, presented in this chapter, is one example of a suitable approach. With the help of the combined subdivision scheme, we have not only described an interpolation technique with an arbitrary curve network, but have also provided

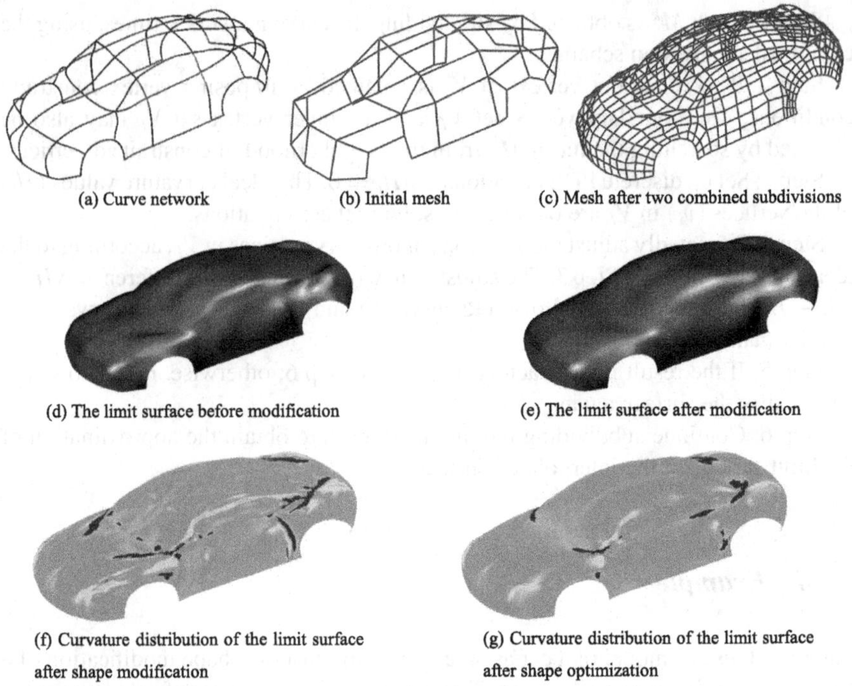

(a) Curve network (b) Initial mesh (c) Mesh after two combined subdivisions

(d) The limit surface before modification (e) The limit surface after modification

(f) Curvature distribution of the limit surface after shape modification (g) Curvature distribution of the limit surface after shape optimization

Fig. 9.16 An example of shape modification for combined subdivision surfaces

a modification method based on curve constraints and an approach for fairing of combined subdivision surfaces based on discrete PDE.

Exercises

Give a concise process for the combined subdivision scheme according to the discussion of this chapter.

Chapter 10
Fitting Unstructured Triangle Meshes Using Subdivision Surfaces

Approximation, interpolation, and fitting are three fundamental topics in CAGD. Fitting may be regarded as a process that constructs a curve or surface whose shape is as close as possible to given points. In reverse engineering, fitting plays a very important role in the construction of such curves or surfaces within a specified tolerance. The data points are always obtained by 3D measurement methods. In this chapter, we assume that these data points form an unstructured triangle mesh and investigate fitting methods based on subdivision surfaces. These fitting surfaces are subdivision surfaces or meshes with subdivision connectivity. Meshes with subdivision connectivity are important because they are necessary for multiresolution analysis, data compression, and multiresolution editing. The majority of existing fitting methods provide meshes with subdivision connectivity, in a process known as remeshing. The meshes to be fitted will have been preprocessed using techniques such as mesh de-noising, simplification, matching, and hole filling; this may be achieved using software such as Geometric. Preprocessing of mesh models is not discussed in this chapter, and readers interested in this area are referred to relevant papers in the literature for more details [191].

10.1 A Simplified Shrink-Wrapping Approach for Remeshing

10.1.1 Introduction to the Shrink-Wrapping Approach

The shrink-wrapping approach was first described by Kobblet [60]. Bertram [65] used this method to construct a mesh with C-C subdivision connectivity in order that wavelet transformations were executed. Zhou [119] has also used this method to construct a mesh with Loop subdivision connectivity. The approach of Sweldens was

W. Liao et al., *Subdivision Surface Modeling Technology*,
DOI 10.1007/978-981-10-3515-9_10

similar to that of Bertram [65] and Zhou [119], which involved construction of base meshes by applying a mesh decimation algorithm to the original mesh. Based on these descriptions [60, 65, 119], Liu [117] has presented an approach to construct a mesh with Catmull–Clark subdivision connectivity. Unlike previous work [65, 119], Liu [117] did not distinguish between sharp vertices and sharp edges, and consequently, this technique did not employ special subdivision rules for sharp features. This technique is therefore unlikely to generate satisfactory results for certain complicated objects, as pointed out by Kobblet [60], who provided a method for solving this problem. Although the approach of Liu [117] may be regarded as a simplification of that of Kobblet [60], in our experience, this method [117] can nonetheless generate good results for many objects. Even if mesh surfaces are obtained with artifacts and distortions, it is possible to improve the fairness by adjusting the values of various factors. The methods described in this section can also be considered to be a simplified version of that of Kobblet [60]. In the view of Kobblet [60], shrink-relaxing does not change the unstable character of projection, and because of this, in their shrink-wrapping algorithm they used a sphere parameterization method in several early steps of the iterative process. Since sphere parameterization is not employed in our approach, the method presented here is termed a simplified shrink-wrapping approach.

10.1.2 Construction of Base Meshes

Base meshes are the start of the shrink-wrapping approach. These may be obtained using the method described in Sect. 6.1 or by applying a mesh decimation algorithm to the original mesh. There are also reverse engineering software packages, such as Geomagic, which can decimate unstructured meshes. By uniting triangles on the decimated meshes, we obtain base meshes. Figure 10.1 illustrates the process for obtaining a base mesh. The uniting algorithm will be discussed in Sect. 10.1.2.

The closer the shape of the base mesh to the original mesh, the smaller the error between the fitting mesh and the original mesh. However, this must be balanced against the requirement to keep the base mesh as simple as possible. The degree to

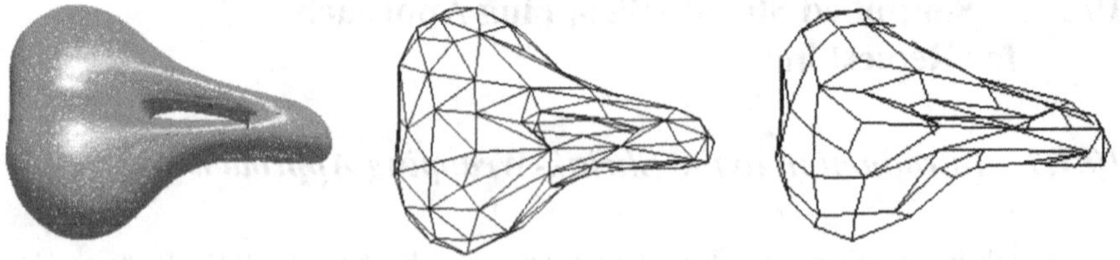

(a) Original mesh (b) Decimated mesh (c) Base mesh

Fig. 10.1 Construction of base mesh

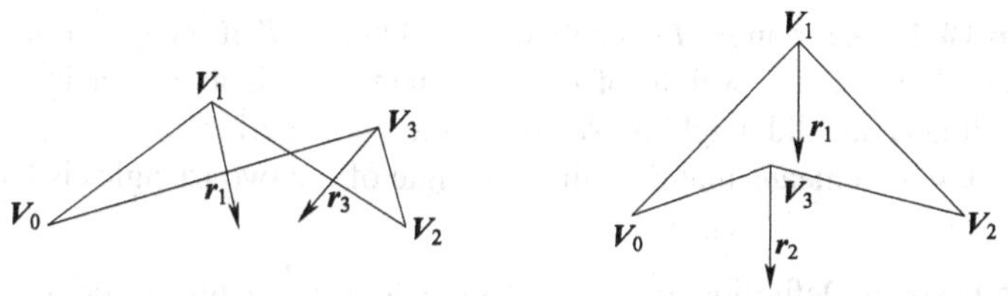

(a) Convex quadrilateral (b) Concave quadrilateral

Fig. 10.2 Convex and concave quadrilaterals in three-dimensional space

which the original mesh should be decimated is therefore probably best decided on an individual basis, according to the actual situation.

10.1.3 Converting a Triangle Mesh to a Quadrilateral Mesh

Zhang [118] provides an algorithm for converting a triangle mesh to a quadrilateral mesh. In order to minimize distortion of the quadrilaterals and avoid deleting sharp edges, Liu [117] modified the algorithm of Zhang [118] using two constraints: the flatness degree and the convexity-preserving flatness. We will first give the definitions of a convex quadrilateral and a concave quadrilateral:

Definition 10.1 For a quadrilateral $\boldsymbol{V}_0\boldsymbol{V}_1\boldsymbol{V}_2\boldsymbol{V}_3$ in three-dimensional space, let $\boldsymbol{r}_i$ be the angle bisector of an angle of less than 180° whose vertex is $\boldsymbol{V}_i$. If $\boldsymbol{r}_0 \cdot \boldsymbol{r}_2 \leq 0$, $\boldsymbol{r}_1 \cdot \boldsymbol{r}_3 \leq 0$, then the quadrilateral is convex, otherwise it is concave, as shown in Fig. 10.2.

We describe the flatness degree of a quadrilateral by the dihedral angle of two triangles with a common edge; the quadrilateral is obtained by uniting the two triangles. Rather than using the distortion energy as described by Zhang [118], we describe the flatness degree by the distortion energy $E_{harm}(h)$ that is defined by Formula (A1.1) when we map the quadrilateral region to a square region with a harmonic map. $-E_{harm}(h)$ is taken as the weight of the common edge. We use the dihedral angle so that the algorithm is simple and intuitionistic. We then use the following procedure to evaluate the weight for every edge of M^0:

Step 1: If $e_{i,j}$ is a boundary edge, set its weight as 1. Otherwise, proceed to step 2.
Step 2: If a concave quadrilateral is obtained by uniting the two triangles with the common edge $e_{i,j}$, set its weight as 1. Otherwise, proceed to step 3.
Step 3: Compute the cosine of the dihedral angle of the two triangles with the common edge $e_{i,j}$, and set its weight as the cosine.

Based on the weights of the edges, we give the following definition:

Definition 10.2 The triangle T_j is called a neighbor of T_i if the two triangles have a common edge e. If the weight of e is less than s, T_j is a valid neighbor of T_i; otherwise, it is an invalid neighbor. According to our experience, it is appropriate to make $s = 0$, which means that the dihedral angle of the two triangles is larger than $\pi/2$.

On the basis of Definition 10.2, we can create a procedure for conversion of a triangle mesh to a quadrilateral mesh:

Step 1: Compute the weight for every edge of the triangle mesh.

Step 2: If all triangles have been united, the procedure finishes. If there are triangles that have not been united but do not have valid neighbors, the procedure finishes. Otherwise, select the triangle T_i with the minimum number of valid neighbors. Of all the valid neighbors of T_i, T_j is the neighbor with minimum weight. Proceed to Step 3.

Step 3: Unite T_i and T_j. Set the weight of the common edge of the two triangles as 1.

Using this procedure, we obtain a mesh M_0' from the triangle mesh M^0. M_0' is likely to still have triangle faces; however, this has no effect on resampling when we use the topology rules of C-C subdivision. We may resample using M^0 as a base mesh, since C-C subdivision can be executed on any two-manifold mesh. However, M^J will have more extraordinary vertices than M'^J. For C-C subdivision, the geometrically uniform resampling rules would have to be redesigned, which would likely be a difficult task.

10.1.4 Adjusting the Positions of the Vertices of Subdivided Meshes

Assume that M is the original mesh, also known as the original surface. In the subsequent discussion, the base mesh is denoted by M^0. Let M^k be a mesh obtained by subdividing $M^0 k$ times. We will now investigate the approaches for adjusting the positions of the vertices of M^k to make the shape of M^k as close to M as possible.

For any inner vertex $\boldsymbol{v}_i$, we first estimate its normal, as shown in Fig. 10.3. Let $\boldsymbol{p}_j (j = 0, \ldots, n$, n is the valence of $\boldsymbol{v}_i)$ be the 1-neighborhood edge vertices of $\boldsymbol{v}_i$. The normal $\boldsymbol{N}_i$ of $\boldsymbol{v}_i$ may be calculated by the following formula:

$$\boldsymbol{N}_i = \sum_{j=0}^{n-1} \frac{d_{j,j+1}}{d_{i,j} + d_{i,j+1} + d_{j,j+1}} \boldsymbol{n}_j, \tag{10.1}$$

where $\boldsymbol{n}_j$ is the unit normal of $\Delta \boldsymbol{v}_i \boldsymbol{p}_j \boldsymbol{p}_{j+1}$, $d_{i,j}$ is the distance between $\boldsymbol{v}_i$ and $\boldsymbol{p}_j$, and $d_{i,n} = d_{i,0}$.

The direction of $\boldsymbol{N}_i$ is regarded as the direction of the attractive force. In order to determine the magnitude of the attractive force, we project $\boldsymbol{v}_i$ to the original mesh

Fig. 10.3 Calculation of the normal at a vertex

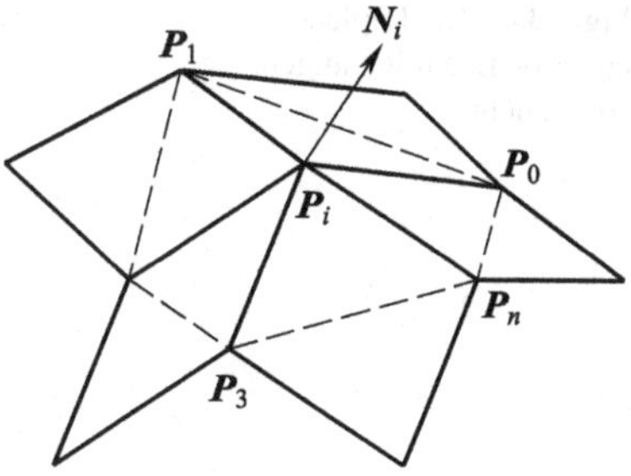

surface M in the direction of N_i and obtain a projection point q_i. The new position of v_i is then calculated as follows:

$$v_i' = v_i + \lambda(q_i - v_i), \quad \lambda \in [0, 1], \tag{10.2}$$

where λ is a factor that can be adjusted. In our experience, λ is best taken as 0.5, because there is a possibility that $q_i = q_j$ for two different vertices v_i and v_j. $\lambda(q_i - v_i)$ represents the magnitude of the attractive force. Since the estimation for N_i may be non-reasonable, we set a threshold value for $\|q_i - v_i\|$. N_i and v_i are considered to be non-reasonable if $\|q_i - v_i\|$ is larger than the threshold. In this case, we choose a vertex of M that is closer to v_i than q_i. In order to determine the threshold value, we compute the bounding box of M: The length d of the diagonal of the bounding box is known, and so τd is used as the threshold value. τ is a factor that may be adjusted according to the user's experience; it can initially take a value of 0.01.

This procedure may be regarded as an attracting procedure, whereby the control vertices are pulled toward the original surface. We now use a relaxing procedure to minimize distortions of M^k. The relaxing procedure is in fact a fairing process that makes the vertices of M^k uniformly distributed. As described in Chap. 4, the limit point of v_i may be written as follows:

$$v_i^{\infty} = \frac{n^2 v_i + 4\sum_j e_j + \sum_j f_j}{n(n+5)} = v_i + \frac{4\sum_j \left(e_j - v_i\right)}{n(n+5)} + \frac{\sum_j \left(f_j - v_i\right)}{n(n+5)}, \tag{10.3}$$

where e_j and f_j are 1-neighborhood edge points and face points, respectively, of v_i. We can then define the Laplacian operator of v_i as follows:

$$L(v_i) = \frac{4}{n}\sum_j \left(e_j - v_i\right) + \frac{1}{n}\sum_j \left(f_j - v_i\right).$$

Since the tangent component of $L(v_i)$ smooths M^k, while the normal component shrinks M^k[122], we take the tangent component of $L(v_i)$ as the relaxing force:

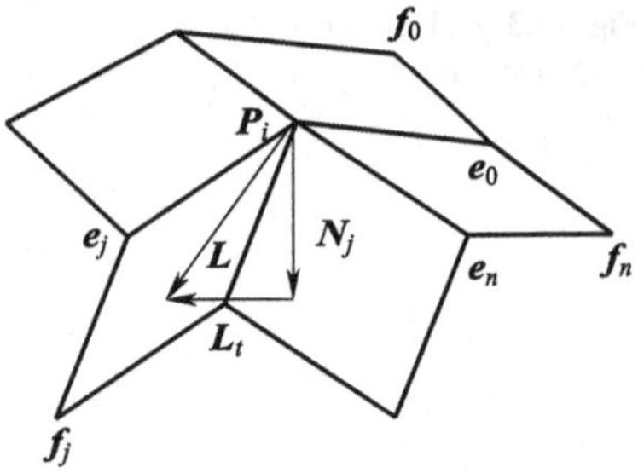

Fig. 10.4 The Laplace operator L and its tangent component

$$L_t(v_i) = L(v_i) - (L(v_i) \cdot N_i)N_i$$

as shown in Fig. 10.4. v_i should therefore be pulled into a new position under the relaxing force:

$$v_i' = p_i + \mu L_t(p_i) \quad \mu \in [0, 1],$$

where μ is a factor that we can take as having a value of 0.2.

For a boundary vertex v_i, we find the closest vertex on the boundary of M to q_i and then shrink it using Formula (10.2). In the relaxing process, we first find two adjacent vertices of v_i, v_{i-1} and v_{i+1}, on the boundary of M^k, and then relax v_i using the following formula:

$$v_i' = (v_{i-1} + 6v_i + v_{i+1})/8.$$

As described in Chap. 2, this is a subdivision formula for a uniform cubic B-spline curve. At every subdivision level k, the attracting-relaxing procedure can be executed several times. In general, we can use the attracting–relaxing–attracting procedure to adjust the vertices of M^k.

10.1.5 Choosing Subdivision Schemes

If we use only the Catmull–Clark subdivision, there are certain cases that are not satisfied: For example, sharp features may be smoothed, or M^k may have edges that cross, as shown in Fig. 10.5a. It is easy to understand why these cases arise. The Catmull–Clark subdivision is a process that smoothes M^k ($k = 0, 1, 2, 3, \ldots$): If the sharp features of M^0 are smoothed, a direct consequence is that these sharp features will be absent on M^k. Hence, a proper projection q_i will probably not be found, due to the large difference between $M^k (k = 1, 2, 3, \ldots)$ and M at sharp feature positions. Since there may be large differences in shape between the parts of M^0 with sharp features and the corresponding parts of M^1, we can use a push-back C-C subdivision

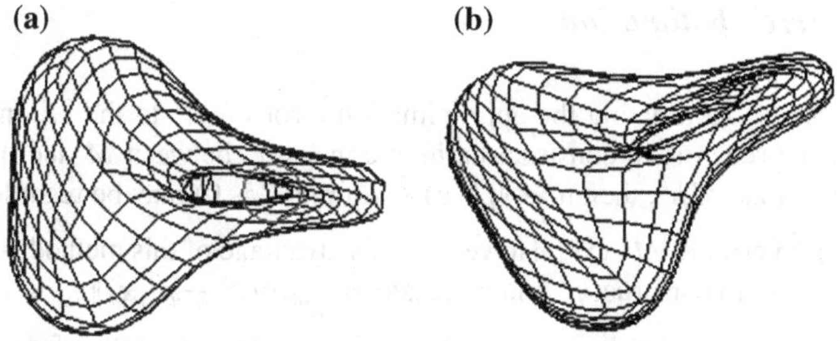

Fig. 10.5 Fitting surfaces and subdivision. **a** Sharp features are smoothed only when Catmull–Clark subdivision is used and **b** distortions appear only when the push-back Catmull–Clark subdivision is used

[47] to obtain M^1 from M^0. Consequently, these sharp features on M^0 will not be smoothed.

The push-back C-C subdivision is described by the following steps:

(1) Subdivide M^k once by the Catmull–Clark scheme, and obtain new vertex points $\{\boldsymbol{v}_i'^k\}$, new edge points $\{\boldsymbol{e}_i'^k\}$, and new face points $\{\boldsymbol{f}_i'^k\}$.

(2) Calculate the difference between vertex $\boldsymbol{v}_i^k$ and its new vertex point $\boldsymbol{v}_i'^k$: $\Delta \boldsymbol{v}_i^k = \delta(\boldsymbol{v}_i^k - \boldsymbol{v}_i'^k)$, where δ is the push factor.

(3) According to $\Delta \boldsymbol{v}_i^k$, make a push-back for new edge point $\boldsymbol{e}_i'^k$: $\boldsymbol{e}_i'^k = \boldsymbol{e}_i'^k + \omega(\alpha)(\Delta \boldsymbol{v}_1^k + \Delta \boldsymbol{v}_2^k)/2$, where $\boldsymbol{v}_1^k$ and $\boldsymbol{v}_2^k$ are two endpoints of the old edge in M^k corresponding to $\boldsymbol{e}_i'^k$, and $\omega(\alpha)$ is the back function.

(4) According to $\Delta \boldsymbol{v}_i^k$, make a push-back for new face point $f_i'^k$:

$$f_i' = f_i' + \omega^2(\alpha) \left(\sum_{v_i \in Corners(f_i')} \Delta \boldsymbol{v}_i \right) /n,$$

where n is the number of vertices in the face of M^k corresponding to $f_i'^k$.

(5) $\{\boldsymbol{v}_i^k\}$, $\{\boldsymbol{e}_i'^k\}$, and $\{\boldsymbol{f}_i'^k\}$ form the new vertices $\{\boldsymbol{v}_i^{k+1}\}$ of M^{k+1}.

In our applications, we use values of $\delta = 1$ and $\omega(\alpha) = 1$. However, it is not feasible to use only push-back C-C subdivision for remeshing. Distortions appear easily because old vertices are not moved during the process of subdivision, a known disadvantage of interpolating subdivision schemes that is discussed in Chap. 5. Figure 10.5b gives an example of this problem.

Let us consider the effects of subdivisions on boundaries. If the push-back C-C subdivision is used, we preserve old vertices. If the C-C subdivision is used, boundary vertices with valence 2 are preserved, while the other vertex points are calculated using Formula (6.8). In both cases, new edge points are calculated from the averages of the endpoints of edges.

10.1.6 Error Estimation

A key problem is estimating the approximation error of M^k to the original M. A reasonable approach is to construct harmonic maps h^k between M and M^k. If ρ^k is the inverse map of h^k, then $\max_{x \in KO} \|\rho^k(\boldsymbol{x}) - \rho^0(\boldsymbol{x})\| (k > 0)$ may be regarded as the approximation error of M^k [59]. However, a disadvantage of this method is that it is time-consuming to construct ρ^k and calculate $\max_{x \in KO} \|\rho^k(\boldsymbol{x}) - \rho^0(\boldsymbol{x})\|$.

Note that we always pull control vertices of M^k to the original surface M in the shrink-wrapping process. The maximum error of the mesh surface M^k should be the distance between a point in a quadrilateral face of M^k to M. Let $\boldsymbol{p}_i$ be a point in a quadrilateral face and $\boldsymbol{q}_i$ be the projection of $\boldsymbol{p}_i$ on M: The distance may then be denoted by $\|\boldsymbol{q}_i - \boldsymbol{p}_i\|$, as shown in Fig. 10.6.

Although it would be feasible to pick several points on every quadrilateral face to calculate $\|\boldsymbol{q}_i - \boldsymbol{p}_i\|$, we can reduce computational time by selecting only the barycenter of each quadrilateral face. Therefore, the following procedure is used:

Step 1: For every quadrilateral face, calculate its barycenter $\boldsymbol{p}_i$ and normal n_i.
Step 2: Project $\boldsymbol{p}_i$ to M in the direction n_i and obtain $\boldsymbol{q}_i$.
Step 3: If $\|q_i - \boldsymbol{p}_i\|$ is larger than the threshold value, choose as $\boldsymbol{q}_i$ a vertex of M that is closest to $\boldsymbol{p}_i$.
Step 4: For all quadrilateral faces, choose $\max_i \|\boldsymbol{q}_i - \boldsymbol{p}_i\|$ as the fitting error of M^k.

Note that $\max_i \|\boldsymbol{q}_i - \boldsymbol{p}_i\|$ is an absolute error that depends on the magnitude of the bounding box of the original surface M. Let d be the diagonal of the bounding box; we can then use a relative fitting error, $\max_i \|\boldsymbol{q}_i - \boldsymbol{p}_i\|/\boldsymbol{d}$.

Previously published studies [65] have used special relaxing rules for sharp features. Relaxation for a vertex adjacent to two sharp edges only weights the adjacent vertices on those edges; sharp vertices are not relaxed. If we were to use special attracting/relaxing rules for sharp features, special subdivision rules for sharp features would probably be required.

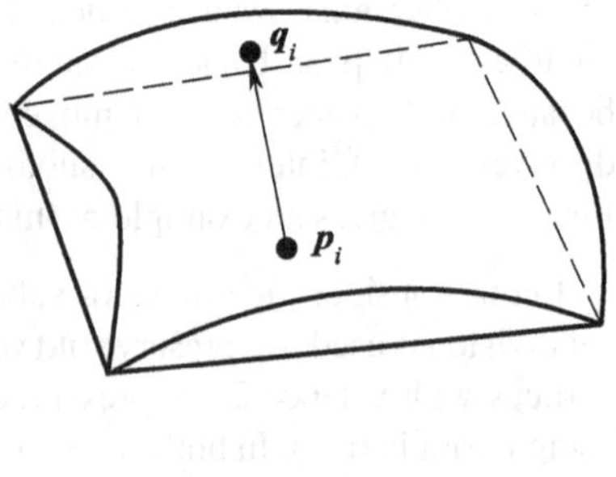

Fig. 10.6 Error in a quadrilateral face relative to the original surface

10.2 Surface Fitting by SLP (Subdivision Limit Position)

The basis of this approach, as presented by Suzuki [61], is to use SLP to adapt the control mesh of the subdivision surface to the data points. This method is not a time-consuming process involving global optimization, but it does fail to capture the local characteristics of data points. Consequently, although this method is not suitable for generating a surface that precisely interpolates the data points, it is useful for rapidly generating a surface that captures the overall shape constituted by the data points. This section introduces the work of Suzuki [61], but uses an algorithm based on C-C subdivision as the example, rather than one based on Loop subdivision as originally described [61]. It should be noted that the method discussed here generates subdivision surfaces, while the techniques described in Sect. 10.1 only generate meshes with subdivision connectivity.

10.2.1 The Error Function and Its Minimization

Let M^k be a mesh obtained by subdividing $M^0 k$ times. We will adjust the vertices of M^k in order that the limit surface S of M^k is an approximation of the original surface M.

Let $\boldsymbol{v}_i^k$ be a vertex of M^k, $\boldsymbol{v}_i^\infty$ be its SLP, and $\boldsymbol{q}_i$ be its nearest point on the original surface M. $\boldsymbol{q}_i$ may be obtained using the approach described in Sect. 6.2. N is the number of vertices of M^k. The error function is defined as follows:

$$E = \sum_{i=1}^{n} \left|\boldsymbol{v}_i^\infty - \boldsymbol{q}_i\right|^2 / N. \tag{10.4}$$

In Eq. (10.4), $\boldsymbol{v}_i^\infty$ is a variable, while $\boldsymbol{q}_i$ may be considered a constant. If we minimize the error function, we can obtain a linear system by differentiating E with respect to $\boldsymbol{v}_i^\infty$:

$$\boldsymbol{v}_i^\infty = \boldsymbol{q}_i. \tag{10.5}$$

Since Eq. (10.3) indicates that $\boldsymbol{v}_i^\infty$ is a linear combination of some vertices of M^k, Eq. (10.5) may be expressed as a large linear system:

$$\boldsymbol{v}_i^\infty = \boldsymbol{v}_i^k + \frac{4\sum_j \left(\boldsymbol{e}_j^k - \boldsymbol{v}_i^k\right)}{n(n+5)} + \frac{\sum_j \left(\boldsymbol{f}_j^k - \boldsymbol{v}_i^k\right)}{n(n+5)} = \boldsymbol{q}_i. \tag{10.6}$$

Here, we would rather not solve such a large equation. Instead, we use an iterative approximation method. From Eq. (10.6), we get:

$$q_i - v_i^k = \frac{4\sum_j \left(e_j^k - v_i^k\right)}{n(n+5)} + \frac{\sum_j \left(f_j^k - v_i^k\right)}{n(n+5)}. \tag{10.7}$$

Referring to Fig. 5.6, this equation is considered to represent a state of equilibrium; the left side of this equation represents the external force from v_i^k, and the right side represents the spring forces from e_j^k and f_j^k. The point v_i^k is attracted toward the resultant force:

$$r_i = \frac{4\sum_j \left(e_j^k - v_i^k\right)}{n(n+5)} + \frac{\sum_j \left(f_j^k - v_i^k\right)}{n(n+5)}.$$

Here, we assume that the vertices e_j^k and f_j^k are fixed. The new position v_i^k is computed by the following equation:

$$v_i'^k = v_i^k + \lambda r_i, \tag{10.8}$$

where λ is an acceleration factor and takes a value between 0 and 1. The convergence speed is determined by λr_i. The convergence speed of the asymptotic loop is increased as the value of λ approaches 1.0, and the asymptotic loop requires only a few iterations when λ is in the range [0.6, 1.0]. Although convergence speed is maximized for $\lambda = 1.0$, the points sometimes move so quickly as to cause undesired undulations in the resulting shape. Therefore, Suzuki [61] has suggested a value of $\lambda = 0.8$. The asymptotic loop is given by the pseudo-code below:

Step 1: Compute $E = \sum_{i=1}^{n} \left|v_i^\infty - q_i\right|^2 / N$; if $E < \varepsilon$, stop; else proceed to step 2.
Step 2: For each v_i^k, compute and evaluate: $v_i^k = v_i^k + \lambda r_i$
Step 3: Return to step 1.

In the above procedure, ε is a threshold value defined by the user.

10.2.2 The Base Mesh and Subdivision

Suzuki [61] suggests that the vertices of M are normalized to fit within the unit cube. Such normalization is probably advantageous to set the threshold value ε for the error function. However, it is also feasible to not normalize the vertices of M, by modifying the error function as follows:

$$E = \sum_{i=1}^{n} \left| \boldsymbol{v}_i^{\infty} - \boldsymbol{q}_i \right|^2 / (dn),$$

where d is the length of the diagonal of the bounding box of M.

The shape of the base mesh M^0 is important for good fitting. Suzuki [61] suggested two approaches: a bounding box and a manual definition. We construct the base mesh using the approaches discussed in Sect. 6.2.1.

After constructing a base mesh M^0, we subdivide it and use the asymptotic loop at every subdivision level. When a given subdivision level is reached, subdivision stops. In the report of Suzuki [61], the subdivision loop is termed the outer loop, and the asymptotic loop is called the inner loop.

10.3 Subdivision Surface Fitting with a Squared Distance Minimization Method

10.3.1 Construction of the Initial Control Mesh

First, all boundary edges with only one associated face are found, and all crease edges whose dihedral angles are smaller than a provided tolerance are identified. Then, the principle direction, principle curvature, and normal direction of each mesh node are evaluated by locally constructing a quadratic fitting surface [213]. According to the marked feature vertices, feature edges, and their differential geometrical attributes, we partition the original mesh into many quadrilateral regions. Here, we provide some basic concepts for region faces, which are similar to those for meshes.

A corner of a region is the intersection point of its two neighboring boundaries. If a boundary is associated with only one region, i.e., it is composed of only boundary edges, it is called a boundary of the whole model. Corners on such boundaries are termed boundary corners, while others are called inner corners. If an inner corner of a region has only 4 associated regions, it is termed a regular corner, otherwise it is called an irregular corner. A boundary corner with 2 or 3 associated regions is called a regular vertex, while boundary corners are called irregular corners. If a region has no irregular corners, it is called a regular region; otherwise, it is termed an irregular region.

For the sake of parameterization, the model should split once if some irregular regions have more than one irregular vertex. The subdivision method for regions is similar to that for meshes. For each boundary, its "midpoint" is taken as the new E-Vertex; for each region, its "barycenter" is taken as the new F-Vertex; for each region corner, its new V-Vertex remains unchanged. Topologically, the splitting method is the same as for the C-C subdivision. All new V-Vertices and new F-Vertices are linked with their neighboring new E-Vertices. Note that the midpoint of a boundary is the point corresponding to the middle parameter of the chord length parameterization for the boundary and that the barycenter of a region is the projection of the barycenter of the corners of the region to the mesh. The boundary link of two vertices is also calculated using projection. It is possible that new vertices are on an edge of the original mesh or on the interior of a triangle face. The initial control meshes of the fitting subdivision surfaces may be constructed using the corners of all regions.

10.3.2 Parameterization of Mesh Vertices

1. Collection of mesh vertices for each region

First, find and mark all triangle faces that are spanned by boundaries, and store these in an array. For each vertex of the marked triangle faces, judge whether or not it is in the region. If it is in the region, mark it and add it to the vertex array. Mark the index of the first triangle added to the face array in each loop step as i. In the next loop step, the search begins from i. Find all untagged neighboring faces and judge whether their vertices have been marked. If there are no untagged faces in some loop, the loop process ends.

For the decision of whether a vertex is in a region, choose two vertices in the middle of each boundary and take their direction as that of the boundary. If a vertex is in the same direction as a boundary of a region, the vertex is regarded as inside the region. This method is correct in most cases, but it may fail if the curvature of one boundary is too large. Figure 10.7 illustrates such a case: Although $\boldsymbol{p}_i$ is located in the region, it is likely that the opposite conclusion will be reached using the method just described. Fortunately, this error can be rectified in the subsequent parameter optimization process.

2. Initial parameterization

A local coordinate system needs to be set up before parameterization. The origin is set as follows: For an irregular region, the origin is associated with the sole irregular

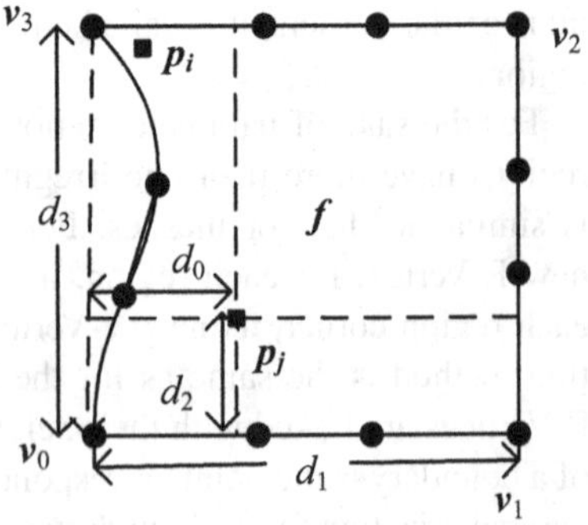

Fig. 10.7 Initial parameterization of the data

vertex; for a feature region, the origin is associated with a feature vertex. Parameters for other vertices may be calculated from the harmonic map [59]. All the local coordinates are normalized in [0,1].

3. Optimization of parameterization

Optimization involves the process of projecting a point onto a base surface. We can use Eq. 8.5 to update the parameters of each vertex. The initial values are the initial parameters calculated as described in the subsection above.

As mentioned previously, points near the boundary may have incorrect parameters (u_i, v_i) since they may be mapped to the wrong regions. In such cases, we turn to the neighboring regions, and the parameters are converted into the local parameters of the neighboring local coordinate system. The loop process continues, and once all the parameters of the vertices on the mesh have been optimized, they are represented in the form of (k_i, u_i, v_i), where k_i is the index of the region, and (u_i, v_i) are local coordinates in that region. It should be noted that not every vertex has a sole parameter, for points on the boundaries have more parameters. For these vertices, any parameter may be used for the process of surface fitting.

10.3.3 Subdivision Surface Fitting Using a Squared Distance Method

The squared distance method was first proposed by Pottmann et al. [214, 215] who used a second-order Taylor approximation of the squared distance function. This approach can produce a second-order convergence speed.

Lemma 10.1 [216] *Let d be the distance between a sample point $\boldsymbol{v}_0$ and the nearest point $\boldsymbol{p}$ on the approximation surface S; the principle directions of $\boldsymbol{p}$ are $\boldsymbol{T}_1$ and $\boldsymbol{T}_2$. If the radii of the corresponding principle curvature are ρ_1 and ρ_2, and the normal vector is $\boldsymbol{N} = \boldsymbol{T}_1 \times \boldsymbol{T}_2$, then the second-order Taylor approximation of the squared distance function is as follows:*

$$F_d(\boldsymbol{v}) = \frac{d}{d+|\rho_1|}[(\boldsymbol{v}-\boldsymbol{p})\cdot \boldsymbol{T}_1]^2 + \frac{d}{d+|\rho_2|}[(\boldsymbol{v}-\boldsymbol{p})\cdot \boldsymbol{T}_2]^2 + [(\boldsymbol{v}-\boldsymbol{p})\cdot \boldsymbol{N}]^2. \tag{10.9}$$

Since precise computation of the principle direction and principle curvature of an arbitrary point of a subdivision surface is complex [217], a simplification is made here. It is economical to take the principle directions and principle curvatures of the mesh nodes, since their projection points on the subdivision surface only need computing once. Considering the fairness of the fitting subdivision surface, an optimization model is constructed as follows:

$$\text{Min} \sum_{i=1}^{N} F_{d_i}(\boldsymbol{v}_i) + \lambda F_s, \tag{10.10}$$

where $F_{d_i}(\boldsymbol{v}_i) = \frac{d_i}{d_i + |\rho_{i,1}|}\left(d\boldsymbol{v}_i \cdot \boldsymbol{T}_{i,1}\right)^2 + \frac{d_i}{d_i + |\rho_{i,2}|}\left(d\boldsymbol{v}_i \cdot \boldsymbol{T}_{i,2}\right)^2 + \left(d\boldsymbol{v}_i \cdot \boldsymbol{N}_i\right)^2$, $F_s = \boldsymbol{X}^{\mathrm{T}}\boldsymbol{Q}\boldsymbol{X}$. λ is a fairness factor and is usually set as 0.01; $\boldsymbol{Q}$ denotes the integral sum of the basis functions. Let $d\boldsymbol{v}_i = \boldsymbol{v}_i - \boldsymbol{S}(t_i) = \boldsymbol{v}_i - \boldsymbol{A}_i(\boldsymbol{V} + \boldsymbol{X})$, where $\boldsymbol{V}$ and $\boldsymbol{X}$ denote the vectors of control vertices and their deviation quantities. $\boldsymbol{A}_i$ is the corresponding coefficient matrix, which is described in Sect. 7.3. Differentiate expression (10.10) about $\boldsymbol{X}$; letting the expression be zero, we can get:

$$\sum_{i=1}^{N} \boldsymbol{A}_i^{\mathrm{T}}\boldsymbol{A}_i\boldsymbol{X}\boldsymbol{C}_i + \lambda\boldsymbol{Q}\boldsymbol{X} = \sum_{i=1}^{N} \boldsymbol{A}_i^{\mathrm{T}}\left(\boldsymbol{A}_i\boldsymbol{V} + \boldsymbol{v}_i\right)\boldsymbol{C}_i, \tag{10.11}$$

where $\boldsymbol{C}_i = \frac{d_i}{d_i + |\rho_{i,1}|}\boldsymbol{T}_{i,1}^{\mathrm{T}}\boldsymbol{T}_{i,1} + \frac{d_i}{d_i + |\rho_{i,2}|}\boldsymbol{T}_{i,2}^{\mathrm{T}}\boldsymbol{T}_{i,2} + \boldsymbol{N}_i^{\mathrm{T}}\boldsymbol{N}_i$.

Equation (10.11) is clearly a matrix equation. From Lemma 10.1, with the Kronecker product of the matrix, Eq. (10.11) can be converted into the following linear system:

$$\overline{\boldsymbol{A}}\,\overline{\boldsymbol{X}} = \overline{\boldsymbol{B}}, \tag{10.12}$$

where $\overline{\boldsymbol{A}} = \sum_{i=1}^{N} \boldsymbol{C}_i^{\mathrm{T}} \otimes (\boldsymbol{A}_i^{\mathrm{T}}\boldsymbol{A}_i) + \lambda\boldsymbol{E}_3 \otimes \boldsymbol{Q}$, $\overline{\boldsymbol{X}} = vec(\boldsymbol{X})$, $\overline{\boldsymbol{B}} = vec\left(\sum_{i=1}^{N} \boldsymbol{A}_i^{\mathrm{T}}(\boldsymbol{A}_i \boldsymbol{V} + \boldsymbol{v}_i)\boldsymbol{C}_i\right)$.

Since feature edges can be computed entirely on the basis of the subdivision rules for curves, data vertices on feature edges may be solved solely by the SDM fitting algorithm for B-spline curves [218]. It should be noted that the computational cost is large when the global linear system of Eq. 7.5 requires solving; we therefore solve the linear system based on the SuperLU algorithm for sparse matrices [219]. To further save time and space, reducing the sampling points on the mesh is recommended. If the error $|d\boldsymbol{v}_i|$ of a point $\boldsymbol{v}_i$ is smaller than a provided tolerance, or its parameter is outside the downsampling interval, $\boldsymbol{v}_i$ is discarded. A downsampling interval is defined as follows:

$$|t - 0.5| < \delta \text{ or } t < \delta \text{ or } t > 1 - \delta.$$

Here, t represents u or v, and δ is set at a value of about 0.1. The gray parts in Fig. 10.8 represent the downsampling intervals. The downsampling technique can ensure the stability of a linear system (10.12). Examples are shown in Figs. 10.9 and 10.10.

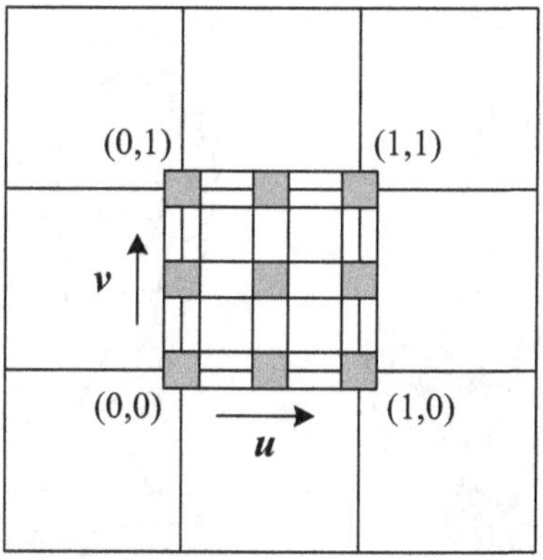

Fig. 10.8 Downsample intervals of the local parameter domain

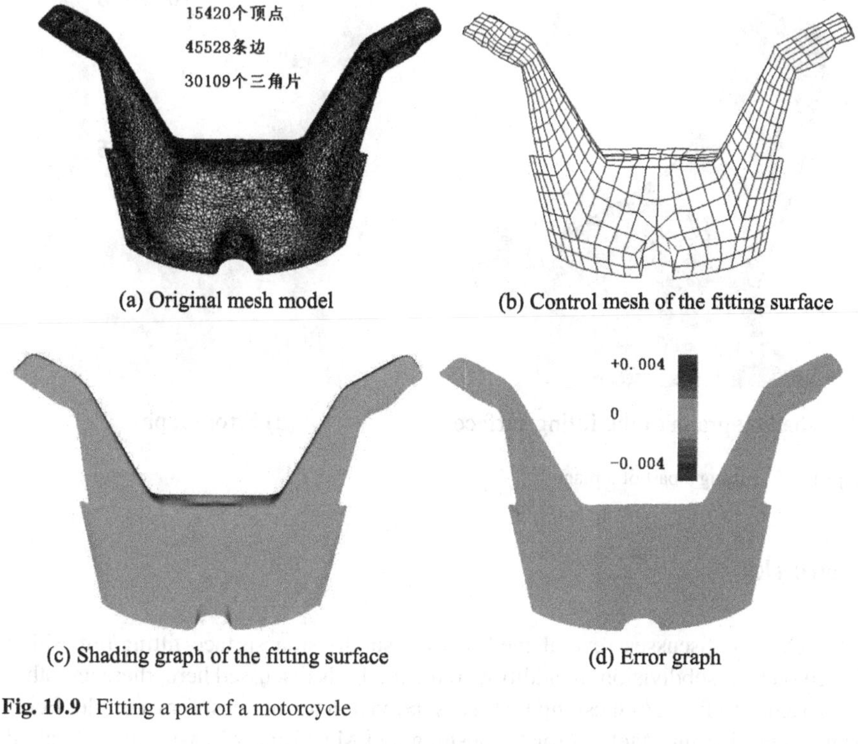

Fig. 10.9 Fitting a part of a motorcycle

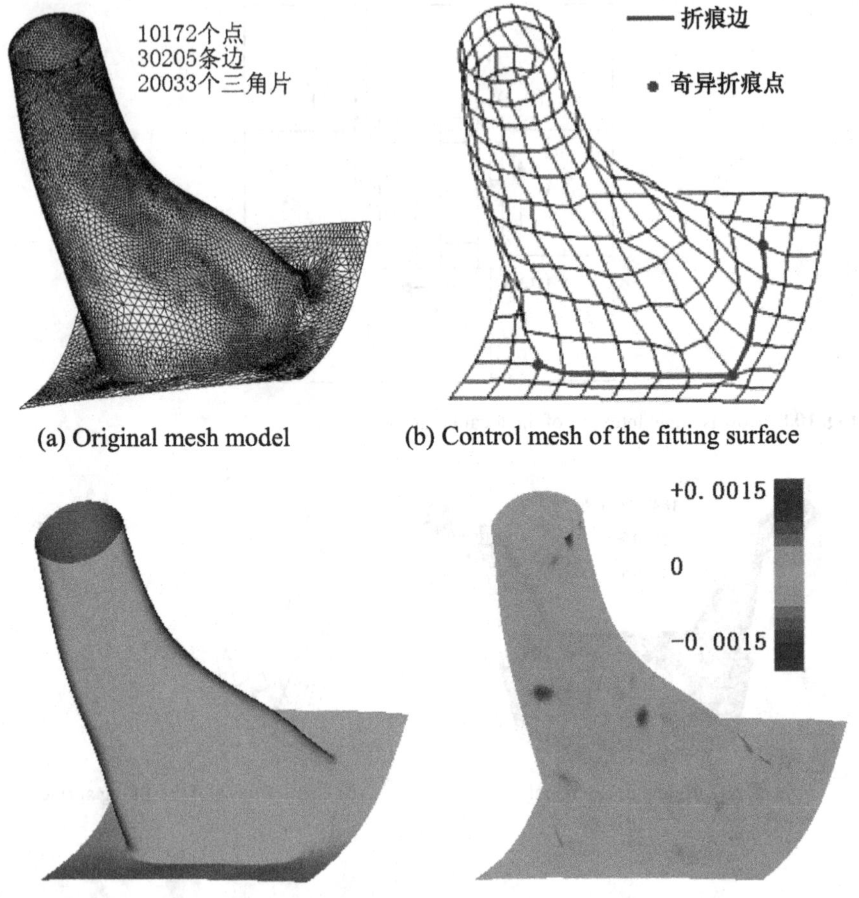

(a) Original mesh model (b) Control mesh of the fitting surface

(c) Shading graph of the fitting surface (d) Error graph

Fig. 10.10 Fitting a part of a plane

Remarks

This chapter discusses several methods for subdivision surface fitting/remeshing based on C-C subdivision. In addition to the methods discussed here, there are other approaches to fitting/remeshing that use subdivision techniques, for example, Eck's Harmonic Mapping Method for Remeshing and Ma's Linear Least Square Method for Surface Fitting. There are also published studies in the literature [59–61, 119, 121] that use triangle subdivision schemes, particularly Loop subdivision schemes. Jeong [121] described construction of a mesh surface with Loop subdivision connectivity directly from unorganized points. Studies [63, 65, 119] have also provided approaches for the construction of sharp features for fitting surfaces and fitting meshes. It is probably important to design a specialized subdivision method for sharp features in order to obtain good fitting effects. For any fitting method, and especially for the iterative methods given in Sects. 10.1 and 10.2, the base mesh is fundamental. Base mesh shape has profound effects on the shape of the fitting surface and fitting mesh. Eck's method is one that is widely applied to partition data points and construct base meshes. Fitting surfaces and fitting meshes of good quality are essential requirements in reverse engineering.

Exercises

1. For a STL original mesh, see Fig. 10.1, (i) decimate the original mesh and obtain a coarse triangle mesh and (ii) convert the coarse triangle mesh to a quadrilateral mesh using the method in Sect. 10.1.3.
2. Write program codes to fit the original mesh using the SLP method.

Chapter 11
Poisson Mesh Editing Guided by Subdivision Surface

This chapter firstly discusses the Poisson mesh editing method guided by subdivision surfaces and then extends it further to the mesh deformation based on surface control. The former constructs the deformation control surface by the subdivision surface determined by the mesh model which was bounded by the mesh to be deformed and converts the corresponding subdivision surface deviation information into the manipulation on the gradient field of the model. The latter uses subdivision surface as a deformation control surface in the specified deformation region and determines the local transformation of the triangle mesh in the deformation region according to the following conditions: the subdivision control surfaces before editing and after editing; the reference curves; and the objective curves. Based on these local transforms, we implement the gradient field manipulation. These two algorithms both realize the intention of the user and then fulfill a particular grid editing task through the manipulation of the differential properties, which are helpful for the effective maintenance of geometric details in the editing process.

11.1 Traditional Mesh Editing Method

Traditional free-form deformation method [237, 251, 252] embeds the mesh into the deformation space composed of control meshes, control curves, control points or other simple elements. And then the method indirectly implements the model deformation by editing the shape of the deformation space. The method has such advantages: (1) Algorithms are simple and (2) having good geometric intuition. However, because deformation is a direct manipulation of vertex coordinates for every vertex in turn, the method is lack of the ability to describe local geometry feature. So some rich geometric details in the deformed mesh cannot be maintained when editing complex models by using this method. Sometimes, unexpected defects may appear, such as small-scale distortions and local self-intersection.

W. Liao et al., *Subdivision Surface Modeling Technology*,
DOI 10.1007/978-981-10-3515-9_11

Compared with the traditional free-form deformation methods, the multiresolution editing techniques and differential domain methods provide an effective solution for the retainment of mesh details in the deformation process. Multiresolution editing method separates the basic shape and surface detail of the mesh model. It sets up a hierarchical model. The user can deform the mesh in a level with fewer details and then rebuild detailed features of the model. The current multiresolution editing techniques [253, 254] provide a hierarchy representation for mesh models, but the editing operation still takes the vertex coordinates as the direct manipulation objects, so it can only maintain parts of details. Differential domain method [255, 256] does not operate on the mesh vertices directly, but resorts to the adjustment of differential attributes of the mesh surface to achieve the editing goals. Compared with vertex coordinates, differential attributes quantify the geometric details and can depict three-dimensional geometric models essentially. At the same time, the definitions of differential attributes build a bridge between spatial coordinates of the vertices and can produce the inherent constraints of the vertices. Taking the differential attributes, constraint as the optimization objective can maintain the geometric details in the editing process, and the energy optimization process defined by the Laplace or Poisson equation can diffuse the error, which will effectively avoid the local defects and ensure a good deformation effect.

Subdivision surfaces have the advantages of arbitrary topology adaptation, locality, and multihierarchy. So they have a natural relation with multiresolution editing techniques. Literature [257] puts subdivision surfaces into the mesh deformation and provides a multiresolution space deformation method based on subdivision surfaces. In this method, an intermediate deformation space is formed by a subdivision surface and its normals. Users can select the control mesh in different resolutions to perform a global or local deformation, which is simple and intuitive. But this deformation is still a direct manipulation of the vertex coordinates in fact and the necessary constraints among vertices are lack, which leads to the disagreement of displacement of neighboring deformed vertices. So when we treat the three-dimensional geometric model with rich details, it cannot ensure the retainment of the features.

11.2 Poisson Mesh Editing

Firstly, let us give the definition of Poisson equation on discrete triangle meshes. Let

$$f(v) = \Sigma f_i \phi_i(\boldsymbol{v})$$

be a piecewise linear scalar field defined on a discrete triangle mesh, where f_i denotes the function value at a vertex $\boldsymbol{v}_i$, and $\phi_i(\boldsymbol{v})$ is a piecewise linear basis function whose value is 1 at the vertex $\boldsymbol{v}_i$ and 0 at the other vertices. The gradient corresponding to the scalar field f is as follows:

$$\nabla f(\boldsymbol{v}) = \Sigma f_i \nabla \phi_i(\boldsymbol{v}). \tag{11.1}$$

Fig. 11.1 Definition of Laplacian operator

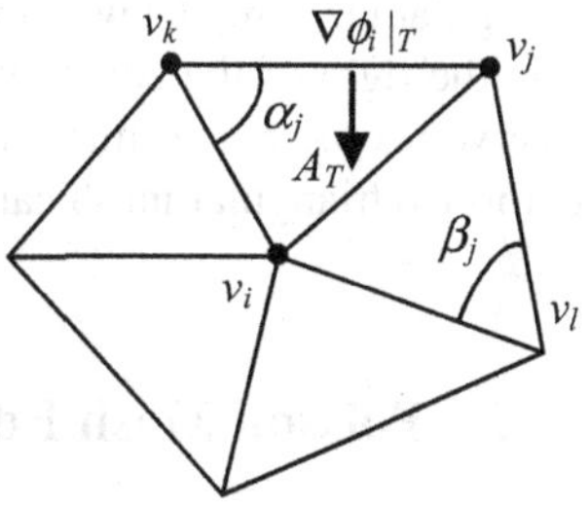

Let $\boldsymbol{w}$ be a piecewise constant vector field on a mesh, then the divergence of $\boldsymbol{w}$ at a vertex $\boldsymbol{v}_i$ is as follows:

$$(div\boldsymbol{w})(\boldsymbol{v}_i) = \sum_{T \in N_T(\boldsymbol{v})} \boldsymbol{w}(T)\nabla\phi_i|_T A_r \tag{11.2}$$

where $N_T(\boldsymbol{v}_i)$s denote a collection of neighbor triangles of the vertex $\boldsymbol{v}_i$, A_T is the area of the triangle T, $\nabla\phi_i|_T = R^{90}(\boldsymbol{v}_k - \boldsymbol{v}_j)/(2A_T)$, and R^{90} said rotated 90° around the triangle normal vector $\boldsymbol{n}$.

The definition of Laplacian operator of the scalar field f at the vertex $\boldsymbol{v}_i$ can be easily deduced by combining the definitions of gradient and divergence in Eqs. (11.1) and (11.2)

$$\Delta f(\boldsymbol{v}_i) = \frac{1}{2} \sum_{v_j \in N_v(\boldsymbol{v}_i)} (\cot\alpha_j + \cot\beta_j)(f_i - f_j). \tag{11.3}$$

α_j and β_j in equation (11.3) are two opposite angles of the edge $v_i v_j$ in the two triangles $v_i v_j v_k$ and $v_i v_j v_l$ as shown in Fig. 11.1.

The discrete Poisson equations can be represented as $\Delta f = \text{Div}(\nabla f) = \text{Div}\boldsymbol{w}$. In the process of numerical calculation, they actually form a sparse linear system:

$$\boldsymbol{A}\boldsymbol{x} = \boldsymbol{b} \tag{11.4}$$

in which the coefficient matrix $\boldsymbol{A}$ is only associated with the triangular mesh of the definition in the Eq. (11.3), while the constant vector $\boldsymbol{b}$ in the right is related to the divergence of $\boldsymbol{w}$ and the selected boundary conditions.

In actual operation, the vertices in the triangle mesh are divided into two classes: constraint vertices and free vertices, among which constraint vertices have fixed locations and form the boundary conditions of the Poisson equations; free vertices have free locations to be determined. According to the classification of constrained vertices and free vertices, triangle meshes whose three vertices are not all constraint vertices form the free deformation regions. In order to impose the Poisson equations on the triangle meshes, the three coordinates x, y, and z of the vertices on each triangular are regarded as three scalar fields, respectively. A new gradient field is obtained by, respectively, imposing local transformation on each triangle in the free deformation region. Since the gradient fields are invariant under transformation, the local transformation contains only rotation and scaling. A discontinuous deformed triangular mesh model will appear after the deformation on each triangle. The vector b in the right of the equation $\boldsymbol{Ax} = \boldsymbol{b}$ is set by calculating the divergence of each free vertex after deformation, and the matrix $\boldsymbol{A}$ remains unchanged, and then, the deformed triangular mesh can be obtained by solving the linear system.

11.3 Poisson Mesh Editing Based on Subdivision Surfaces

The particular process of Poisson mesh editing method based on subdivision surface is shown in Fig. 11.2. Firstly, define the bounding mesh of the mesh to be deformed. Then, the bounding mesh is subdivided with an ordinary subdivision scheme, where C-C subdivision method is used. Since the mesh after subdivision is all quads, in order to set up the relationship between the mesh to be deformed and the subdivision surface, triangularization operation is needed for the subdivided mesh. For the choice of the subdivision times, on the one hand, the more times subdivision is done, the smoother the subdivided mesh will be, and hence, the better the effect of the deformation will be. On the other hand, because of the increasing of the number of subdivision, the storage and computation cost will increase correspondingly, which will inevitably the deformation speed. In order to balance the two aspects, subdividing the bounding mesh twice or thrice will meet the deformation requirements in most cases. After getting the subdivision surface corresponding to the bounding mesh, we can construct the relationship between elements (the triangles and the vertices) in the mesh and the subdivision surfaces by using the nearest point principle. Then, the bounding mesh is edited and modified. The rotation and scaling transformations are determined according to the change of the subdivision surface. The Poisson gradient field of the deformed mesh model is set up, and the mesh model is rebuilt.

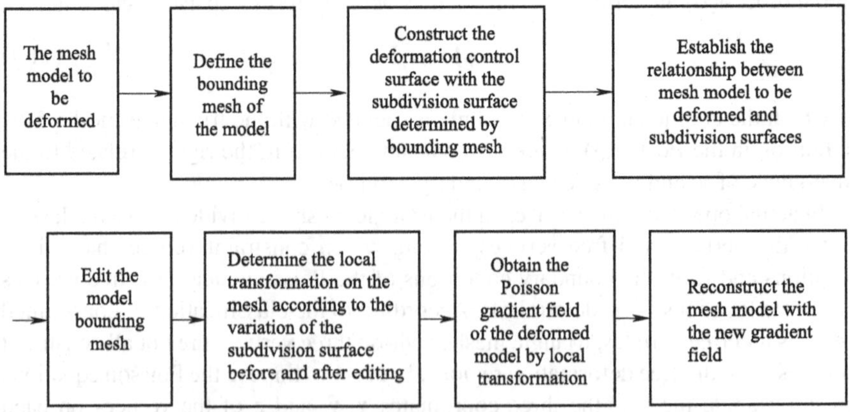

Fig. 11.2 Flowchart of the Poisson mesh editing method based on subdivision surfaces

11.3.1 Construction of the Model Bounding Mesh

Construction of the model bounding mesh is the first step to perform Poisson mesh editing method based on subdivision surfaces, which directly affects the construction of the relationship between the mesh model and subdivision surfaces. Construction of the model bounding mesh also affects the final deformation effect. There are mainly three ways at present to construct the bounding mesh:

(1) build the bounding mesh interactively;
(2) build the control mesh with skeleton extraction technology [258];
(3) simplify the original mesh to get the bounding mesh [259]. The first method can construct an ideal bounding mesh, but needs too much human–computer interaction, which is not helpful for the realization of the automation. The second method is not effective for objects that do not have obvious axis curves, and the skeleton extraction algorithm is costly and not robust. The third method can produce a bounding mesh with consistent topology and similar shape as the original mesh. It also has higher efficiency and degree of automation. However, there are still some gaps between the simplified mesh and the shape demand for the bounding mesh.

Using the Reeb graph technology, Literature [257] could automatically generate a control mesh that has similar shape and consistent topological structure with the deformation object. For the construction of bounding mesh, this chapter mainly uses the third method. However, the bounding mesh is not the simplest mesh, but the one displaced of the original mesh. Generally, the displaced mesh needs simplifying and editing. For some mesh models with regular shape, the bounding mesh can generally be created by interactive manner.

11.3.2 Relationship Between the Mesh Model and Subdivision Surfaces

There are mainly two aspects needing to be treated for the mesh and the corresponding subdivision surfaces: For each mesh vertex, calculate its corresponding position on the subdivision surface, and for each triangle, calculate the corresponding position of its centroid on the subdivision surface. Here, the corresponding location means the nearest point to the subdivision surface. The corresponding location may not necessarily be the mesh vertices of the subdivision mesh, and it can be on an edge or in a face of the subdivision mesh. Since the methods to solve these two kinds of corresponding locations are consistent, here we take the mesh vertex as an example to demonstrate the particular process.

For an arbitrary vertex $\boldsymbol{v}$ on the mesh model, the nearest triangle ΔT_s to $\boldsymbol{v}$ is found first, and then, the corresponding location $\boldsymbol{v}_{ds}$ of $\boldsymbol{v}$ on the subdivision mesh can be expressed as a linear combination of three vertices on ΔT_s. Traditional methods to find the nearest triangle from the mesh by investigating triangles in the subdivision mesh one by one is a very time-consuming process, and its computational complexity is $O(n^2)$, which will inevitably impact the deformation efficiency. In order to speed up the searching speed of ΔT_s, the algorithm is divided into the following two steps:

Step 1: Search all vertices of the subdivided mesh to find the nearest vertex $\boldsymbol{v}_s$ to the vertex $\boldsymbol{v}$;

Step 2: Extract all the triangles adjacent to the vertex $\boldsymbol{v}_s$ and calculate the distance from $\boldsymbol{v}$ to the triangles, and pick the nearest one as ΔT_s.

Since the number of vertices is much smaller than that of faces in most cases, the above search algorithm can greatly reduce the computational complexity and improve the search speed than traditional methods. So for any mesh vertex $\boldsymbol{v}$ on the model, the nearest triangle ΔT_s on the subdivision surface can be obtained by distance comparison. At the same time, the nearest point $\boldsymbol{v}_d$ from $\boldsymbol{v}$ on ΔT_s is obtained, which is also the corresponding location of the mesh vertices $\boldsymbol{v}$ on the subdivision mesh as.

However, it should be noted that this search algorithm for ΔT_s is not entirely accurate, and in some special cases, ΔT_s is not necessarily located within the adjacent triangles of the vertex $\boldsymbol{v}_s$. Figure 11.3 gives a two-dimensional interpretation of the situation, where the plotline in the figure can be considered as the subdivision mesh. The broken lines on both sides of a control vertex represent the 1-ring neighborhood triangles. In this case, the nearest triangle ΔT_s is located below $\boldsymbol{v}$ and is not within the 1-ring neighborhood triangles of the nearest vertex $\boldsymbol{v}_s$ which is the nearest subdivision grid vertex from $\boldsymbol{v}$. As shown in Fig. 11.3, this situation will only occur when a triangle in the subdivision surface has a very long edge or a very large area, which will appear only in this particular case. So such slim or large triangles need to be picked out and made special consideration.

Set the corresponding location of the centroid of an arbitrary triangle of the mesh model on the subdivision mesh be

$$\boldsymbol{v}_{ds} = a\boldsymbol{v}_{s0} + b\boldsymbol{v}_{s1} + c\boldsymbol{v}_{s2}$$

where $\boldsymbol{v}_{s0}$, $\boldsymbol{v}_{s1}$, and $\boldsymbol{v}_{s2}$ are three vertices of the triangular. After editing the bounding mesh, three vertices are changed into $\boldsymbol{v}'_{s0}$, $\boldsymbol{v}'_{s1}$, and $\boldsymbol{v}'_{s2}$, respectively. Then, the corresponding location $\boldsymbol{v}_{ds}$ is changed into

$$\boldsymbol{v}'_{ds} = a\boldsymbol{v}'_{s0} + b\boldsymbol{v}'_{s1} + c\boldsymbol{v}'_{s2}.$$

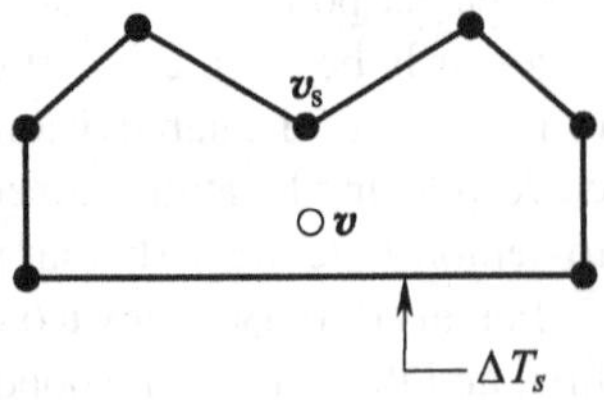

Fig. 11.3 Special circumstances of ΔT_s to speed up the search for the error

11.3.3 Boundary Conditions of Poisson Equations and Mesh Local Transformation

For the mesh surface, first give the definitions of 1-ring neighborhood of vertices, edges, and faces. The 1-ring neighborhood of a mesh vertex means the mesh composed of all the edges and faces that are associated with the vertex. The 1-ring neighborhood of an edge is composed of its two end vertices and all the three vertices' (or edges') 1-ring neighborhoods forming a face. The following is the judgment method of free vertices and constraint vertices.

As mentioned in Sect. 11.3.2, for any vertex $\boldsymbol{v}$ on the model to be deformed, we can obtain its corresponding location on the deformation control mesh, that is, the corresponding location on the subdivision mesh. According to whether the corresponding location is at a vertex or on an edge or within a face of a subdivision surface, we can extract a 1-ring structure of the corresponding vertex, edge, or face and get a local mesh that is denoted by M_s. Then, compare the differences in the local mesh M_s before editing and after editing. If all the vertices in M_s do not change, such a mesh vertex $\boldsymbol{v}$ is looked on as a constraint vertex; otherwise, it is thought to be a free vertex.

As we divide all the vertices of the mesh model to be deformed into constraint vertices and free vertices, the free deformation region of the mesh will also be determined, which is a collection of triangles whose three vertices are not all the constraint vertices. The local rotation and scaling transformation operations are all based on these triangles.

For any triangle ΔT in the free deformation region, first of all we need to calculate the normals $\boldsymbol{n}_{ds}$ and $\boldsymbol{n}'_{ds}$ on the corresponding locations $\boldsymbol{v}_{ds}$ and $\boldsymbol{v}'_{ds}$ of the centroid of the triangle on the mesh before and after subdivision. Then, we will take $\boldsymbol{n}_{ds}$ as an example to illustrate the solving process. Consider the situation when $\boldsymbol{v}_{ds}$ is located within a triangle face of the subdivision mesh. If $\boldsymbol{v}_{ds}$ is at a vertex or on an edge of the subdivision mesh, it can be managed in a similar way. As shown in Fig. 11.4, the average normals $\boldsymbol{n}_{as0}$, $\boldsymbol{n}_{as1}$, and $\boldsymbol{n}_{as2}$ of these three vertices $\boldsymbol{v}_{s0}$, $\boldsymbol{v}_{s1}$, and $\boldsymbol{v}_{s2}$ of the triangle which $\boldsymbol{v}_{ds}$ belongs to are calculated are an average of the unit normals weighted by the areas of the surrounding neighbor triangular of the vertex. All the average normals are normalized. Suppose

$$\boldsymbol{v}_{ds} = a\boldsymbol{v}_{s0} + b\boldsymbol{v}_{s1} + c\boldsymbol{v}_{s2}$$

then

$$\boldsymbol{n}_{ds} = a\boldsymbol{n}_{s0} + b\boldsymbol{n}_{s1} + c\boldsymbol{n}_{s2}.$$

Then, the local Descartes coordinate system is defined at the corresponding position on the subdivision mesh. One direction of the local coordinate system is above the normal of the corresponding position on subdivision mesh, and the other two are within the tangent plane of the corresponding position, one of which is given by the projection point of an adjacent vertex of the corresponding location in the

Fig. 11.4 Calculation of the average normal vector $\boldsymbol{n}_{ds}$

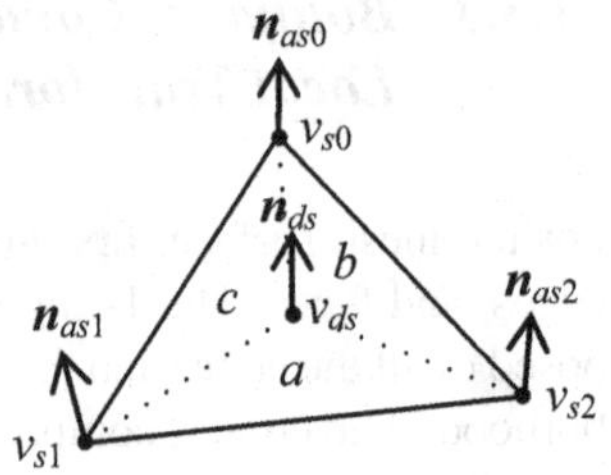

Fig. 11.5 Definition of local orthogonal frame

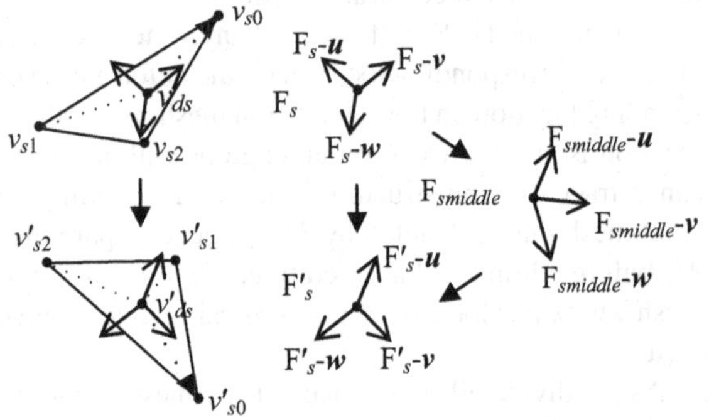

tangent plane. The last one is the cross product of the first two directions. As shown in Fig. 11.5, suppose $\boldsymbol{v}_{ds}$ is located in the triangle face of the subdivision mesh, the local coordinate systems F_s and F'_s are calculated at the two corresponding locations $\boldsymbol{v}_{ds}$ and $\boldsymbol{v}'_{ds}$ on the subdivision mesh before editing and after editing, respectively. In the figure, $\mathrm{F}_s - \boldsymbol{u}$ and $\mathrm{F}'_s - \boldsymbol{u}$ are normals of the corresponding locations $\boldsymbol{v}_{ds}$ and $\boldsymbol{v}'_{ds}$, and $\mathrm{F}_s - \boldsymbol{v}$ and $\mathrm{F}'_s - \boldsymbol{v}$ are the corresponding projection directions of the vectors $\boldsymbol{v}_{ds}\boldsymbol{v}_{s0}$ and $\boldsymbol{v}'_{ds}\boldsymbol{v}'_{s0}$ on the tangent plane.

Thus, for any triangle ΔT in the free deformation region, we can get two local coordinate systems F_s and F'_s corresponding to two corresponding locations $\boldsymbol{v}_{ds}$ and $\boldsymbol{v}'_{ds}$. By using these two local coordinate systems, we can construct a rotation transformation matrix between them. Since the gradient field is independent of the translation transform, the rotation transformation can be determined by the above method. For each triangle in free deformation region, we only need to decide the scaling transformation. For the scaling transformation, here we only consider iso-scaling transformation, which can be determined by the circumferences of the triangular near the corresponding location on the subdivision meshes before editing and after editing. The specific solution process of the rotation and scaling transform can be given below.

1. Rotation transforms

As shown in Fig. 11.5, for any triangle ΔT in the free deformation region, we can get two local coordinate systems F_s and F'_s corresponding to the two corresponding locations $\boldsymbol{v}_{ds}$ and $\boldsymbol{v}'_{ds}$ on subdivision meshes before and after editing, and the solution of the rotation transformation can be divided into two steps:

Step 1: Take the normals of subdivision meshes at the two corresponding locations, namely the $\boldsymbol{u}$ direction of the local coordinate system $\mathrm{F}_s - \boldsymbol{u}$ and $\mathrm{F}_s' - \boldsymbol{u}$, and obtain the first rotation axis *rollaxis_1* by cross product and the first rotation angle *rollangle_1* by the dot product of them.

Step 2: Taking the corresponding location $\boldsymbol{v}_{ds}$ as the rotation center, rotate the local frame F_s around the first axis *rollaxis_1* by an angle *rollangle_1* to get a middle coordinate system, denoted as $\mathrm{F}_{smiddle}$ (shown in Fig. 11.5). And then taking the $\boldsymbol{v}$ direction of the two local coordinate systems $\mathrm{F}_{smiddle}$ and F_s', namely $\mathrm{F}_{smiddle}$-$\boldsymbol{v}$ and F_s'-$\boldsymbol{v}$, make their cross product as the second rotation axis *rollaxis_2* and dot product as the second rotation angle.

For any triangle in the free deformation region, taking its centroid as the rotation center, the rotation transformation can be constructed by rotating *rollangle_1* around *rollaxis_1* first and then rotating *rollangle_2* around *rollaxis_2*.

2. Scaling transform

The definition of scalar value length is introduced at the corresponding location first. As shown in Fig. 11.4, we just consider the situation that the corresponding location lies inside the triangle. Define the average scalar value lengths of l_{as0}, l_{as1}, and l_{as2} of three vertices $\boldsymbol{v}_{s0}$, $\boldsymbol{v}_{s1}$, and $\boldsymbol{v}_{s2}$ as follows:

$$l_{asi} = \sum_j A_{sj} \cdot l_{sj} \Big/ \sum_j A_{sj},$$

where l_{sj} is the jth triangle perimeter within the 1-ring neighborhood of the vertex $v_{si}(i = 0, 1, 2)$, and A_{sj} represents the jth area of the triangle. Suppose

$$\boldsymbol{v}_{ds} = a\boldsymbol{v}_{s0} + b\boldsymbol{v}_{s1} + c\boldsymbol{v}_{s2},$$

where a, b, and c are the combination coefficients, then the length scalar value l_{ds} can be calculated by the equation

$$l_{ds} = al_{as0} + bl_{as1} + cl_{as2}.$$

The scale factor of the scaling transform corresponding to ΔT is determined by the ratio of l_{ds} and l'_{ds}.

11.3.4 Realization of Poisson Reconstruction Deformation

As the bounding mesh of the model to be deformed is edited according to the deformation intent of the user, the related subdivision mesh will accordingly deform. The free deformation region of the mesh is determined automatically according to the variation of the subdivision mesh. At the same time, for each triangle in the free deformation region, a compound local transformation can be constructed. By taking

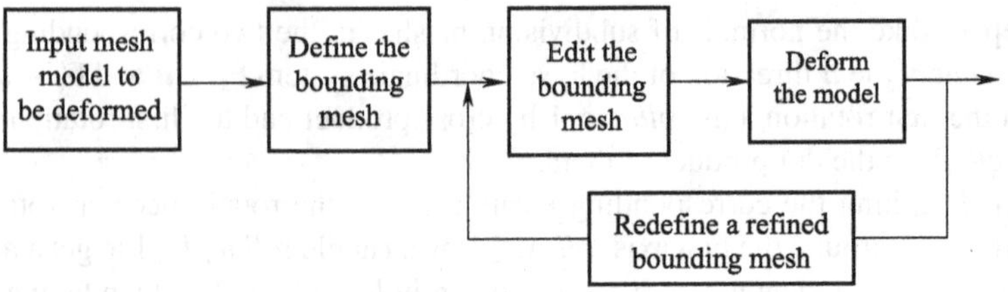

Fig. 11.6 Multiresolution-based deformation process

the triangle centroid as the origin of the local transformation, we can get a gradient field which reflects the deformation information and is put into the Poisson equation. We can get the target deformation mesh model by using the least-squares method. In the deformation process, the constraint vertices remain unchanged and play the role of boundary conditions.

11.3.5 Multiresolution Deformation

The Poisson mesh editing method discussed in this chapter takes the subdivision surface determined by the bounding mesh as the deformation control surface. The locality of subdivision surface leads to the locality of the deformation method. However, the locality is limited. For the mesh model to be deformed, once the bounding mesh and the subdivision rules are established, the local deformation region is determined when a vertex of the bounding mesh is moved. When the users want to edit the mesh model in a smaller scope, it cannot meet the deformation requirements to rely solely on the bounding mesh modifications. In order to get a richer mesh editing effect and enhance the local modeling ability, the following supplies the multiresolution-based deformation method.

The specific process of the multiresolution-based deformation is shown in Fig. 11.6. At first, the initial bounding mesh of the mesh to be deformed is relatively coarse, and the deformation under the action of the initial bounding mesh can be regarded as a local mesh deformation or called a "primary deformation." Then, the modified bounding mesh is subdivided once, and a new bounding mesh should be redefined based on the old bounding mesh. Compared with the primary deformation, the mesh model under the action of the new bounding has better locality. The above process can be repeated. According to the user's requirements on the model deformation, we can define new and refined bounding mesh. On this basis, satisfying deformation effect can be gotten.

In the deformation process, the relationship between the mesh model and the subdivision surface corresponding to the new bounding mesh is unchanged. The relationship between the mesh model and the subdivision surface only needs to be

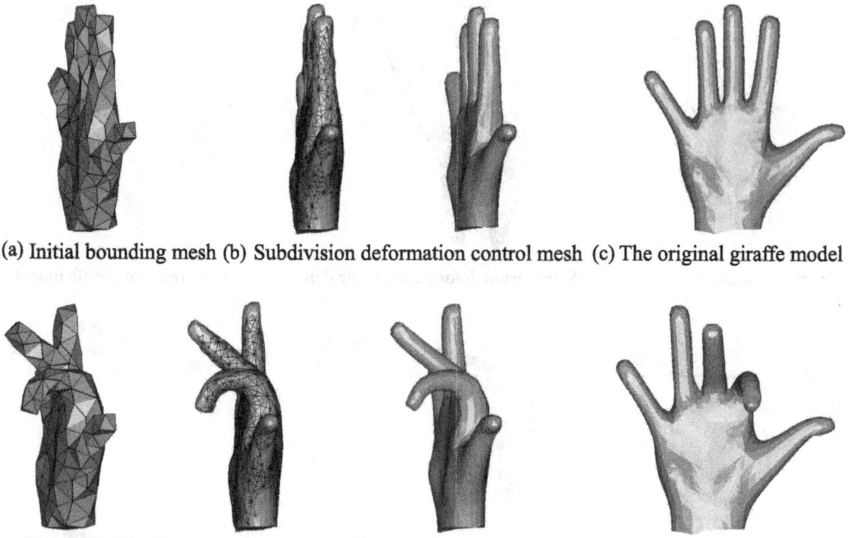

(a) Initial bounding mesh (b) Subdivision deformation control mesh (c) The original giraffe model

(d) Edit the initial bounding mesh in (a) (e) Modify the subdivision mesh (b) (f) Deformed model

Fig. 11.7 Deformation of the hand model

determined in the primary deformation stage, and it does not have to be recalculated in the later deformation process.

11.3.6 Deformation Instance Analysis

Figures 11.7 and 11.8 give deformation examples of the hand and giraffe model, respectively, the bounding meshes of the model before and after editing, the C-C subdivision deformation control mesh corresponding to the bounding mesh (the mesh has been triangularized after subdivision), and the mesh models before and after deformation.

11.4 Mesh Deformation Based on Surface Control

As an important complement to the FFD deformation, the deformation through surface control [242, 243] is more intuitive to describe the deformation about some of the objects associated with the contact surface, such as a snake crawling on the uneven ground or an soft object placed on a rock. The deformation method under surface control [260, 261] is still to use the parametric spline surface as the control surface whose control mesh is a rectangular grid and it is difficult to fit the shape of objects with

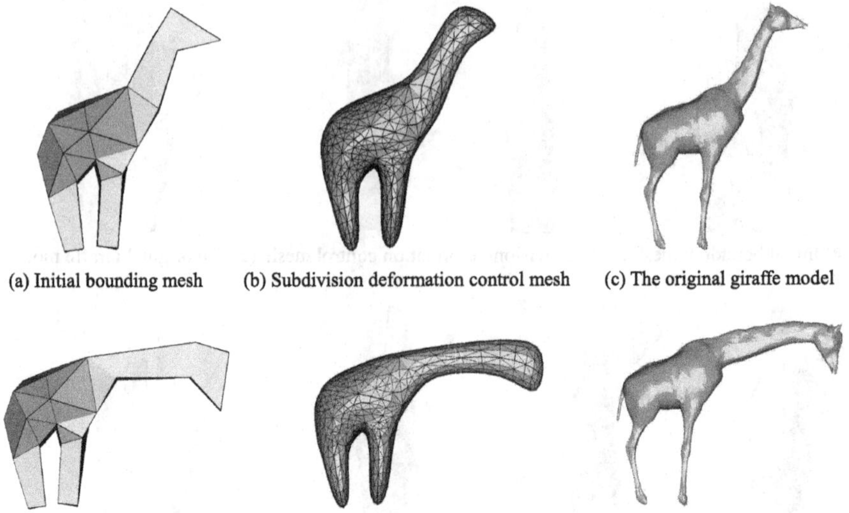

(a) Initial bounding mesh (b) Subdivision deformation control mesh (c) The original giraffe model

(d) Edit the initial bounding mesh in (a) (e) Modify the subdivision mesh (b) (f) Deformed model

Fig. 11.8 Deformation of the giraffe model

arbitrary topology. Moreover, in the deformation process, the deformation method based on surface control is still a direct manipulation of vertex coordinates in turn, which is lacking the ability to depict the local geometry. So these deformation methods are difficult to maintain minutiae and probably produce unnatural deformation results and cause self-intersection phenomena sometimes, especially when the shape to be edited has rich geometric details.

In Sect. 11.1, we give the Poisson mesh editing method based on subdivision surfaces. On this basis, we will further extend it to the mesh deformation based on surface control. The basic steps of deformation process are as follows:

Step 1. Specify the scope of the mesh deformation region interactively.

Step 2. Subdivide the mesh in the deformation region and rebuild the information of vertices, edges, and faces in the model.

Step 3. Design a subdivision surface as a deformation control surface in the deformation region.

Step 4. For all the triangles in the deformation region, calculate their corresponding locations in the subdivision control surfaces and classify the triangles accordingly.

Step 5. Modify the control mesh vertices of the subdivision surface and design the corresponding control curve. The local transformation for the triangle in the deformation region is determined by the variation of the subdivision surface and the control curve.

Step 6. The modified gradient field is gotten from the local transformation, and the deformed mesh model is obtained with the new gradient field.

The above surfaces based on mesh deformation method in Step 5 are fit for general situations. Since the boundaries of the subdivision surface have changed after editing, the local transformation for the triangle in the deformation region needs the participation of the control curves. Due to the local nature of subdivision surface, to move a vertex of the control mesh only affects the vertices in the 2-ring neighborhood. So when we modify a control vertex of the subdivision surface, if we barely allow the inner vertices to be moved freely and confine the boundary vertices unchanged, the boundaries of the subdivision surface will remain unchanged after editing. At this time, the deformation region is no longer in the scope the user predesignated, but all the inner triangles in the free deformation region are in. The local transform of triangles in the deformation region is completely determined by the variation of subdivision surface, and there is no need to design the control curve.

In the following, we will specify each step of the mesh deformation based on the surface controls.

11.4.1 Determination of the Deformation Region

The deformation region is completely determined by manual interactions. The boundary curves of the deformation region are obtained firstly by picking up some vertices on the mesh and connecting them with the Dijkstra's algorithm. The vertices on the boundary curves form constraint vertex set. And then specify an arbitrary vertex in the deformation region as a seed vertex and put the vertex into a free vertex set. According to the topology connection relationship of the mesh model, we make a search starting from this seed vertex. If the neighbor vertices of the seed are not on the boundary curves, put them into the free vertex set. Take the neighbor vertices as new seed vertices and do the same neighborhood searching task until all the seeds are all on the boundary curves. Then, we can get the whole free vertex set. At the same time, we can get the free deformation region of the mesh indeed, which is composed of the triangles associated with free vertices.

In the model deformation process, all the vertices in the constraint vertex set are fixed, which constitute the boundary conditions of Poisson equation. The vertices in the free vertex set are the ones to be sought. The resolution of local transformation for the mesh is only for the triangles in the deformation region.

11.4.2 Refinement of the Deformation Region

In the model editing process, if the mesh in the deformation region is very coarse and the deformation is sharp, we need to subdivide the triangles in the deformation region. Otherwise, the visual effect of deformation will be bad, and the deformed model will not look smooth. Here, we take a given edge length threshold as the guide whether to split a triangle or not. If the edge length of a triangle is beyond

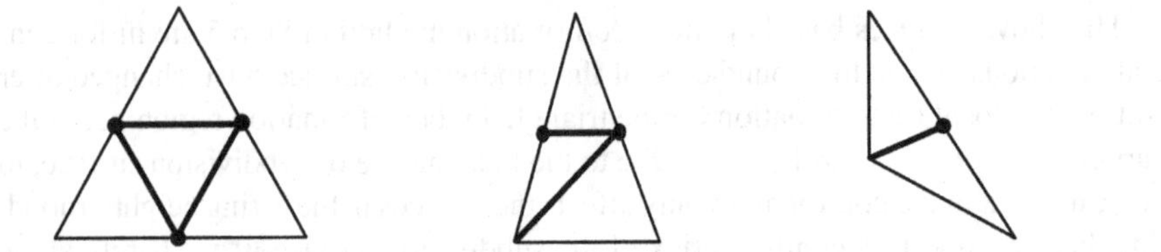

(a) Split into four new triangles (b) Split into three new triangles (c) Split into two new triangles

Fig. 11.9 Splitting of the triangle mesh

the threshold, a new vertex is placed at the midpoint of the edge. As a result, there are three kinds of situations when splitting the triangle, as shown in Fig. 11.9. After splitting the triangle, the information of vertices, edges, and faces of the mesh in the deformation region needs to be rebuilt.

11.4.3 Design of Subdivision Control Surfaces

Here, we mainly consider two kinds of subdivision models: Loop subdivision and C-C subdivision. Firstly, the user picks up some mesh vertices on the mesh to be deformed by interaction and then links these vertices according to certain topological relation, which is based on the control surface shape requirements wanted by the user. The mesh generated in the interactive manner is the mesh M whose vertices are to be inserted. According to the method in Literature [262], we construct the subdivision surface in order to insert all vertices into the mesh to be inserted. The construction process is as follows:

(1) Copy the mesh M to be inserted and denote the copy mesh as P. Set the energy parameter E to be sufficiently large;
(2) For each vertex $\boldsymbol{v}_{pi}$ of P, calculate the corresponding subdivision limit point $\boldsymbol{v}_{pi}^{\infty}$, and the difference between $\boldsymbol{v}_{pi}^{\infty}$ and the corresponding vertex to be inserted $\boldsymbol{v}_{mi}$ in M: $\boldsymbol{r}_i = \boldsymbol{v}_{mi} - \boldsymbol{v}_{pi}^{\infty}$;
(3) Update the energy parameter $E = \sum_i |\boldsymbol{r}_i|$; if E is less than any given value ε, the algorithm ends; otherwise, turn to (4);
(4) Update the vertex in P by $\boldsymbol{v}_{pi} = \boldsymbol{v}_{pi} + \lambda \boldsymbol{r}_i$ with the parameter λ equals 0.8, and go to step (2).

After the above iterative computation, we can get the mesh P that is a topological isomorphic of the mesh M. The deviation between each subdivision limit point in P and the corresponding vertex to be inserted is within the given error range. So we can get the subdivision control surface by subdividing the mesh P. It should be noted that the preselected vertices to be inserted are usually not located in the subdivision control surface.

11.4.4 The Construction of Relationship Between The Deformation Mesh and the Subdivision Control Surface

The construction of relationship between the deformation mesh and the subdivision control surface means that we search the nearest point of the centroid of each triangle in the deformation region on the subdivision surface. The nearest point is called the corresponding position of the triangle on the subdivision surface. The corresponding position is not necessarily at a mesh vertex, it can be on an edge or in a triangle of subdivision mesh. The specific solution process was given in Sect. 11.3.2.

According to whether the corresponding location is located in the interior of subdivision control surfaces, the triangles in the deformation region are divided into two categories: Class I triangle (the corresponding position located within the subdivision control surface) and Class II triangle (the corresponding position located on boundaries of the subdivision control surface).

11.4.5 Design of the Control Curve

Since the local transformation of Class II triangles in the deformation region is related to the control curve, we give the definition of the control curve in the following.

As shown in Fig. 11.10, for any Class II triangle in the deformation region, set its centroid be $\boldsymbol{v}_c$, and then, the corresponding positions $\boldsymbol{v}_{ds}$ and $\boldsymbol{v}'_{ds}$ of $\boldsymbol{v}_c$ on the subdivision control surface before and after editing are on a boundary of the control surface. Set the point $\boldsymbol{v}_b$ be the intersection point of the plane Σ and the boundary curve of the deformation region. The plane Σ passes the points $\boldsymbol{v}_c$, $\boldsymbol{v}_{ds}$, and $\boldsymbol{v}'_{ds}$. $\boldsymbol{m}_0$ and $\boldsymbol{m}_1$ are the tangent vectors at the points $\boldsymbol{v}_b$ and $\boldsymbol{v}'_{ds}$, respectively. The straight line $\boldsymbol{v}_b\boldsymbol{v}_{ds}$ is defined as the reference control curve. The target control curve is determined by the points $\boldsymbol{v}_b$, $\boldsymbol{v}'_{ds}$ and the tangent vector $\boldsymbol{m}_0$ and $\boldsymbol{m}_1$:

$$\boldsymbol{v}(t) = (2t^3 - 3t^2 + 1)v_b + (t^3 - 2t^2 + t)\boldsymbol{m}_0 + (t^3 - t^2)\boldsymbol{m}_1 + (-2t^3 + 3t^2)v'_{ds},$$

where $t \in [0,1]$. In order to ensure the continuity of the boundary in the deformation process, the tangent vectors $\boldsymbol{m}_0$ and the direction $\boldsymbol{v}_b\boldsymbol{v}_{ds}$ of the reference control curve should be the same. Set $\boldsymbol{\alpha}$ as the unit vector of the direction $\boldsymbol{v}_b\boldsymbol{v}_{ds}$, then

$$\boldsymbol{m}_0 = k_0|\boldsymbol{v}_b\boldsymbol{v}_{ds}|\boldsymbol{\alpha}.$$

For the determination of the tangent vector $\boldsymbol{m}_1$, we need to ensure the continuity of the mesh in the transition region of Class I and Class II triangles; that is, we have to ensure the consistency under rotation transformation. The local orthogonal coordinate systems F_s and F'_s are established at the corresponding positions $\boldsymbol{v}_{ds}$ and $\boldsymbol{v}'_{ds}$ of the subdivision control surface before and after editing. A unique rotation transformation matrix $\boldsymbol{R}$ is determined between the two local coordinate systems. At the same time, the corresponding positions $\boldsymbol{v}_{ds}$ and $\boldsymbol{v}'_{ds}$ are also the ends of the reference curve and the objective control curve. For the tangent direction of the two ends, the same rotation transformation matrix is required. Therefore, the tangent vector $\boldsymbol{m}_1$ can be defined as follows:

$$\boldsymbol{m}_1 = k_1 |\boldsymbol{v}_{ds}\boldsymbol{v}'_{ds}| \boldsymbol{\beta}|,$$

where the unit vector $\boldsymbol{\beta}$ is obtained from $\boldsymbol{\alpha}$ multiplied by $\boldsymbol{R}$. From the definitions of the tangent vectors $\boldsymbol{m}_0$ and $\boldsymbol{m}_1$, it can be seen that the target control curve is a space curve, and k_0 and k_1 are two adjustment parameters, which affect the shape of the curve.

For any Class II triangle in the deformation region, we can obtain an associated reference curve and objective control curve, so the local transformation of the triangle can be determined by them.

11.4.6 Local Transformation of the Triangle in The Deformation Region

For the determination of the local rotation transformation and scaling transformation of Class I triangles, a reader may refer to Sect. 11.3.3. In the following, we will give the local transformation method of Class II triangles.

The local rotation transformation and the scaling transformation of Class II triangles are related to the reference curve and the objective control curve discussed in Sect. 11.3.5. The specific construction process is as follows:

(1) Calculate the projection point of the centroid of the Class II triangle on the reference control curve and define the corresponding parameter t. As depicted in Fig. 11.10, the point $\boldsymbol{v}_{cc}$ is the projection point of the point $\boldsymbol{v}_c$ on the straight line $\boldsymbol{v}_b\boldsymbol{v}_{ds}$, and the corresponding parameter t of the point v_c is $t = |\boldsymbol{v}_b\boldsymbol{v}_{cc}|/|\boldsymbol{v}_b\boldsymbol{v}_{ds}|$.

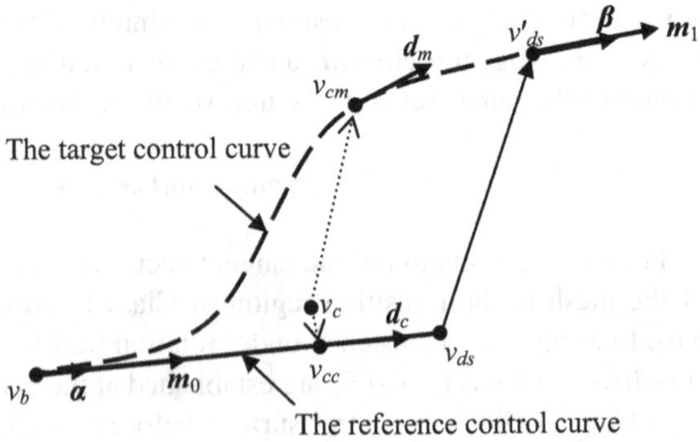

Fig. 11.10 Definition of the control curve

(2) Substituting the value t obtained in (1) into the parametric equations of the target control curves, then the point $\boldsymbol{v}_{cm}$ is obtained;
(3) Calculate the tangent direction $\boldsymbol{d}_m$ at the point $\boldsymbol{v}_{cm}$ on the target control curve. Get the tangent direction $\boldsymbol{d}_c$ at the point $\boldsymbol{v}_{cc}$ on the control curve, namely $\boldsymbol{v}_b\boldsymbol{v}_{ds}$. We can obtain the rotation axis and rotation angle by cross product and dot production on the vectors $\boldsymbol{d}_m$ and $\boldsymbol{d}$;
(4) The scaling factor of the scaling transformation is determined by the ratio of the total length of the reference control curve and the target control curve.

11.4.7 Realization of the Poisson Reconstruction Deformation

According to the deformation intent, the user manually determines the deformation region and designs the corresponding subdivision control surface. By the edition for the initial control mesh with fewer vertexes, the subdivision control surface has consequent shape change. If the control surface boundaries do not change before and after editing, the local transformation of each triangle in the deformation region is completely determined by the variation of the control surface. Otherwise, a control curve should be designed between the control surface and the boundary curve of the deformation region. The local transformations of all the triangles in the deformation region are determined by the two conditions: (1) the variation of the subdivision control surface before and after editing and (2) the shapes of the reference curves and objective curves. By taking the centroids of the triangles as the local origins, the gradient fields reflecting the deformation information can be constructed by the local rotation transformation and scaling transformation. By substituting the gradient information into the Poisson equation and solve Poisson equation with the least-squares method, we can obtain the target deformation mesh model. In the solving process, the constraint points keep unchanged and play the role of boundary conditions.

11.4.8 Deformation Instances

Figure 11.11 gives two plane-carved pattern deformation examples of "the cross flowers" and "five-star," which reflect the arbitrary topology adaptivity advantage of using subdivision surfaces as deformation control surfaces. Figure 11.11 shows the plane mesh before carving deformation where the crimson mesh lines identify the interactively designated deformation region(left), the plane mesh after splitting and the subdivision control surface which is shown in blue before and after editing(middle), and lighting effect after carving deformation(right). In these two plane-carved pattern deformation instances, the editions of the subdivision control surfaces all enhance a

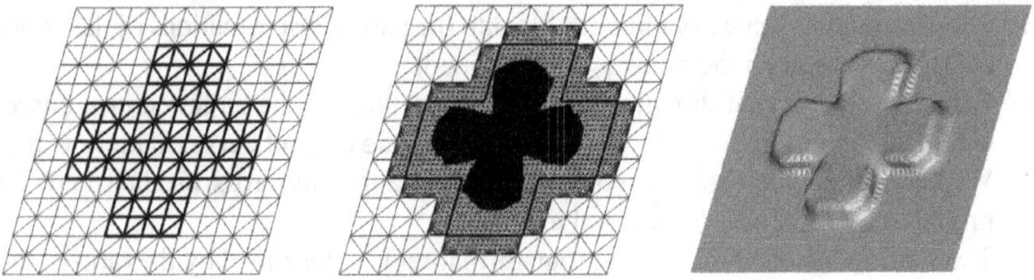

(a) Cross flower pattern engraving

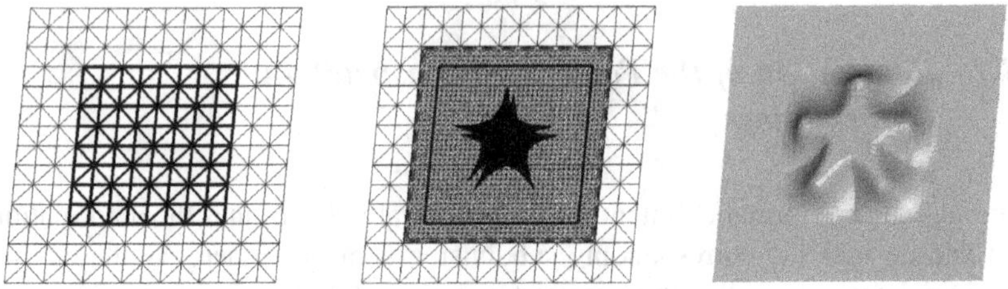

(b) "Five-pointed star" pattern engraving

Fig. 11.11 Plane-carved pattern deformation examples

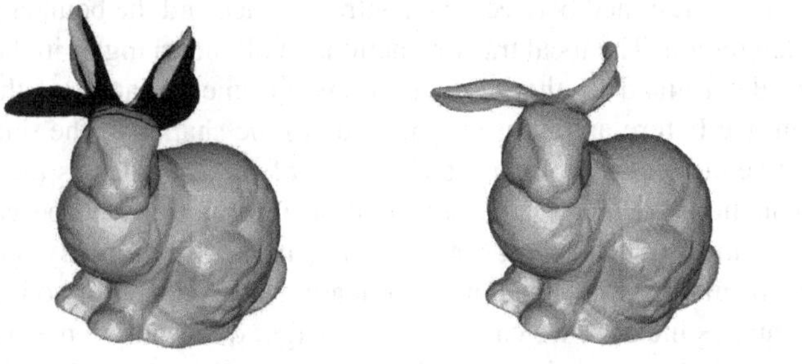

(a) The rabbit model before deformation (b) The rabbit model after deformation

Fig. 11.12 Rabbit ears deformation

displacement along the normal. Because the boundaries of the subdivision control surfaces changed before and after editing, the determination of the local transformation on the triangles in the designated deformation region needs the participation of control curves. It can be seen from the lighting effect of these two carved patterns that the deformation results maintain a very good continuity, whether on the boundaries of the deformation region or in the border regions of Class I and Class II triangles.

Figure 11.12 gives an example of deformation of the rabbit ears. Figure 11.12a shows the model before deformation, and the blue surface is the subdivision control surface before and after editing; Fig. 11.12b shows the model after deformation.

Remarks

The Poisson mesh editing algorithm based on subdivision surface combines the advantages of both the spatial deformation method based on subdivision surfaces and the Poisson mesh editing algorithm. So, users can get high-quality edited results in a simple and intuitive interaction way. By taking subdivision surfaces as deformation control surfaces, the algorithm has the advantage of multiresolution editing that enhances the ability of the local shaping. The introduction of the local gradient field for three-dimensional meshes makes the deformation results better maintain the original model details. At present, the construction of bounding meshes of complex models mainly resorts to offset and simplification operation of the mesh model. There is often a gap between the target of the mesh simplification and the shape demand of bounding mesh, which will inevitably affect the final deformation results. How to automatically create a bounding mesh with the similar shape as the mesh model to be deformed is worth of further studying.

Furthermore, the Poisson mesh editing algorithm based on subdivision surface is extended and applied in mesh deformation based on surface controls. Instead of traditional parametric spline surfaces, subdivision surfaces overcome the defects that rectangular mesh is difficult to model the shape of arbitrary topology. The deviation of the Poisson gradient field of a mesh is determined by the change of the control curve and subdivision control surfaces. The mesh model is rebuilt according to the revised Poisson gradient field, which makes the mesh deformation algorithm based on surface controls has more remarkable advantages in editing mesh models with rich detail features.

Exercises

1. Give the basic mathematic principles of the Poisson mesh editing method.
2. Give a concise procedure for the Poisson mesh editing method according to the discussions in this chapter.

References

1. Warren Joe, Weimer Henrik. Subdivision Methods for Geometric Design: A Constructive Approach. San Fransisco: Morgan Kaufmann, 2001.
2. Wang Guojin. Status and Development Trend of Surface Modeling. China Computer World, 1998, 36 (D):D12–D13.
3. Shi Fa Zhong. CAGD&NURBS. Beijing: Binjing University of Aeronautics and Astronautics Press, 1994.
4. Les Piegl, Wayne Tiller. The NURBS Book. New York: Springer-Verlag, 1997.
5. Zhu Xinxiong. Free Curve and Surface Sculpt Technology. Beijing: Science Press, 2000.
6. Wang Guo Jin, Wang Guo Zhao, Zheng Jian Min. Computer Aided Geometric Design. Beijing: High Education Press, 2001.
7. T. DeRose, Kass M,Truong T. Subdivision surfaces in character animation. In: Computer Graphics, Annual Conference Series(ACM SIGGRAPH). New York: ACM Press, 1998:85–96.
8. Zorin D, Schröder P. Subdivision for Modeling and Animation. In: Computer Graphics, Annual Conference Series (ACM SIGGRAPH). SIGGRAPH 99 Course Notes, 1999.
9. Chaikin. An algorithm for high speed curve generation. Computer Graphics and Image Processing, 1974, 3:346–349.
10. Catmull E, Clark J. Recursively generated B-splines Surfaces on Arbitrary topological meshes. Computer Aided Design, 1978, 10(6):350–355.
11. Doo D, Sabin M. Behaviour of recursive division surfaces near extraordinary points. Computer Aided Design, 1978, 10(6):356–360.
12. Loop C. Smooth subdivision surfaces based on triangles[Master thesis]. Utah:University of Utah, 1987.
13. Dyn Nira, Levin David, Gregory John A. A 4-point interpolatory subdivision scheme for curve design. Computer Aided Geometric Design, 1987, 4:257–268.
14. Dyn Nira, Levin David, Gregory John A. A butterfly subdivision scheme for surface interpolatory with tension control. ACM Transactions on Graphics, 1990, 9(2):160–169.
15. Ball A.A., Storry D.J.T.. Conditions for tangent plane continuity over recursively generated B-spline surfaces. ACM Transactions on Graphics. 1988, 7(2):83–102.
16. Sabin M, Recent Progress in Subdivision: a Survey. In: Advances in Multiresolution for Geometric Modelling, Springer Publisher, 2006:203–230.
17. Reif U, A unified approach to subdivision algorithms near extraordinary vertices. Computer Aided Geometric Design, 1995, 12(2):153–174.
18. U. Reif 1996 A degree estimate for subdivision surfaces of higher regularity. Proc AMS 124,7, 2167–2174.

W. Liao et al., *Subdivision Surface Modeling Technology*,
DOI 10.1007/978-981-10-3515-9

19. Zorin Denis N. Stationary subdivision and multiresolution surface representation[Ph.D Thesis]. California: California Institute of Technology, 1998.
20. Prautzsch Hartmut. Analysis of Gk-subdivision surfaces at extraordinary points. In:Oberwolfach conference, Germany, 1995.
21. Jörg Peters, Reif Ulrich. Shape characterization of subdivision surfaces——basic principles. Computer Aided Geometric Design, 2004:21(6):585–599.
22. Karčiauskas K, Jörg Peters, Reif Ulrich. Shape characterization of subdivision surfaces——case studies. Computer Aided Geometric Design, 2004, 21(6):601–614.
23. Leif P. Kobbelt. Interpolatory subdivision on open quadrilateral nets with arbitrary topology, in: Proceedings of Eurographics 96, Computer Graphics Forum, 1996:409–420.
24. Jörg Peters, Ulrich Reif. The Simplest subdivision scheme for smoothing polygon. ACM Transactions on Graphics, 1997, 16(4):420–431.
25. Sederberg T, Zheng J, Swell D, Sabin M.. Non-Uniform recursive Subdivision Surfaces. In:Computer Graphics, Annual Conference Series(ACM SIGGRAPH). New York: ACM Press, 1998:387–394.
26. Kobbelt Leif. $\sqrt{3}$-Subdivision. In:Computer Graphics, Annual Conference Series (ACM SIGGRAPH). New York: ACM Press, 2000:103–112.
27. Velho Luiz, Zorin Denis. 4-8 Subdivision. Computer Aided Geometric Design, 2001, Special Issue on Subdivision Techniques, 1–31.
28. Li G, Ma W, Bao H. $\sqrt{2}$ Subdivision for quadrilateral meshes. Visual Computer, 2004, 20(2–3):180–198.
29. Dyn, N., Levin, D., Liu, D. Interpolatory convexity preserving subdivision schemes for curves and surfaces. Computer-Aided Design, 1992, 24 (4):211–216.
30. Zhang Hongxin, Wang Guojin. Honeycomb subdivision. Journal of Software, 2002, 13(7): 1199–1208.
31. Denis Zorin, Peter Schröder and Wim Sweldens. Interpolating subdivision for meshes with arbitrary topology. In Computer Graphics Proceedings, Annual Conference Series, ACM SIGGRAPH, 1996:189–192.
32. Prautzsch Hartmut. Freeform splines. Computer Aided Geometric Design, 1997, 14(3):201–206.
33. Prautzsch H., Umlauf G. A G^2-subdivision algorithm. In: G. Farin *et al*, ed. Geometric Modeling. New York: Springer-Verlag, 1998:217–224.
34. Prautzsch Hartmut, Umlauf Georg. A G^1 and a G^2 subdivision scheme for triangular nets. International Journal of Shape Modeling, 2000, 6(1):21–35.
35. Ivrissimtzis I.P., Dodgson N.A., Sabin M.A. A generative classification of mesh refinement rules with lattice transformations. Computer Aided Geometric Design, 2004, 21(1):99–109.
36. Alexa, M. Split operators for triangular refinement. Computer Aided Geometric Design, 2002, 19 (3):169–172.
37. Stam J. On subdivision schemes generalizing uniform B-spline surfaces of arbitrary degree. Computer Aided Geometric Design, 2001, 18(5):383–396.
38. Zorin D., Schröder P. A unified framework for primal/dual quadrilateral subdivision scheme. Computer Aided Geometric Design, 2001, 18(5):429–454.
39. Oswald P., Schröder P. Composite Primal/Dual $\sqrt{3}$-Subdivision Schemes. Computer Aided Geometric Design, 2003, 20(3):135–164.
40. Stam Jos. Exact evaluation of Catmull-Clark subdivision surfaces at arbitrary parameter values. In: Computer Graphics, Annual Conference Series (ACM SIGGRAPH). New York: ACM Press, 1998:395–404.
41. Stam Jos. Evaluation of Loop subdivision surfaces. In: Zorin D. and Schröder P. eds. Subdivision or Modeling and Animation. New York: ACM Press, 1999.
42. Wang Huawei, Qin Kaihuai, Ron Kikinis. Exact evaluation of NURSS at arbitrary parameter values. In: Proceedings of the IASTED International Conference on Computer Graphics and Imaging. USA: Las Vegas. 169–174.
43. Shuhua Lai, Fuhua (Frank) Cheng. Parametrization of General Catmull-Clark Subdivision Surfaces and Its Applications Computer Aided Design & Applications, 2006, 3 (1–4):513–522.

44. Halstead Mark, Kass Michael, DeRose Tony. Efficient, fair interpolation using Catmull-Clark surfaces. In:Computer Graphics, Annual Conference Series (ACM SIGGRAPH). New York: ACM Press, 1993:35–44.
45. Labisk U., Greiner G. Interpolatory $\sqrt{3}$-Subdivision. Computer Graphics Forum, 2000, 19(3):131–138.
46. Li G, Ma W, Bao H. Interpolatory $\sqrt{2}$-subdivision surfaces. Proceedings of GMP: geometric modelling and processing—theory and applications. California: IEEE Computer Press, 2004:185–94.
47. Zhang Hongxin and Wang Guojin. Semi-stationary subdivision operators and their applications in geometric modeling. Progress in Natural Science, 2002, 12(10): 772–776.
48. Ahmad H. Nasri. Polyhedral subdivision methods for free-form surfaces. ACM Transactions on Graphics, 1987, 6(1):29–73.
49. Ahmad H. Nasri. Surface interpolation on irregular networks with normal conditions, CAGD, 1991, 8(1):89–96.
50. Li Gui Qing, LI Hua. Vertex and Normal Interpolation of Surfaces Based on Control Net Generated by Mixed Subdivisions. Journal of Computer Aided Design & Computer Graphics, 2001, 13(6):537–544.
51. Zheng Jianmin, Cai Yiyu. Interpolation Over Arbitrary Topology Meshes Using a Two-Phase Subdivision Scheme. IEEE Transactions on Visualization and Computer Graphic, 2006, 12(3):301–310.
52. Lai Shuhua, Cheng Fuhua. Similarity based Interpolation using Catmull-Clark Subdivision Surfaces. The Visual Computer: International Journal of Computer Graphics, 2006, 22(9): 865–873.
53. Levin Adi. Interpolating nets of curves by smooth subdivision surfaces. In: Computer Graphics, Annual Conference Series (ACM SIGGRAPH). New York: ACM Press, 1998:57–64.
54. Litke N, Levin A, Schröder P. Trimming for subdivision surfaces. Computer Aided Geometry Design, 2001, 18(5):463–481.
55. Adi Levin. Levin Adi. Filling an N-sided hole using combined subdivision schemes. In: Proceedings of Curves & Surfaces. Saint-Malo(France): Vanderbilt University Press, 1999: 221–228.
56. Levin Adi. Combined Subdivision Scheme. [Ph.D Thesis]. Israel(France): Tel-Aviv University, 2000.
57. Nasri A. Recursive subdivision of polygonal complexes and its applications in computer aided geometric design. Computer Aided Geometric Design, 2000, 17(4):595–619.
58. Zhang Jingqiao, Wang Guojin, Zhen Jianmin. Curve Interpolation Based on Non-Uniform Catmull-Clark Subdivision Scheme. Journal Of Software, 2003, 14(12):2082–2091.
59. Eck Matthias, DeRose Tony, Duchamp Tom *et al.* Multiresolution Analysis of Arbitrary Meshes In: Proceedings of the 22nd annual conference on Computer graphics and interactive techniques New York: ACM Press, 1995:173–182.
60. Kobbelt Leif P, Vorsatz Jens, Labsik Ulf *et al.* A Shrink Wrapping Approach to Remeshing Polygonal Surfaces. Computer Graphics Forum, 1999, 18(1):209–237.
61. Suzuki H, Takeuchoi S, Kanai T. Subdivision surface fitting to a range of points. In: Proceedings of the 7th Pacific Conference on Computer Graphics and Applications Washington:IEEE Computer Society, 1999:158–167.
62. Ma Weiyin. Catmull-Clark Surface Fitting for Reverse Engineering Applications.
63. Hoppe Hugues, DeRose Tony, Duchampy Tom, *et al.* Piecewise Smooth Surface Reconstruction. In Computer Graphics Proceedings, ACM SIGGRAPH, 1994:295–302.
64. Lounsbery Michael, DeRose Tony, Warren Joe. Multiresolution analysis for surfaces of arbitrary topological type. ACM Transactions on Graphics, 1997, 16(1):34–73.
65. Bertram M., Duchaineau M.A., Hamann B., *et al.* Bicubic subdivision-surface wavelets for large-scale isosurface representation and visualization. In: IEEE Visualization. New York: ACM Press, 2000:279–284.
66. Bertram M., Duchaineau M.A., Hamann B., *et al.* Generalized B-spline Subdivision-surface wavelets for Geometry Compression. In: IEEE Transactions on Visualization and Computer Graphics. New York: ACM Press, 2004:326–338.

67. Zorin Denis, Peter Schröder, Wim Sweldens. Interactive Multiresolution Mesh Editing. In: Proceedings of the 24th annual conference on Computer graphics and interactive techniques. New York: ACM Press, 1997:615–625.
68. Biermann H, Kristjansson D, Zorin D. Approximate Boolean operations on free-form solids. Proceedings of ACM SIGGRAPH computer graphics, 2001: 185–194.
69. Hui K.C., Lai Y.H. Smooth blending of subdivision surface. Computer Aided Design, 2006, 38(7):786–799.
70. Zhou Hai. Research on Modeling Technology of Subdivision Surface. [Ph.D Thesis]. Nanjing University of Aeronautics and Astronautics, 2004.
71. Bolz J., Schröder P. Rapid evaluation oc Catmull-Clark subdivision surfaces. In: Proceedings of the Web3D 2002 Symposium, 11–18.
72. Li Guiqing, Wu Zhuangzhi, Ma Weiyin. Research Advances in Adaptive Subdivision Techniques. Journal of Computer-Aided Design & Computer Graphics, 2002, 18(12):1789–1798.
73. Thomas W. Sederberg, Jianmin Zheng. T-splines and T-NURCCs. In: International Conference on Computer Graphics and Interactive Techniques. New York: ACM Press, 2003:477–484.
74. Stephan Bischoff, Leif P. Kobbelt, Hans-Peter Seidel. Towards Hardware Implementation Of Loop Subdivision. In: Proceedings of the ACM SIGGRAPH/EUROGRAPHICS workshop on Graphics hardware. New York: ACM Press, 2000:41–50.
75. Shiue L.J., Jones I, Peters J. A realtime GPU subdivision kernel[J]. ACM Transactions on Graphics, 2005, 24(3):1010–1015.
76. Kun Zhou Xin Huang Weiwei Xu. Direct Manipulation of Subdivision Surfaces on GPUs.
77. Jörg Peters. Patching Catmull-Clark Meshes. In: Computer Graphics, Annual Conference Series (ACM SIGGRAPH). New York: ACM Press, 2000:255–258.
78. Charles Loop, Scott Schaefer. Approximating Catmull-Clark Subdivision Surfaces with Bicubic Patches. ACM Transactions on Graphics, 2008, 27(1):1–11.
79. Ding Jun Yong, Hu Shi Min, Zhou Deng Wen. Offset Approximation of Loop Subdivision Surfaces. Chinese Journal of Computers, 2003, 26(7):789–795.
80. Zhou Hai, Zhou Lai Shui, Wang Zhan Dong, Zhong Da Ping. An Algor ithm for Generating Offset Loop Subdivision Surface. Mechanical Science and Technology, 2003, 22(6):1016–1020.
81. Yang Junqing, Zhou Min, Zhang Lining. Offset Approximation Algorithm for Subdivision Surfaces. In: 8-th ACIS International Conference on Software Engineering, Artificial Intelligence, Networing, and Parallel/Distributed Computing. Washington: IEEE Computer Society, 2007:216–221.
82. Kobbelt Leif. A variational approach subdivision. Computer Aided Geometric Design, 1996, 13(8):743–761.
83. Leif Kobbelt, and Schröder. A multiresolution framework for variational subdivision. ACM Transactions on Graphics, 1998, 17(4):209–237.
84. Jörg Peters, Umlauf Georg. Gaussian and mean curvature of subdivision surfaces [Technical Report]. Florida: University of Florida, 2000.
85. Prautzsch Hartmut. Freeform splines. Computer Aided Geometric Design, 1997, 14(3):201–206.
86. Ma Weiyin. Subdivision surfaces for CAD—an overview. Computer Aided Desgin, 2005, 37(7):693–709.
87. Liao Wenhe, Liu Hao. Research Situation and Application Prospect of Subdivision Surface, 2007.
88. Gerald Farin. Curves and Surfaces for CAGD. Fifth Edition. 2001.
89. P. Alfeld. A case study of multivariate piecewise polynomials. In G. Farin, editor. Geometric Modeling: Algorithms and New Trends, 149–160. SIAM, Philadelphia, 1987.
90. Zhang Hongxin and Wang Guojin, Semi-stationary subdivision operators and their applications in geometric modeling, Progress in Natural Science, 2002, 12(10): 772–776.
91. Wolfgang Dahmen Subdivision algorithm for the generation of box spline surface. Computer Aided geometric design, 1984:115–129.

92. Hartmut Prautzsch. Generalized subdivision and convergence. Computer Aided geometric design, 1985:69–75.
93. Wolfgang Boemhm. Triangular spline algorithms. Computer Aided geometric design, 1985: 61–67.
94. Hartmut prautzsch. Box Splines. In: Computer Aided geometric design, Hoschek, Kim(eds),2005.
95. Kobbelt Leif. Stable Evalution of Box Splines. Numerical Algorithms, 1997, 14(4): 377–382.
96. Luiz Velho. Semi-Regular 4-8 Refinement and Box Spline Surfaces. In: Proceedings of the 13th Brazilian Symposium on Computer Graphics and Image Processing. Washington: IEEE Computer Society, 2000:131–138.
97. Luiz Velho. Generalizing the C^4 Four-directional Box Spline to Surfaces of Arbitrary Topology. Mathematical Methods for Curves and Surfaces: Oslo 2000, 507–516.
98. Ying Lexing, Zorin Denis. Non-manifold subdivision. In: Proceedings of IEEE Visualization, California, 2001.
99. Peters Jörg. C^2 Free-Form Surfaces of degree (3,5). Computer Aided Geometric Design, 2003, 19(2):113–126.
100. Qin Kaihuai, Wang Huawei. Non-uniform subdivision surface modeling and its continuity analysis. Science in China(Series E), 2000, 30(3):271–281.
101. Lane, J.M., Riesenfeld, R.F. A theoretical development for the computer generation and display of piecewise polynomial surfaces. IEEE Transactions on Pattern Analysis and Machine Intelligence, 1980, 2 (1):35–46.
102. Hartmut Prautzsch. Smoothness of subdivision surfaces at extraordinary points. Advances in Computational Mathematics, 1998 (9):377–389.
103. Denis Zorin. C^k Continuity of Subdivision Surfaces. Technical Report, Report 00000152. California: California Institute of Technology, 1996.
104. Umlauf Georg. Analyzing the characteristic map of triangular subdivision schemes. Constr. Approx., 2000, 16(1):145–155.
105. Barthe Loc, Kobbelt Leif.Subdivision scheme tuning around extraordinary vertices.Computer Aided Geometric Design, 2004, 21(6):561–583.
106. Levin Adi Modifid subdivision surfaces with continuous curvature. In: Computer Graphics, Annual Conference (ACM sIGGRAPH) New York: ACM Press, 2006:1035–1040.
107. George Celniker, Dave Gossard. Deformable curve and surface finite elements for free-form shape design. In: Proceeding of SIGGRAPH'91, 257–265.
108. Wang Xuefu, Cheng Fuhua, Barsky B.A, Energy and B-spline Interpolation. Computer Aided Design, 1997, 29(7):485–496.
109. Adi Levin. Analysis of combined subdivision schemes for the interpolation of curves.
110. Ahmad H. Nasri. Designing Catmull-Clark subdivision surfaces with curve interpolation constraints, 2000.
111. Zhang Jingqiao. Curve Interpolation based on Catmull-Clark Subdivision Scheme.
112. Zhang Jingqiao. Research Surface on Generating Subdivion and Applying to Surfaee Modeling [Ph.D Thesis]. Hangzhou: ZhejiangUniversity, 2003.
113. Olga A. karpenko, John F. Hughes. Smooth Sketch: 3D free-form shapes from complex sketches. In: Computer Graphics, Annual Conference Series (ACM SIGGRAPH). New York: ACM Press, 2006:589–598.
114. Andrew Nealen, Takeo Igarashi, Olga Sorkine, *et al.* FiberMesh: Design Freeform Surface with 3D Curves. In: Computer Graphics, Annual Conference Series (ACM SIGGRAPH). New York: ACM Press, 2007:1–8.
115. Nasri A. Interpolation of open B-spline curves by recursive subdivision surfaces. Mathematics of Surfaces VII, 1997:173–188.
116. Ahmad H Nasri, Tae-wan Kim, Kunwoo Leey. Fairing Recursive Subdivision Surfaces With Curve Interpolation Constraints. In: Shape Modeling International Proceedings of the International Conference on Shape Modeling & Application. Washington: IEEE Computer Society, 2001:49–60.

117. Liu Hao, Liao Wenhe. Surface reconstruction based on Catmull-Clark subdivision. Journal of the Graduate School of the Chinese Academy of Sciences, 2007, 24(3):307–315.
118. Zhang Liyan. Research on the key techniques of model reconstruction in reverse engineering. [Ph.D Thesis]. Nanjing University of Aeronautics and Astronautics, 2001.
119. Zhou Hai, Zhou Laishui, Wang Zhandong, Zhong Daping. Surface Reconstruction from Unorganized Points Using Subdivision Techniques. Journal of Computer Aided Design & Computer Graphics, 2003, 15(10): 1287–1292.
120. Desbrn M., Meyer M., Schröder P., *et al.* An isotropic feature preserving denoising of height fields and bivariare. In:Proceeding of Graphics Interface 2000. Washington: IEEE Computer Society, 2000:145–152.
121. Won-Ki Jeong. Direct reconstruction of a displaced subdivision surface from unorganized points. Graphical Models, 2002, 64(2): 78–93.
122. Wim Sweldens. The Lifting Scheme:A New Philosophy in Biorthogonal Wavelet Constructions. In: Wavelet Applications In Signal and Image Processing III, A.F.Lain and M.Unser, Eds., proceeding of SPIE, 1995, Vol. 2569:68–79.
123. Huawei Wang, Kaihuai Qin, Kai Tang. Efficient wavelet construction with Catmull–Clark subdivision. The Visual Computer, 2006, 22(9–11):874–884.
124. Mark A. Duchaineau, Martin Bertram, Serban Porumbescu.et.al. Interactive Display of Surfaces Using Subdivision Surfaces and Wavelets. In: Proceedings of the 17th Spring conference on Computer graphics, Washington: IEEE Computer Society, 2001:117–118.
125. Nathan Litke, Adi Levin, Peter Schröder. Fitting Subdivision Surfaces. In: IEEE Visualization 2001. Washington: IEEE Computer Society, 2001:319–324.
126. Sandrine Lanquetin. Reverse Catmull-Clark Subdivision. http://wscg.zcu.cz/WSCG2006/Papers_2006/Full/B89-full.pdf.
127. Valette, S., Prost, R.: Wavelet-based multiresolution analysis of irregular surface meshes. IEEE TVCG, 2004, 10(2):113–122.
128. Valette, S., Prost, R. Wavelet-based progressive compression scheme for triangle meshes: Wavemesh. IEEE TVCG, 2004, 10(2):123–129.
129. Bertram, M. Biorthogonal loop-subdivision wavelets. Computing, 2004, 72(1–2):29–39.
130. Li, D., Qin, K., Sun, H. Unlifted Loop subdivision wavelets. In: Proceedings of Pacific Graphics '04, Seoul, Korea, 2004, 25–33.
131. Wu, J., Amaratunga, K. Wavelet triangulated irregular networks. Int. J. Geograph. Inform. Sci. 2003, 17(3):273–289.
132. Samavati, F.F., Bartels, R.H. Multiresolution curve and surface editing: reversing subdivision rules by least-squares data fitting. Comput. Graph. Forum, 1999, 18(2):97–119.
133. Samavati, F.F., Mahdavi-Amiri, N., Bartels,R.H. Multiresolution surfaces having arbitrary topologies by a reverse Doo subdivision method. Comput. Graph. Forum, 2002, 21(2): 121–136.
134. Jun Hai Yong, Fuhua Cheng. Adaptive Subdivision of Catmull-Clark Subdivision Surfaces.
135. Junfu Dai, Wang Huawei, Qin Kaihuai. Parallel Generation of NC Tool Paths for Subdivision Surfaces. International Journal of CAD/CAM, 2004, 4(10):1–9.
136. Li Sheng, Huang Xin, Wang Guoping. View-Dependent Adaptive Subdivision on GPU. Journal of Computer Aided Design & Computer Graphics, 2007, 19(4):409–414.
137. Cheng F, Yong J. Subdivision Depth Computation for Catmull-Clark Subdivision Surfaces, Computer Aided Design & Applications, 2006, 3:1–4.
138. Fuhua Frank Cheng, Gang Chen, and Jun-Hai Yong. Subdivision Depth Computation for Extra-Ordinary Catmull-Clark Subdivision Surface Patches. Advances in Computer Graphics (Lecture Notes in Computer Science), 2006:404–416.
139. Thomas W. Sederberg, David L. Cardon. T-spline Simplification and Local Refinement. ACM Transactions on Graphics, 2004, 23(3):276–283.
140. Amresh A., Farin G. Adaptive subdivision schemes for triangular meshes. Hierarchical and geometrical methods in scientific visualization, 2002:319–326.
141. Liu W, Kondo K1 An adaptive scheme for subdivision surfaces based on triangular meshes. Journal for Geometry and Graphics, 2004, 8 (1):69–80.

142. Doggett M., Hirche J. Adaptive view dependent tessellation of displacement maps. In: Proceedings of the ACM SIGGRAPH/EUROGRAPHICS workshop on Graphics hardware. New York: ACM Press, 2000:59–66.
143. Wu X, Peters J. An accurate error measure for adaptive subdivision surfaces. In: Proceedings of the International Conference on Shape Modeling and Applications 2005. Washington: IEEE Computer Society, 2005:51–56.
144. Wang, H. and Qin, K. Estimating subdivision depth of Catmull-Clark Surfaces, Journal of Computer Science and Technology, 2004, 19(5):657–664.
145. Xu Z, Kondo K1 Local subdivision process with Doo-Sabin subdivision surfaces [C] Proceedings of Shape Modeling International, Banff, Alberta, 2002:7–12.
146. Shuhua Lai, Fuhua(Frank) Cheng. Inscribed Approximation based Adaptive Tessellation of Catmull-Clark Subdivision Surfaces. International Journal of CAD/CAM, 2006, 6(1):1–16.
147. Zhong Daping , Zhou Laishui , Zhou Hai. Research of an adaptive algorithm for subdivision of hybrid meshes. Mechanical Science and Technology, 2004, 23 (9):1090–1092.
148. Ying He, Kexiang Wang, Hongyu Wang, Xianfeng Gu, Hong Qin. Manifold T-spline. Geometric modeling and processing (Lecture notes in computer science), 2006:409–422.
149. Tong Weihua Feng Yuyu Chen Falai. Tong Weihua, Feng Yuyu, Chen Falai. A Surface Reconstruction Algorithm Based on implicit T-Spline Surface. Journal of Computer Aided Design & Computer Graphics, 18(3):358–365.
150. Yimin Wang, Jianmin Zheng, Hock Soon Seah. Conversion between T-spline and Hierarchic B-splines. Computer Graphics and Imaging, 2005:8–13.
151. Wanchiu Li, Nicolas Ray, Bruno Lévy. Automatic and Interactive Mesh to T-spline Conversion. In: Proceedings of the fourth Eurographics symposium on Geometry processing. Switzerland: Eurographics Association, 2006:191–200.
152. Li Gui qing, LU Bing, LI Xian min, LI Hua. Generation of Sharp Features for Subdivision Surfaces. Journal Of Software, 2000, 11 (9):1189–1195.
153. Wei Guo fu, Chen Fa lai. Sharp Features of Subdivision Surfaces over Triangular Meshes. Journal of University of Science and Technology of China, 2002, 32(2):140–146.
154. Ayman Habib, Joe Warren. Edge and Vertex Insertion for a Class of C1 Subdivision Surfaces. Computer Aided Design, 1978, 10(6):350–355.
155. Mueller H, JaeschkeR. Adaptive subdivision curves and surfaces. In: Proceeding computer graphics international'98, 1998:48–58.
156. Amresh A, Farin G, Razdan Z. Adaptive subdivision schemes for triangular meshes.In:Farin G, Hagen H, Hamann B, editors. Hierarchical and geometric methods in scientific visualization. Springer-Verlag, 2003:319–327.
157. Qin Hong. Dynamic Catmull-Clark Subdivision Surface. IEEE Transactions on Visualization and Computer Graphics, 1998, 4(3):215–229.
158. Qin Kaihuai, Wang Huawei, Li D, *et al.* Physics-based Subdivision Surfaces Modeling for Medical Imaging and Simulation. In: Proceedings of the International Workshop on Medical Imaging and Augmented Reality. Hong Kong, 2001:10–12.
159. Qin Kaihuai, Chang Zhengyi, Wang, Huawei *et al.* Physics-Based Loop Surface Modeling, Journal of Computer Science and Technology, 2002, 17(6):851–858.
160. Huawei Wang, Hanqiu Sun, Kaihuai Qin. Deformable surface modeling based on dual subdivision, Progress in Natural Science, 2005, 15(1):81–88.
161. Feng Jieqing, Jin Shao, Qunsheng Peng, A. Robin Forrest. Multiresolution free-form deformation with subdivision surface of arbitrary topology. The Visual Computer, 2006, 22(1): 28–42.
162. Boler Martin I., Ronfard R., Bernardini F. Detail-preserving variational surface design with multiresolution constraints. In: Proceedings of the Shape Modeling International 2004, 2004:119–128.
163. Lien, S.L., Shantz, M., Pratt, V. Adaptive Forward Differencing for Rendering Curves and Surfaces. Computer Graphics (Proceedings of SIGGRAPH 87) 21, 4 (1987), 111–118.
164. Alex Vlachos, Jörg Peters, Chas Boyd *et al.* Curved PN Triangles. ACM, New York:: Symposium on Interactive 3D Graphics Proceedings of the 2001 symposium on Interactive 3D graphics, 2001:159–166.

165. Bolz Jeff, Schröder Peter. Evaluation of Subdivision Surfaces on Programmable Graphics Hardware. http://www.multires.caltech.edu/pubs/GPUSubD.pdf.
166. Mikaël Bourges-Sévenier. An Introduction to MPEG-4 Animation Framework Extension(AFX).
167. Yannick Maret. Implementation of MPEG-4's Subdivision Surfaces Tools[Master's Thesis]. Lausanne: Ecole Polytechnique de Lausanne, 2003.
168. Francisco Morán, Patrick Gioia, Michael Steliaros. Subdivision Surfaces in MPEG-4. In: International Conference on Image Processing. Washington: IEEE Computer Society, 2002, vol. 3: III-5–III-8.
169. ISO/IEC 14496-16, Information technology-Coding of audio-visual objects-Part 16: Animation Framework eXtension(AFX). Second edition, 2006.
170. Peng Wu, Hiromasa Suzuki, Joe Kuragano, et al. Three-axis NC Cutter Path Generation for Subdivision Surface. In: Proceedings of GMP(04), 2004:1–6.
171. Terzopoulos, D., Platt, J. and Barr, A. Elastically Deformable Models. ACM, Computer Graphics, 1987, 24(4):205–214.
172. Liu Hao, Liao Wenhe. Modeling G-2 Continuous Free-Form Surfaces by C-C Subdivision Method and Manifold Method. Journal of Computer Aided Design & Computer Graphics, 2005(04).
173. Greiner, G. Surface Construction Based on Variational Princuples. In Wavelets, Images and surface fitting, P.J. Laurent, A.Le. Mehaute and L.L. Schumaker, Eds, 277–286.
174. Guoliang Xu, Qin Zhang, G2 surface modeling using minimal mean-curvature-variation flow. Computer Aided Design.
175. Guoliang Xu, Qin Zhang. A general framework for surface modeling using geometric partial differential equations. Computer Aided Geometric Design.
176. Halstead Mark, Kass Michael, DeRose Tony. Efficient, fair interpolation using Catmull-Clark surfaces. In: Computer Graphics, Annual Conference Series (ACM SIGGRAPH). New York: ACM Press, 1993:35–44.
177. Vida H., Matin R. R., Varady T. A survey of blending methods that use parametric surfaces. Computer Aided Design, 1994, 16(5):341–365.
178. Li Guiqing, Li Hua. Blending Parametric Patches with Subdivision Surfaces.Journal Computer Science & Technology, 2002, 17(4):498–506.
179. Cheng Jinsan. Blending Quadric Surfaces via Base Curve Method. Mathematics Mechanization Research, 2002, 21(3):15–22.
180. Levin Adi. Filling an n-sided hole using combined subdivision schemes. In:Proceedings of Curves & Surfaces. Saint-Malo, France: Association Française d'Approximation, 1999:221–228.
181. Hwang Wei-Chung, Chuang Jung-Hong. N-Sided Hole Filling and Vertex Blending Using Subdivision Surfaces. Journal of Information Science and Engineering, 2003, 19(6):745–763.
182. Zhu Xinxiong. Free Curve and Surface Sculpt Technology. Beijing: Science Press, 2000.
183. Jörg Peters, Umlauf Georg. Gaussian and mean curvature of subdivision surfaces [Technical Report]. Florida: University of Florida, 2000.
184. Lai Shuhua, Cheng Fuhua. Similarity based Interpolation using Catmull-Clark Subdivision Surfaces. The Visual Computer: International Journal of Computer Graphics, 2006, 22(9):865–873.
185. Zhang Xiangyu, Liao Wenhe, Liu Hao. Simple geometric constrained deformation and shape editing of Subdivision surfaces. Journal of Jiangsu University (Natural Science Edition), 2009, 30(2):118–123.
186. Zhang Xiangyu. Research on Deformation Technology of Curves and Surfaces Based on Subdivision, [Ph.D Thesis]. NUAA, 2010.
187. Li Tao,Zhou Laishui,Zhang Weizhong. Interactive Shape Modification of C-C Subdivision SurfacesJournal of Image and Graphics, 2008, 13(1):170-175
188. LiTao. Key Technology Research on Approximation Subdivision Surfaces Based on Quad Meshes. [Ph.D Thesis]. NUAA, 2007.

189. Wyvill B, Wyvill G. Field functions for implicit surfaces. Visual Computer, 1989, 5(1–2):75–82.
190. Wang X.F., Cheng F.H., Barsky B.A. Energy and B-spline interproximation. Computer-Aided Design, 1997, 29(7):485–496.
191. Linda M. Wills, Philip Newcomb. Reverse engineering. Springer, 1996.
192. Assem S. Deif. Advanced Matrix Theory for Scientists and Engineers. Abacus Press, Tunbridge Wells & London, 1982.
193. Nasri A Polyhedral subdivision methods for free form surfaces ACM Transactions on graphics, 1987, 6(1):29–73.
194. Lanquetin S, Foufou S, Kheddouci H, *et al.* A Graph Based Algorithm for Intersection of Subdivision Surfaces http://sandrine.Lanquetin.free.fr/Articles/LANQUETIN20CGGM202003.pdf.
195. Zhu X P, Hu S M, Tai C L, *et al.* A marching method for computing intersection curves of two subdivision solid http://cg.cs.tsinghua.edu.cn/-shimin/pdf/mos_2005_sub.pdf.
196. Habib A, Warren J Edge and vertex insertion for a class of C1 subdivision surfaces Computer Aided Geometric Design, 1999, 16(4):223–247.
197. Biermann H, Martin I M, Zorin D, *et al.* Sharp features on multiresolution subdivision surfaces Graphical Models, 2002, 64(2):61–77.
198. Litke N, Levin A, Schröder P Trimming for subdivision surfaces Computer Aided Geometric Design, 2001, 18(5):463–481.
199. Reif U. TURBS—topologically unrestricted rational B-splines. Constructive Approximation, 1998, 14(1):57–77.
200. Dai H. Matrix theory. Science Presser, Beijing, China, 2001.
201. Cotrina J., Pla N., Vigo M. Towards free form surfaces[Technical Report]. Politècnica de Catalunya: Universitat: Politècnica de Catalunya, 2001.
202. Zhu X P, Hu S M, Tai C L, *et al.* A marching method for computing intersection curves of two subdivision solid http://cg.cs.tsinghua.edu.cn/-shimin/pdf/mos_2005_sub.pdf.
203. Levin A. Interpolating nets of curves by smooth subdivision surfaces. Computer Graphics Proceedings. SIGGRAPH 99. ACM, 1999:57–64.
204. Levin A. Combined subdivision scheme[Ph.D. Thesis]. Israel: Tel-Aviv University, 2000.
205. He Gang. Fundamental Technology Research on Subdivision Surface Modeling Based on Geometric Constraints, [Ph.D. Thesis]. Nanjing: Nanjing University of Aeronautics and Astronautics, 2008.
206. Kuragano J, Suzuki H, Takarada Y, *et al.* Subdivision surface generation from a set of unconnected characteristic curves. Journal of the Japan Society of Precision Engineering, 2003, 69(9):1264–1269.
207. Litke N, Levin A, Schroder P. Trimming for subdivision surfaces. Computer-Aided Geometric Design, 2001, 18(5):463–481.
208. Schneider R, Kobbelt L. Mesh fairing based on an intrinsic PDE approach. Computer-Aided Design, 2001, 33(11):767–777.
209. Schneider R, Kobbelt L. Geometric fairing of irregular meshes for free-form surface design. Computer-Aided Geometric Design, 2001, 18(4):359–379.
210. Moreton H. Minimum Curvature Variation Curves, Networks, and Surfaces for Fair Free-Form Shape Design[Doctor of Philosophy Thesis]. UNIVERSITY of CALIFORNIA at BERKELEY, 1992.
211. Desbrun M, Meyer M, Schroder P, *et al.* Implicit fairing of irregular meshes using diffusion and curvature flow. Proceedings of the ACM SIGGRAPH Conference on Computer Graphics, 1999:317–324.
212. Li Guiqing, Li Hua. Blending Parametric Patches with Subdivision Surfaces.Journal Computer Science & Technology, 2002, 17(4):498–506.
213. Liu Shenlan. Research on Key Technology in Reconstruction of Free-Form & Regular Surfaces in Reverse Engineering,[Ph.D. Thesis]. NUAA, 2004.
214. Pottmann H, Leopoldseder S A Concept for Parametric Surface Fitting which avoids the Parametrization Problem Computer Aided Geometric Design, 2003, 20(6):343–362.

215. Pottmann H, Hofer M Geometry of the Squared Distance Function to Curves and Surfaces Visualization and Mathematics III, Hege H, Polthier K, Eds New York, Springer Presser, 2003.
216. Cheng K S, Wang W P, Qin H, *et al.* Fitting subdivision surfaces to unorganized point data using SDM In: Proceedings of the 12th Pacific Conference on Computer Graphics and Applications, Seoul 2004:16–24.
217. Peters J, Umlauf G Gaussian and mean curvature of subdivision surfaces[Technical Report], Florida: the University of Florida, 2002.
218. Wang W P, Pottmann H, Liu Y Fitting B-Spline Curves to Point Clouds by Squared Distance Minimization ACM Transactions on Graphics, 2006, 25(2):214–238.
219. Sherry Li, Jim Demmel, John Gilbert SuperLU, Version 3.0 http://crd.lblgov/-xiaoye/SuperLU/.
220. Wang Lazhu, Zhu Xinxiong. Construction of N-sided Patches.Computer Aided Drafting, Design and Manufacturing, 1995, 5 (1): 26–32.
221. Cotrina J., Pla N., Vigo M. n-sided Patches with B-Spline Boundaries[Technical Report]. Politècnica de Catalunya: Universitat Politècnica de Catalunya, 2001.
222. Cotrina J., Pla N. Modeling surfaces from planar irregular meshes. Computer Aided Geometric Design, 2000, 17(1):1–15.
223. Guiqing Li. Modeling and Application of Subdivision Surfaces. [Ph.D. Thesis]. Institute of Computing Technology Chinese Academy of Sciences, 2000.
224. Wei-Chung Hwang, Jung-Hong Chuang. N-Sided Hole Filling and Vertex Blending Using Subdivision Surfaces. Journal of Information Science and Engineering, 2003, 19(6):745–763.
225. Levin Adi. Filling an n-sided hole using combined subdivision schemes. In: Proceedings of Curves & Surfaces. Saint-Malo, France: Association Française d'Approximation, 1999:221–228.
226. Prautzsch H., Umlauf G. A G^2-subdivision algorithm. In: G. Farin *et al.*, ed. Geometric Modeling. New York: Springer-Verlag, 1998:217–224.
227. Li Guiqing, Li Xianmin, Li Hua. G^2 Filling of N-Sided Holes with Subdivision Surfaces. In: Seventh International Conference on Computer Aided Design and Computer Graphics. Kunming, 2001.
228. Sederberg T, Zheng J, Swell D, Sabin M. Non-Uniform recursive Subdivision Surfaces. In: Computer Graphics, Annual Conference Series (ACM SIGGRAPH). New York: ACM Press, 1998:387–394.
229. Reif U. A Degree Estimate for Subdivision Surfaces of Higher Regularity. In: Proceedings of the American mathematical Society, New York, 1996:2167–2174.
230. Prautzsch Hartmut. Freeform splines. Computer Aided Geometric Design, 1997, 14(3): 201–206.
231. Piegl L Modifying the shape of rational B-spline, Part 2:surfaces Computer-Aided Design, 1989, 21(9):538–546.
232. Au C K, Yuen M M F Unified approach to NURBS curve shape modification Computer-Aided Design, 1995, 27(2):85–93.
233. Hu S. M., Ju T., Li Y. F., *et al.* Modifying the shape of NURBS surfaces with geometric constraints Computer-Aided Design, 2001, 33(12):903–912.
234. Zhang Lenian, Zhou Laishui, Zhou Rurong. Research on controllably modeifying the shape of NURBS curve and surface. Journal of Engineering Graphics, 1995, 16(1):21–27.
235. Wang Zhiguo. Research on key technology in shape modification and deformation of curves and surfaces[Ph.D. Thesis]. Nanjing: Nanjing University of Aeronautics and Astronautics, 2006.
236. Kobbelt L, Schröder P A multiresolution framework for variational subdivision ACM Transactions on Graphics, 1998, 17(4):209–237.
237. Sederberg T W, Parry S R. Free-Form deformation of solid geometric models. Computer Graphics, 1986, 20(4): 151–160.
238. Coquillart S. Extended free-form deformation: A sculpting tool for 3D geometric modeling. Computer Graphics, 1990, 24(4):187–196.

239. Kalra P, Mangil A, Thalmann N M, et al. Simulation of facial muscle action based on rational free-form deformation. Computer Graphics Forum, 1992, 11(3):59–69.
240. Lamousin H J, Waggenspack W N. NURBS-based free-form deformation. IEEE Computer Graphics and Applications, 1994, 14(6):59–65.
241. MacCracken R, Joy K I. Free-form deformations with lattices of arbitrary topology. Proceedings of the 23rd annual conference on Computer graphics and interactive techniques, New York: ACM SIGGRAPH, 1996:181–188.
242. Feng J. Q., Ma L. Z., Peng Q. S. A new free-form deformation through the control of parametric surfaces. Computers & Graphics, 1996, 20(4):531–539.
243. Feng Jieqing, Peng Qunsheng, Ma Lizhuang. Homogeneous deformation following the fhape of a parametric surface. Journal of software, 1998, 9(4): 263–267.
244. Lazarus F, Coquillart S, Jancene P. Axial deformation: an intuitive deformationtechnique. Computer-Aided Design, 1994, 26(8):607–613.
245. Chang Y K, Rockwood A P. A generalized de casteljua approach to 3D free-form deformation. Proceedings of 21st International SIGGRAPH Conference on Computer Graphics, New York: ACM SIGGRAPH, 1994:257–260.
246. Borrel P, Rappoport A. Simple constrained deformation for geometric modeling and interactive design. ACM Transaction on Graphics, 1994, 13(2):137–155.
247. Raffin R, Neveu M, Jaar F. Curvilinear displacement of free-form-based deformation. Visual Computer, 2000, 16(1):38–46.
248. Jin X G, Li Y F, Peng Q S. General constrained deformation based on generalized meatballs. Computers & Graphics, 2000, 24(2):219–231.
249. Jin Xiaogang, Peng Qunsheng. General constrained deformationsbased on generalized metaballs. Journal of software, 1998, 9(9):677–682.
250. Li Leilei. Geometric Primitives Based deformation techniques for arbitrary Meshes[Master Thesis]. Hangzhou: Zhejiang University, 2004.
251. Hus W M, Hughes J F, Kaufman H. Direct manipulation of freeform deformation. Computer Graphics, 1992, 26(2):177–184.
252. Singh K, Fiume E. Wires: A geometric deformation technique. Computer Graphics, 1998, 32(3):405–414.
253. Sauvage B., Hahmann S., Bonneau G. P. Volume preservation of multiresolution meshes. Computer Graphics Forum, 2007, 26(3):275–283.
254. Marinov M, Botsch M, Kobbelt L. GPU-based multiresolution deformation using approximate normal reconstruction. Journal of Graphics Tools, 2007, 12(1):27–46.
255. Xu Weiwei, Zhou Kun, Yu Yizhou, et al. Gradient domain editing of deforming mesh sequences. ACM Transactions on Graphics, 2007, 26(3):Article No.84.
256. Wang Jun, Zhang Hongxin, Xu Dong, Bao Hujun. Sketch-based poisson mesh editing. Journal of Computer-Aided Design & Computer Graphics, 2006, 18(11):1723–1729.
257. Shao Jin. Subdivision-based space deformation method and generation of control meshes[Master Thesis]. Hangzhou: Zhejiang University, 2004.
258. Yoshizawa S, Belyaev A G, Seidel H P. Free-form skeleton-driven mesh deformations. Proceedings of the eighth ACM Symposium on Solid Modeling and Applications, New York: ACM SIGGRAPH, 2003:247–253.
259. Dai Ning, Zhou Yongyao, Liao Wenhe, *et al.* Research and implementation of local deformation design technology on dental restoration model. Chinese Journal of Biomedical Engineering, 2008, 27(3): 378–382.
260. Zhou Tingfang, Feng Jieqing, Xiao Chunxia, *et al.* Mesh Deformation with hierarchical B-splines. Journal of Computer-Aided Design & Computer Graphics, 2006, 18(3):443–450.
261. Wu Jinzhong, Liu Xuehui, Wu Enhua. Spline-based mesh editing. Journal of Computer-Aided Design & Computer Graphics, 2007, 19(7):907–912.
262. Liu Hao. Fundamental technology research on subdivision surface modeling based on quad meshes[Ph. D. Thesis]. Nanjing University of Aeronautics and Astronautics, 2005.

Printed in the United States
By Bookmasters